Complex and Hypercomplex Analytic Signals

Theory and Applications

Complex and Hypercomplex Analytic Signals

Theory and Applications

Stefan L. Hahn
Kajetana M. Snopek

ARTECH
HOUSE

BOSTON | LONDON
artechhouse.com

Library of Congress Cataloging-in-Publication Data
A catalog record for this book is available from the U.S. Library of Congress

British Library Cataloguing in Publication Data
A catalog record for this book is available from the British Library.

ISBN-13: 978-1-63081-132-7

Cover design by John Gomes

© 2017 Artech House
685 Canton St.
Norwood, MA

Contents

Preface   **xi**

1   Introduction and Historical Background   **1**

1.1   Introduction   1
       1.1.1   The Signal Domain Method   3
       1.1.2   The Frequency Domain Method   4
1.2   A Historical Survey   4
       References   7

2   Survey of Chosen Hypercomplex Algebras   **9**

2.1   Cayley-Dickson Algebras   9
       2.1.1   The Cayley-Dickson Construction   11
       2.1.2   The Cayley-Dickson Algebra of
                Quaternions   15
       2.1.3   The Cayley-Dickson Algebra of
                Octonions   19
2.2   Selected Clifford Algebras   23
       2.2.1   The Clifford Algebra of Biquaternions   24

	2.2.2	The Clifford Algebra of Bioctonions	28
2.3		Comparison of Algebras	31
2.4		Applications of Hypercomplex Algebras in Signal Processing	31
2.5		Summary	36
		References	37

3   Orthants of the n–Dimensional Cartesian Space and Single-Orthant Operators   43

3.1	The Notion of an Orthant	43
3.2	Single-Orthant Operators	45
3.3	Decomposition of Real Functions into Even and Odd Terms	47
	References	52

4   Fourier Transformation in Analysis of n-Dimensional Signals   53

4.1		Complex n-D Fourier Transformation	54
	4.1.1	Spectrum of a 1-D Real Signal in Terms of its Even and Odd Components	59
	4.1.2	Spectrum of a 2-D Real Signal in Terms of its Even and Odd Components	61
	4.1.3	Spectrum of a 3-D Real Signal in Terms of its Even and Odd Components	67
4.2		Cayley-Dickson Fourier Transformation	69
	4.2.1	General Formulas	69
	4.2.2	Quaternion Fourier Spectrum in Terms of its Even and Odd Components	71
	4.2.3	Octonion Fourier Spectrum in Terms of its Even and Odd Components	73
4.3		Relations Between Complex and Hypercomplex Fourier Transforms	77
	4.3.1	Relation Between QFT and 2-D FT	78
	4.3.2	Relation Between OFT and 3-D FT	79
4.4		Survey of Applications of Complex and Hypercomplex Fourier Transformations	81
	4.4.1	Applications in the Domain of Analytic Signals	83

| 4.5 | Summary | 83 |
| | References | 84 |

| **5** | **Complex and Hypercomplex Analytic Signals** | **87** |

5.1	1-D Analytic Signals as Boundary Distributions of 1-D Analytic Functions	88
5.2	The n-D Analytic Signal	94
	5.2.1 The 2-D Complex Analytic Signals	94
	5.2.2 3-D Complex Analytic Signals	99
5.3	Hypercomplex n-D Analytic Signals	109
	5.3.1 2-D Quaternion Signals	110
	5.3.2 3-D Hypercomplex Analytic Signals	112
5.4	Monogenic 2-D Signals	113
5.5	A Short Survey of the Notions of Analytic Signals with Single Orthant Spectra	117
5.6	Survey of Application of n-D Analytic Signals	120
	5.6.1 Applications Presented in Other Chapters of this Book	120
	5.6.2 Applications Described in Hahn's Book on Hilbert Transforms	121
	5.6.3 Selected Applications	121
	References	121

| **6** | **Ranking of Analytic Signals** | **125** |

6.1	Definition of a Suborthant	126
	6.1.1 Subquadrants in 2-D	128
	6.1.2 Suboctants in 3-D	128
6.2	Ranking of Complex Analytic Signals	129
	6.2.1 Ranking of 2-D Complex Analytic Signals	129
	6.2.2 Ranking of 3-D Complex Analytic Signals	131
6.3	Ranking of Hypercomplex Analytic Signals	135
	6.3.1 Ranking of 2-D Cayley-Dickson Analytic Signals	135
	6.3.2 Ranking of 3-D Cayley-Dickson Analytic Signals	138

6.4	Summary	141
	References	141

7   Polar Representation of Analytic Signals   143

7.1	Introduction	143
7.2	Polar Representation of Complex Numbers	144
7.3	Polar Representation of 1-D Analytic Signals	144
7.3.1	Representation of the Instantaneous Complex Frequency using the Wigner Distribution	148
7.4	Polar Representation of 2-D Analytic Signals	148
7.4.1	2-D Complex Analytic Signals with Single-Quadrant Spectra	148
7.4.2	2-D Hypercomplex Quaternion Analytic Signals with Single-Quadrant Spectra	150
7.4.3	Relations between the Analytic and Quaternion 2-D Phase Functions	151
7.4.4	Polar Representation of the Monogenic 2-D Signal	152
7.4.5	Common Examples for 2-D Polar Representations of Analytic, Quaternion, and Monogenic Signals	153
7.5	Polar Representation of 3-D Analytic Signals	165
7.5.1	3-D Complex Signals with Single Octant Spectra	165
7.5.2	3-D Octonion Signals with Single Octant Spectra	167
	References	173

8   Quasi-Analytic Signals   175

8.1	Definition of a Quasi-Analytic Signal	175
8.2	The 1-D Quasi-Analytic Signals	176
8.3	Phase Signals	181
8.4	The n-D Quasi-Analytic Signals	183
	References	185

9   Space-Frequency Representations of *n*-D Complex and Hypercomplex Analytic Signals   187

9.1   Wigner Distributions and Woodward Ambiguity Functions of Complex Analytic Signals   187
    9.1.1   WDs and AFs of 1-D Signals   190
    9.1.2   WDs and AFs of 2-D Complex Signals   200
    9.1.3   WDs and AFs of 2-D Complex Analytic Signals   203
9.2   Wigner Distributions and Woodward Ambiguity Functions of Quaternion and Monogenic Signals   211
    9.2.1   The WDs of Quaternion Signals   211
    9.2.2   AFs of Quaternion Signals   214
    9.2.3   WDs of Monogenic Signals   219
    9.2.4   AFs of Monogenic Signals   220
9.3   Double-Dimensional Wigner Distributions   223
9.4   Applications of Space-Frequency Distributions in Signal Processing   227
    9.4.1   Wigner Distribution in Noise Analysis   227
    9.4.2   Wigner Distribution in Image Processing   232
    References   235

10   Causality of Signals   241

10.1   Kramers-Kronig Relations   241
10.2   Extension of the Notion of Causality to Higher Dimensions   243
    10.2.1   Derivation of the Dispersion Relations   244
10.3   Summary   246
    References   248

11   Summary   249

    References   251

Appendix A   Table of Properties of 1-D Fourier Transformation   **253**

Selected Bibliography   254

Appendix B   Table of Chosen 1-D Fourier Pairs   **255**

Selected Bibliography   257

Appendix C   Table of Properties of 2-D Fourier Transformations   **259**

Selected Bibliography   261

Appendix D   Chosen 2-D Fourier Pairs   **263**

Selected Bibliography   264

Appendix E   Table of Properties of Quaternion Fourier Transformation of Real Signals   **265**

Selected Bibliography   266

Appendix F   Properties of 1-D Hilbert Transformations   **267**

Appendix G   1-D Hilbert Pairs   **269**

Appendix H   2-D Hilbert Quadruples   **273**

List of Symbols   **277**

About the Authors   **283**

Index   **285**

Preface

The content of this book is the result of the authors' 20 years worth of research in the domain of complex and hypercomplex multidimensional signals. The starting point was the theory of multidimensional analytic signals with single-orthant spectra defined in 1992 by Stefan L. Hahn in the *Proceedings of IEEE*. Four years later, Artech House published *Hilbert Transforms in Signal Processing*. The theory and applications of the hypercomplex analytic signals have been presented in the dissertation of D.Sc. Kajetana M. Snopek, *Studies of Complex and Hypercomplex Multidimensional Analytic Signals*.

During their research, the authors often put forth the following questions:

Complex or hypercomplex? Which approach is better? Are they both equivalent?

This book is an attempt to answer these questions. The reader will find lengthy descriptions of the theory of multidimensional complex and hypercomplex signals illustrated with numerous examples and followed by practical applications. Of course, some problems are still open and can serve as inspiration for postgraduate or doctoral students in their research. For example, the field of octonion signals is still only being superficially explored, and the problem of their polar form has not yet been solved.

The majority of the figures presented in this book were created in a MATLAB™ environment and are the result of programming in C++. The authors have made every possible effort to avoid errors in presented formulas and derivations. However, if errors should occur, the authors appreciate any remarks from their readers and are open to any discussions.

The authors express their deep gratitude to all those who have contributed to this book and supported it, especially to the reviewer, Dr. Stephen Sanguine, for his suggestions and remarks, which significantly improved the quality of this work. To a large extent, the problems presented in this book are result of many years of research done by the authors at the Institute of Radioelectronics and Multimedia Technology of the Warsaw University of Technology (originally known as the Politechnika Warszawska). The authors would like to thank the director of the Institute, Professor Józef Modelski, and the scientific deputy director, Professor Wiesław Winiecki, for their support and encouragement at different stages of the research. This book was initiated by a letter from William M. Bazzy from Artech House to Professor Hahn with suggestions to present new results.

1

Introduction and Historical Background

1.1 Introduction

There are two domains in theoretical descriptions of n-dimensional (n-D) signals: the signal domain (in 1-D usually the time domain) and the frequency domain. In the signal domain, signals are functions of the Cartesian coordinates $\boldsymbol{x} = (x_1, x_2, \ldots, x_n)$. Note that the signal u is assumed to be a real-valued (not complex-valued) function of $\boldsymbol{x}$. On the other hand, analytic signals are complex or hypercomplex functions of $\boldsymbol{x}$. The name *analytic* is used because analytic signals are *boundary distributions* of complex analytic functions of an n-D complex variable $\boldsymbol{z} = (z_1, z_2, \ldots, z_n)$, where $z_k = x_k + jy_k$. (for 1-D time signals: $z = t + j\tau$). Details defining the boundary distribution are given in Chapter 5. The second domain of analysis of n-D signals is the frequency domain $\boldsymbol{f} = (f_1, f_2, \ldots, f_n)$. For a given real signal $u(\boldsymbol{x})$, we can define a pair of functions:

$$u(\boldsymbol{x}) \overset{\text{n-F}}{\Longleftrightarrow} U(\boldsymbol{f}) \tag{1.1}$$

where $U(\boldsymbol{f})$ is called the n-D Fourier spectrum (notation n-F) of $u(\boldsymbol{x})$. In general, $U(\boldsymbol{f})$ is a complex (or hypercomplex) function of frequency $\boldsymbol{f}$.

As will be described in Chapter 3, the real signal $u(\boldsymbol{x})$ may be expressed as a sum of 2^n terms with even and/or odd parity. The 1-D signal may be a union of two 1-D terms:

$$u(t) = u_e + u_o \tag{1.2}$$

The 2-D real signal is a union of four terms:

$$u(x_1, x_2) = u_{ee} + u_{oe} + u_{eo} + u_{oo} \tag{1.3}$$

and the 3-D real signal may be a sum of eight 3D terms:

$$u(x_1, x_2, x_3) = u_{eee} + u_{oee} + u_{eoe} + u_{ooe} + u_{eeo} + u_{oeo} + u_{eoo} + u_{ooo} \tag{1.4}$$

where the subscripts are written using the reverse notation of binary numbers: e (even) = binary "0" and o (odd) = binary "1." Examples of the reverse notation: eee = 000, oee = 001,,,,,.

In Chapter 4, we will present the theory of complex and hypercomplex Fourier transforms. For 1-D signals, we have only a complex Fourier transform (real for even functions). In 2-D, we have a choice between complex or hypercomplex (quaternion) Fourier transforms. For 3-D signals, we have a choice between the complex and more than one hypercomplex Fourier transforms. They differ by the applied algebra of the imaginary units (see Chapter 4). The corresponding complex and hypercomplex analytic signals are described in Chapter 5.

In 1-D, the complex Fourier spectrum has the form:

$$U(f) = \int_{-\infty}^{\infty} u(t)e^{-j2\pi ft}\, dt$$

$$= \int_{-\infty}^{\infty} u_e(t)\cos(2\pi ft)\, dt - j\int_{-\infty}^{\infty} u_o(t)\sin(2\pi ft)\, dt = U_e(f) - jU_o(f) \tag{1.5}$$

We observe that, in general, the spectrum of a real signal is a complex-valued function. However for even signals, it is a real function.

As mentioned previously, the complex and hypercomplex spectra of 2-D signals differ. The complex spectrum of a 2-D real signal is calculated as the 2-D complex Fourier transform (see Chapter 4),

$$U(f_1, f_2) = U_{ee} - U_{oo} - j(U_{oe} + U_{eo}) \tag{1.6}$$

is a real-valued function, if the terms indexed *oe* and *eo* equal zero. The hyper-complex 2-D Fourier transform will be described in Chapter 4. In this book, we refer to it as the *Quaternion Fourier Transform* (QFT) given by

$$U_q\left(f_1, f_2\right) = U_{ee} - U_{oe} \cdot e_1 - U_{eo} \cdot e_2 + U_{oo} \cdot e_3 \qquad (1.7)$$

It is a real function, only if $u(x_1, x_2)$ is an even-even function. The 3-D complex Fourier spectrum is

$$U\left(f_1, f_2, f_3\right) = U_{eee} - U_{ooe} - U_{oeo} - U_{eoo} - j\left(U_{oee} + U_{eoe} + U_{eeo} - U_{ooo}\right) \qquad (1.8)$$

while the corresponding spectrum defined by the 3-D hypercomplex Fourier transform (*Octonion Fourier Transform* (OFT)) applying the Cayley-Dickson algebra is

$$
\begin{aligned}
&U_{CD}\left(f_1, f_2, f_3\right) \\
&= U_{eee} - U_{oee} \cdot e_1 - U_{eoe} \cdot e_2 + U_{ooe} \cdot e_3 - U_{eeo} \cdot e_4 + U_{oeo} \cdot e_5 + U_{eoo} \cdot e_6 - U_{ooo} \cdot e_7
\end{aligned}
$$
$$(1.9)$$

Therefore, the spectrum of a real signal u may be a complex/hypercomplex function. The corresponding analytic signal with single-orthant spectrum is always a complex (hypercomplex) function. The analytic signal can be calculated directly in the signal domain x or in the frequency domain f by the inverse Fourier transform of a single-orthant spectrum. Both methods yield exactly the same complex (hypercomplex) signal.

1.1.1 The Signal Domain Method

The n-D analytic signal with a single-orthant spectrum is defined by the n-fold convolution with the n-D complex (hypercomplex) delta distribution. The general formula is

$$
\begin{aligned}
&\psi(x) \\
&= \left\{ \left[\delta(x_1) + \frac{1}{\pi x_1} \cdot e_1 \right] \times \left[\delta(x_2) + \frac{1}{\pi x_2} \cdot e_2 \right] \times \ldots \times \left[\delta\left(x_n + \frac{1}{\pi x_n} \cdot e_n \right) \right] \right\} \underset{n\text{-fold}}{\overset{*}{}} u(x)
\end{aligned}
$$
$$(1.10)$$

Equation (1.10) is written assuming imaginary units $e_1, e_2, \ldots, e_n$ of the Cayley-Dickson algebra (see Chapter 2). In the complex case all $e_i = j$.

1.1.2 The Frequency Domain Method

The n-D analytic signal with a single-orthant spectrum is defined by the inverse Fourier transform of the spectrum $U(f)$ multiplied by the single-orthant operator in the form of a product of the frequency domain unit step functions (distributions). For the first orthant with all positive unit steps, we have

$$\text{Single orthant operator} = 2^n \times \mathbf{1}\big(f_1\big) \times \mathbf{1}\big(f_2\big) \times \ldots \times \mathbf{1}\big(f_n\big) \qquad (1.11)$$

The methods presented in (1.10) and (1.11) yield exactly the same analytic signals for a given algebra. In 1-D, we get the Gabor's analytic signal given by (1.17), whereas in 2-D, the complex analytic signal with the first quadrant spectrum is

$$\psi_1\big(x_1, x_2\big) = u - v + j\big(v_1 + v_2\big) \qquad (1.12)$$

where v is the total Hilbert transform and v_1 and v_2 are the partial Hilbert transforms. The 2-D hypercomplex analytic signal with the spectrum in the first quadrant of the frequency plane has the form of a quaternion

$$\psi_q^1\big(x_1, x_2\big) = u + v_1 \cdot e_1 + v_2 \cdot e_2 + v \cdot e_3 \qquad (1.13)$$

1.2 A Historical Survey

A large part of this survey was been presented in 2007 by Stefan Hahn at a historical session of the EUROCON conference [1]. The imaginary unit j = e_1 = $\sqrt{-1}$ was introduced in the sixteenth century to solve some problems concerning cubic equations. Three centuries later, in *Introductio in Analysim Infinitorum* (1748), Leonard Euler (1707–1783) derived his famous formula:

$$\exp(jx) = \cos x + j \sin x \qquad (1.14)$$

Then, the English mathematician, William Rowan Hamilton (1805–1865), expanded the notion of complex numbers and complex functions defining so-called *quaternions* [2] in a general form

$$q = a \cdot 1 + b \cdot i + c \cdot j + d \cdot k \qquad (1.15)$$

The complete work about quaternions appeared in 1853 under the title *Lectures on Quaternions* [3]. in 1843, John T. Graves (1806–1870)—a close friend of W. R. Hamilton—discovered the eighth-order algebra of hypercomplex numbers, which were called *octonions*. However, the discovery of octonions is ascribed to Arthur Cayley (1821–1895), who in his 1845 paper *On Jacobi's elliptic functions, in reply to the Rev. B. Bronwin: and on quaternions*, described the algebra of octonions; three years later, he presented their interpretation in terms of matrices. It is after Arthur Cayley and the famous American mathematician, Leonard Eugene Dickson (1874–1954), that the Cayley-Dickson construction of hypercomplex numbers was named.

The next step was the 1870 publication of *Linear Associative Algebras* by Benjamin Peirce (1809–1890); this was one of the first systematic studies on hypercomplex numbers. The nineteenth century was a period of other significant discoveries in the field of hypercomplex algebras. In 1873, William Kingdon Clifford (1845–1879) generalized the Hamilton's quaternions into biquaternions, and in 1886, Rudolf O.S. Lipschitz (1832–1903) rediscovered the Clifford algebras and proposed to apply them to study rotation in Euclidean spaces. Some of these algebras will be described in detail in Chapter 2.

In the signal theory, the spectra are always expressed in terms of the Fourier transforms. The real notation of the Fourier transform has been introduced by the French mathematician Jean Baptiste de Fourier (1768–1830) in the form of cosine and sine functions. Today, the Fourier transform is usually written using the complex notation. This notation was first introduced to electrical engineering by the American engineer Charles Proteus Steinmetz (1865–1923). He used the following notation of the Euler's formula (1.14):

$$\exp(j\omega t) = \cos(\omega t) + j\sin(\omega t) \qquad (1.16)$$

This notation was later adapted to the complex Fourier transformation. The representation (1.16) can be classified as the first analytic signal. Hungarian physicist and winner of the Nobel Prize for Physics in 1971, Dennis Gabor (1900–1979), generalized this notion defining the 1-D analytic signal [4] as

$$\psi(t) = u(t) + jv(t) \qquad (1.17)$$

where $u(t)$ and $v(t)$ form a pair of Hilbert transforms (invented by the German mathematician David Hilbert (1862–1943)). The polar notation of the Gabor's analytic signal (1.17),

$$\psi(t) = A(t)\exp\big(j\varphi(t)\big) \qquad (1.18)$$

uniquely defines the instantaneous amplitude and instantaneous frequency (derivative of the phase) of u [5]. Using (1.18), Hahn defined the notion of *instantaneous complex frequency* [6]. The Gabor's analytic signal has found numerous applications—especially in modulation theory (single-sideband modulation), time-frequency analysis (Wigner-Ville distribution), and estimation of the instantaneous parameters (amplitude, frequency or phase) of a given signal [7, 8]. Of course, the notions of the instantaneous amplitude and instantaneous frequency were earlier known to radio engineers developing AM and FM radio systems [9].

Analytic signals are defined as boundary distributions of analytic functions. A prominent cofounder of the theory of analytic functions is the French mathematician Augustus Louis Cauchy (1789–1857). The theory of analytic signals also applies the notions of unit step functions and of the delta pulse. The unit step function $\mathbf{1}(t)$ has been defined by Oliver Heaviside and the delta pulse $\delta(t)$ by Paul Dirac. Their theoretical background was later formulated by Schwartz [10] and Mikusiński [11] using theory of distributions. Schwartz defined distributions using test functions and Mikusiński applied approximating functions (see Chapter 3).

In [12], Hahn introduced the notion of the *complex delta distribution* used later in signal-domain definitions of n-D analytic signals [13]. In his paper, Hahn defined the n-D analytic signals with single-orthant spectra (see Chapter 3) as an extension of the 1-D Gabor's analytic signal to n dimensions. Specifically, he introduced notions of the *partial Hilbert transforms*. The idea of the n-D *hypercomplex delta distribution* was proposed by Hahn to the IEEE Transactions on Signal Processing. However, it was rejected by the associate editor. Later, this notion was applied in Snopek's work [14].

The theory of hypercomplex analytic signals has been developing quickly in the last twenty years and has found numerous applications in different domains. The introduction of the hypercomplex (quaternion) Fourier transformation by Todd A. Ell in 1992 [15] opened new possibilities of applications. The quaternion analytic signal defined by Thomas Bülow in his 1999 Ph.D. thesis [16] has been used first in the analysis of 2-D seismic signal attributes [17]. Further research has resulted in new applications, for example, in color image watermarking and filtering [18–20]. The new discrete versions of hypercomplex Fourier transforms have also appeared [21, 22]. In [23], Sommer and Bülow applied the 2-D quaternionic version of analytic signals. In [24], Sommer and Felsberg defined a 2-D monogenic hypercomplex signal using the Riesz transforms [25] as opposed to the Hilbert transforms. The Riesz transform has been also applied in numerous works of Larkin, et al. in [26].

References

[1]   Hahn, S., "The History of Applications of Analytic Signals in Electrical and Radio Engineering," *Intern. Conf. Computer as a Tool*, EUROCON, 9–12 Sept. 2007, 2007, pp. 2627–2631.

[2]   Hamilton, W. R., "On Quaternions," *Proc. Royal Irish Academy*, Vol. 3, 1847, pp. 1–16.

[3]   Hamilton, W. R., *Lectures on Quaternions*, Dublin: Hodges and Smith, 1853.

[4]   Gabor, D., "Theory of Communications," *Trans. Inst. Electr. Eng.*, Vol. 3, 1946, pp. 429–456.

[5]   Hahn, S. L., "On the uniqueness of the definition of the amplitude and phase of the analytic signal," *Signal Processing*, Vol. 83, 2003, pp. 1815–1820.

[6]   Hahn, S., "Complex variable frequency electric circuit theory," *Proc. IEEE*, Vol. 52, No. 6, 1964, pp. 735–736.

[7]   Hahn, S. L., *Hilbert Transforms in Signal Processing*, Norwood, MA: Artech House, 1996.

[8]   Ville, J., "Théorie et Applications de la Notion de Signal Analytique," *Câbles et Transmission*, Vol. 2A, 1948, pp. 61–74.

[9]   Bedrosian, E., "The Analytic Signal Representation of Modulated Waveforms," *Proc. IRE*, Vol. 50, No. 10, 1962, pp. 2071–2076.

[10]   Schwartz, L., *Méthodes Mathématiques pour les Sciences Physiques*, Paris, France: Hermann, 1965.

[11]   Antosik, P., J. Mikusiński, and P. Sikorski, *Theory of Distributions: The Sequential Approach*, Warsaw, Poland: PWN, 1973.

[12]   Hahn, S. L., "The n-Dimensional complex delta distribution," *IEEE Trans. Sign. Proc.*, Vol. 44, No. 7, 1996, pp. 1833–1837.

[13]   Hahn, S. L., "Multidimensional Complex Signals with Single-Orthant Spectra," *Proc. IEEE*, Vol. 80, No. 8, 1992, pp. 1287–1300.

[14]   Snopek, K. M., "New Hypercomplex Analytic Signals and Fourier Transforms in Cayley-Dickson Algebras," *Electr. Tel. Quarterly*, Vol. 55, No. 3, 2009, pp. 403–419.

[15]   Ell, T. A., "Hypercomplex Spectral Transformations," Ph.D. dissertation, University of Minnesota, Minneapolis, 1992.

[16]   Bülow, T., "Hypercomplex spectral signal representation for the processing and analysis of images," in Bericht Nr. 99–3, Institut für Informatik und Praktische Mathematik, Christian-Albrechts-Universität Kiel, August 1999.

[17]   Le Bihan, N., and J. Mars, "New 2D attributes based on complex and hypercomplex analytic signal," *71st Meeting of Society of Exploration Geophysicists SEG01*, San Antonio, TX, September 2001.

[18]   Bas, P., N. Le Bihan, and J.-M. Chassery, "Color Image Watermarking Using Quaternion Fourier Transform," *Proc. ICASSP*, Hong Kong, 2003.

[19]   Denis, P., P. Carre, and C. Fernandez-Maloigne, "Spatial and spectral quaternionic approaches for colour images," in *Computer Vision and Image Understanding*, Elsevier, Vol. 107, 2007, pp. 74–87.

[20]   Ell, T. A., and S. J. Sangwine, "Hypercomplex Fourier Transforms of Color Images," *IEEE Trans. Image Processing*, Vol. 16, No.1, January 2007, pp. 22–35.

[21]   Sommer, G. (Ed.), *Geometric Computing with Clifford Algebras*, Berlin Heidelberg, Germany: Springer-Verlag, 2001.

[22]   Said, S., N. Le Bihan, and S. J. Sangwine, "Fast Complexified Quaternion Fourier Transform," *IEEE Trans. Signal Proc.*, Vol. 56, No. 4, April 2008, pp. 1522–1531.

[23]   Bülow, T., and G. Sommer, "Hypercomplex Signals: A Novel Extension of the Analytic Signal to the Multidimensional Case," *IEEE Trans. Sign. Proc.*, Vol. 49, No. 11, 2001, pp. 2844–2852.

[24]   Felsberg, M., and G. Sommer, "The Monogenic Signal," *IEEE Trans. Sign. Proc.*, Vol. 49, No. 12, 2001, pp. 3136–3144.

[25]   Riesz, M., "Sur les fonctions conjuguées," *Mathematische Zeitschrift*, Vol. 27, No. 1, 1928, pp. 218–244.

[26]   Larkin, K.G., D. J. Bone, and M. A. Oldfield, "Natural demodulation of two-dimensional fringe patterns. I. General background of the spiral phase quadrature transform," *J. Opt. Soc. Am. A*, Vol. 18, No. 8, August 2001, pp. 1862–1870.

2

Survey of Chosen Hypercomplex Algebras

The goal of this chapter is to give a brief overview of chosen Cayley-Dickson and Clifford algebras of hypercomplex numbers. Bearing in mind their practical applications, we focus on Cayley-Dickson algebras of quaternions and octonions and Clifford algebras of biquaternions and bioctonions. We start with a discussion of their basic properties and then present some exemplary applications in the different fields. We also point out the differences between considered algebras. In the final section, we come to general definitions of complex and hypercomplex signals based on the previously presented Cayley-Dickson and Clifford algebras.

2.1 Cayley-Dickson Algebras

The goal of this book is to study the properties of algebras of order 2, 4, and 8. Presently, the higher-order Cayley-Dickson algebras seem to be out of interest from the point of view of possible applications. The Cayley-Dickson algebras are 2^N-order algebras over the real number field $\mathbb{R}$, $N \in \mathbb{N}$. These are the complex numbers, quaternions, octonions, and sedenions; that is, algebras defined using the Cayley-Dickson construction described in Section 2.1.1. It should be noted that only complex numbers, quaternions, and octonions are *normed division algebras*, meaning that every nonzero element has its inverse. This property is very important from the point of view of potential practical applications. Let us start with the known complex numbers algebra.

The complex numbers, denoted with $\mathbb{C}$, form the Cayley-Dickson algebra of order 2 over $\mathbb{R}$. In all other chapters of this book, we will denote any complex number $z \in \mathbb{C}$ as $z = x + y \cdot j$ where $x, y \in \mathbb{R}$. However, bearing in mind the compatible notation used in this chapter, let us write z in the form $z = r_0 + r_1 \cdot e_1$, $r_0, r_1 \in \mathbb{R}$ and e_1 is the imaginary unit usually denoted by mathematicians with i or j, $e_1 = \sqrt{-1}$. The real number r_0 is called the *real part* of the complex number z: $r_0 = \mathrm{Re}\, z$, and r_1 is its *imaginary part*: $r_1 = \mathrm{Im}\, z$.

Let us recall some commonly known notions and definitions concerning the complex numbers. The *reverse* of $z \in \mathbb{C}$ is a complex number $-z$:

$$-z = -r_0 - r_1 \cdot e_1 \tag{2.1}$$

and the *conjugate* of $z \in \mathbb{C}$ is a complex number $z^* \in \mathbb{C}$ defined as

$$z^* = r_0 - r_1 \cdot e_1 \tag{2.2}$$

The norm (*modulus*) of $z \in \mathbb{C}$ is a real nonnegative number

$$\|z\| = |z| = \sqrt{z \cdot z^*} = \sqrt{r_0^2 + r_1^2} = \sqrt{(\mathrm{Re}\, z)^2 + (\mathrm{Im}\, z)^2} \tag{2.3}$$

Remark: Let us note that in this entire book, we express the norm of a complex/hypercomplex number as a square root of a corresponding inner product.

Then, the conjugate divided by the norm or modulus squared defines the *inverse* of $z \in \mathbb{C}$ as follows

$$z^{-1} = \frac{1}{z} = \frac{1}{r_0 + r_1 \cdot e_1} = \frac{r_0 - r_1 \cdot e_1}{r_0^2 + r_1^2} = \frac{z^*}{|z|^2} \tag{2.4}$$

Any complex number z can be written in its polar form

$$z = |z| e^{e_1 \arg z} = |z| \cos \arg z + |z| \sin \arg z \cdot e_1 \tag{2.5}$$

where $\arg z$ is the *angle* of z. Thus, the real and imaginary parts of z (respectively) are expressed as

$$r_0 = |z| \cos \arg z, \quad r_1 = |z| \sin \arg z \tag{2.6}$$

Moreover, we see that the sums and differences of the complex number with its conjugate yield the real and imaginary parts with suitable scaling:

$$r_0 = \frac{z + z^*}{2}, \quad r_1 = \frac{z - z^*}{2e_1} \tag{2.7}$$

Due to the commutativity of summation and multiplication in $\mathbb{R}$, it is evident that these operations in $\mathbb{C}$ are also commutative. So, for any two complex numbers $z_0, z_1 \in \mathbb{C}$: $z_0 = r_0 + r_1 \cdot e_1$, $z_1 = r_2 + r_3 \cdot e_1$ we have:

$$z_0 + z_1 = \left(r_0 + r_2\right) + \left(r_1 + r_3\right) \cdot e_1 = \left(r_2 + r_0\right) + \left(r_3 + r_1\right) \cdot e_1 = z_1 + z_0 \tag{2.8}$$

$$z_0 \cdot z_1 = \left(r_0 + r_1 \cdot e_1\right)\left(r_2 + r_3 \cdot e_1\right) = \left(r_0 r_2 - r_1 r_3\right) + \left(r_0 r_3 + r_1 r_2\right) \cdot e_1$$
$$= \left(r_2 r_0 - r_3 r_1\right) + \left(r_3 r_0 + r_2 r_1\right) \cdot e_1 = z_1 \cdot z_0 \tag{2.9}$$

Let us now express a complex number $z \in \mathbb{C}$: $z = r_0 + r_1 \cdot e_1$ as an ordered pair of two real numbers $r_0, r_1 \in \mathbb{R}$:

$$z = \left(r_0, r_1\right) \tag{2.10}$$

that is, a complex number of order 2 is an ordered pair of two reals (of order 1). We will later show that the idea of forming an ordered pair of elements of lower-order algebra is applied in the Cayley-Dickson process of construction of higher-order algebras of hypercomplex numbers.

The operations (2.1)–(2.4), (2.8)–(2.9) defined in $\mathbb{C}$ can be generalized for quaternions and octonions; this will be shown in Sections 2.1.2 and 2.1.3. In Table 2.1, we show chosen operations on elements of Cayley-Dickson algebras of complex numbers, quaternions and octonions. Let us notice that (2.8) and (2.9) are special cases of general operations of summation and multiplication, if we replace $\left(x_0, x_1\right)$ and $\left(x_2, x_3\right)$ with $\left(r_0, r_1\right)$ and $\left(r_1, r_2\right)$ respectively.

2.1.1 The Cayley-Dickson Construction

The method known as the Cayley-Dickson construction [1] is based on the following step-by-step procedure. Let us consider two complex numbers $z_0, z_1, \in \mathbb{C}$: $z_0 = r_0 + r_1 \cdot e_1$, $z_1 = r_2 + r_3 \cdot e_1$, $r_0, r_1, r_2, r_3, \in \mathbb{R}$. Then, we define the complex ordered pair $\left(z_0, z_1\right)$ and denote it with q, that is

$$q = \left(z_0, z_1\right) = \left(\left(r_0, r_1\right), \left(r_2, r_3\right)\right) \tag{2.11}$$

Table 2.1
Operations in Cayley-Dickson Algebras

Property	Operation
Reversibility	$-(x_0, x_1) = (-x_0, -x_1)$
Conjugation	$(x_0, x_1)^* = (x^*_0, -x_1)$
Norm	$\left\Vert (x_0, x_1) \right\Vert = \sqrt{(x_0, x_1)^* (x_0, x_1)}$ $= \sqrt{(x_0^* x_0 + x_1 x_1^*, x_0 x_1 - x_1 x_0)} = \sqrt{\Vert x_0 \Vert^2 + \Vert x_1 \Vert^2}$
Multiplicative Inverse	$(x_0, x_1)^{-1} = \left(\dfrac{x_0^*}{\Vert x_0 \Vert^2 + \Vert x_1 \Vert^2}, -\dfrac{x_1}{\Vert x_0 \Vert^2 + \Vert x_1 \Vert^2} \right)$
Summation	$(x_0, x_1) + (x_2, x_3) = (x_0 + x_2, x_1 + x_3)$
Additive Identity	$(0, 0)$
Multiplication	$(x_0, x_1)(x_2, x_3) = (x_0 x_2 - x_3^* x_1, x_0 x_3 + x_2^* x_1)$
Multiplicative Identity	$(1, 0)$

where q is called a *quaternion*, being a quadruple of real numbers r_0, r_1, r_2, r_3. Of course, the above procedure demands the introduction of the next imaginary unit e_2, $e_2 = \sqrt{-1}$, since

$$q = z_0 + z_1 \cdot e_2 = \left(r_0 + r_1 \cdot e_1 \right) + \left(r_2 + r_3 \cdot e_1 \right) \cdot e_2 \tag{2.12}$$

In this way, we have just defined the fourth-order Cayley-Dickson algebra of quaternions, denoted with $\mathbb{H}$ in honor of William Rowan Hamilton (1805–1865), their "inventor" [2]. Using one of the Hamiltonian rules of multiplication in $\mathbb{H}$, presented in Table 2.2, that is, $e_1 \cdot e_2 = e_3$, the definition (2.12) gets the form

$$q = r_0 + r_1 \cdot e_1 + r_2 \cdot e_2 + r_3 \cdot e_3 \tag{2.13}$$

The above formula is the most popular definition of a quaternion hypercomplex number (sometimes other letters are used to denote imaginary units, for example i, j, k instead of e_1, e_2, e_3).

Remark: The Hamiltonian rule $e_1 \cdot e_2 = e_3$ used in (2.12) is a matter of choice. The sign of a product can be positive or negative and this leads to

Table 2.2
Multiplication Rules in $\mathbb{H}$

$\times$	1	e_1	e_2	e_3
1	1	e_1	e_2	e_3
e_1	e_1	-1	e_3	$-e_2$
e_2	e_2	$-e_3$	-1	e_1
e_3	e_3	e_2	$-e_1$	-1

other rules of multiplication and in consequence, the multiplication tables different than Table 2.2.

Example: A red, green, and blue (RGB) image can be represented in the form of a pure (reduced) quaternion (with a zero real part): $q = R \cdot e_1 + G \cdot e_2 + B \cdot e_3$, where R, G, B, respectively, are red, green, and blue components of a color image. More exactly, if n_1, n_2 are pixel indices, a color image can be represented by a 2-D array whose elements are pure quaternions $q(n_1, n_2)$.

Following the same procedure as in (2.10) and (2.11), we can define the eighth-order Cayley-Dickson algebra of *octonions* ($\mathbb{O}$), then the sixteenth-order algebra of *sedenions* ($\mathbb{S}$) and so on.

Let us present the algebra of octonions. According to the Cayley-Dickson construction, any octonion $o \in \mathbb{O}$ is a complex ordered pair of quaternions $q_0, q_1 \in \mathbb{H}$.

$$o = \left(q_0, q_1\right) = \left(\left(z_0, z_1\right), \left(z_2, z_3\right)\right) \tag{2.14}$$

that is, the quadruple of complex numbers z_i, $i = 0,1,2,3$, $z_0 = r_0 + r_1 \cdot e_1$, $z_1 = r_2 + r_3 \cdot e_1$, $z_2 = r_4 + r_5 \cdot e_1$, $z_3 = r_6 + r_7 \cdot e_1$ in the form

$$o = q_0 + q_1 \cdot e_4 = \left(z_0 + z_1 \cdot e_2\right) + \left(z_2 + z_3 \cdot e_2\right) \cdot e_4 \tag{2.15}$$

or, equivalently, the 8-tuple of real numbers r_i, $i = 0,1,\ldots,7$:

$$o = \left(r_0 + r_1 \cdot e_1 + r_2 \cdot e_2 + r_3 \cdot e_3\right) + \left(r_4 + r_5 \cdot e_1 + r_6 \cdot e_2 + r_7 \cdot e_3\right) \cdot e_4 \tag{2.16}$$

and finally

$$o = r_0 + \sum_{i=1}^{7} r_i \cdot e_i = r_0 + r_1 \cdot e_1 + r_2 \cdot e_2 + r_3 \cdot e_3 + r_4 \cdot e_4 + r_5 \cdot e_5 + r_6 \cdot e_6 + r_7 \cdot e_7.$$

$$(2.17)$$

It should be noted that in (2.16) we introduced the next imaginary unit e_4, $e_4 = \sqrt{-1}$ and in (2.17) we applied new rules of multiplication of $e_i, i = 0,1,\ldots,7$ in $\mathbb{O}$ presented in Table 2.3: $e_1 \cdot e_4 = e_5$, $e_1 \cdot e_5 = e_6$, $e_1 \cdot e_6 = e_7$. We will come back to these rules and study the basic properties of octonions in Section 2.1.3.1.

It should be noted that Table 2.3 is only one of many possible multiplication tables for the octonions. Evidently, its part for imaginary units from the set $\{1, e_1, e_2, e_3\}$ coincides with Table 2.2. The remaining cells are derived bearing in mind different properties of multiplication in $\mathbb{O}$ described by (2.37)–(2.39).

Of course, the Cayley-Dickson construction can continue to define an infinite set of algebras each of twice the dimension of the previous algebra and each containing all previous algebras as proper subalgebras. However, their properties become worse with each application of the Cayley-Dickson procedure. First, we lose commutativity (quaternions) and associativity (octonions), and finally, we lose the division algebra property (sedenions).

Let us explain what this means for successive Cayley-Dickson algebras. For any two quaternions, $q_1, q_2 \in \mathbb{H}$, we have $q_1 \cdot q_2 \neq q_2 \cdot q_1$ (i.e., the multiplication in $\mathbb{H}$ is noncommutative and the same applies for all subsequent algebras ($\mathbb{O}, \mathbb{S}, \ldots$)). However, the multiplication in $\mathbb{H}$ is still associative because

Table 2.3
Multiplication Rules in $\mathbb{O}$

$\times$	1	e_1	e_2	e_3	e_4	e_5	e_6	e_7
1	1	e_1	e_2	e_3	e_4	e_5	e_6	e_7
e_1	e_1	-1	e_3	$-e_2$	e_5	$-e_4$	$-e_7$	e_6
e_2	e_2	$-e_3$	-1	e_1	e_6	e_7	$-e_4$	$-e_5$
e_3	e_3	e_2	$-e_1$	-1	e_7	$-e_6$	e_5	$-e_4$
e_4	e_4	$-e_5$	$-e_6$	$-e_7$	-1	e_1	e_2	e_3
e_5	e_5	e_4	$-e_7$	e_6	$-e_1$	-1	$-e_3$	e_2
e_6	e_6	e_7	e_4	$-e_5$	$-e_2$	e_3	-1	$-e_1$
e_7	e_7	$-e_6$	e_5	e_4	$-e_3$	$-e_2$	e_1	-1

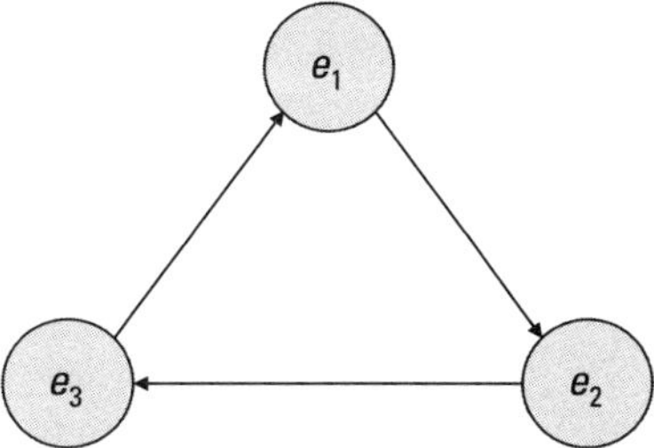

Figure 2.1 Multiplication of imaginary units in $\mathbb{H}$.

for any three quaternions $q_1, q_2, q_3 \in \mathbb{H}$, we have $q_1 \cdot (q_2 \cdot q_3) = (q_1 \cdot q_2) \cdot q_3$. It is no more true in $\mathbb{O}$, where, for any three octonions, $o_1, o_2, o_3 \in \mathbb{O}$: $o_1 \cdot (o_2 \cdot o_3) \neq (o_1 \cdot o_2) \cdot o_3$.

We have already pointed out that only complex numbers, quaternions and octonions are divison algebras. The sixteenth-order algebra of sedenions is not a divison algebra because it contains zero divisors, which means that there exist nonzero sedenions $s_1, s_2 \in \mathbb{S}$, such that $s_1 \cdot s_2 = 0$, for example $(e_1 + e_{13})(e_2 - e_{14}) = 0$. There are many other zero divisors in $\mathbb{S}$ [3].

2.1.2 The Cayley-Dickson Algebra of Quaternions

Let us describe in detail the aforementioned Hamiltonian rules of multiplication in $\mathbb{H}$ presented in Table 2.2. They are clearly visualized on the scheme from Figure 2.1. The first two units, respectively, are a multiplicand and a multiplier, and the third one is their product. If we move clockwise, we get:

$$e_1 \cdot e_2 = e_3, \ e_2 \cdot e_3 = e_1, \ e_3 \cdot e_1 = e_2 \tag{2.18}$$

However, if we multiply the units counterclockwise, their product becomes negative. It means that the multiplication in $\mathbb{H}$ is not commutative and we have:

$$e_2 \cdot e_1 = -e_3, \ e_3 \cdot e_2 = -e_1, \ e_1 \cdot e_3 = -e_2 \tag{2.19}$$

2.1.2.1 Chosen Properties of Quaternions

Reversibility

The quaternion $-q$ *reversible* to $q \in \mathbb{H}$: $q = (z_0, z_1) = r_0 + r_1 \cdot e_1 + r_2 \cdot e_2 + r_3 \cdot e_3$, is defined as

$$-q = -r_0 - r_1 \cdot e_1 - r_2 \cdot e_2 - r_3 \cdot e_3 = \left(-z_0, -z_1\right) \tag{2.20}$$

Conjugation

The *conjugate* of a quaternion $q \in \mathbb{H}$: $q = \left(z_0, z_1\right) = r_0 + r_1 \cdot e_1 + r_2 \cdot e_2 + r_3 \cdot e_3$ is also a quaternion

$$q^* = r_0 - r_1 \cdot e_1 - r_2 \cdot e_2 - r_3 \cdot e_3 = \left(z_0^*, -z_1\right) \tag{2.21}$$

Proof: For $q = \left(z_0, z_1\right)$ and using (2.11), we have: $q^* = \left(z_0, z_1\right)^* = \left(z^*_0, -z_1\right) = (r_0 - r_1 \cdot e_1, -r_2 - r_3 \cdot e_1) = \left(r_0 - r_1 \cdot e_1\right) + \left(-r_2 - r_3 \cdot e_1\right) \cdot e_2$, which is identical to (2.21).

The conjugation is an involution,[1] meaning that the double conjugate of a quaternion q returns the original: $\left(q^*\right)^* = q$.

Norm

The *norm* of a quaternion $q \in \mathbb{H}$: $q = \left(z_0, z_1\right) = r_0 + r_1 \cdot e_1 + r_2 \cdot e_2 + r_3 \cdot e_3$ is

$$\|q\| = \sqrt{q \cdot q^*} = \sqrt{q^* \cdot q} = \sqrt{\|z_0\|^2 + \|z_1\|^2} = \sqrt{r_0^2 + r_1^2 + r_2^2 + r_3^2} \tag{2.22}$$

Proof: The above result is evident because for $q = \left(z_0, z_1\right)$, we have

$$\|q\| = \left\|\left(z_0, z_1\right)\right\| = \sqrt{\left(z_0, z_1\right)^* \left(z_0, z_1\right)} = \sqrt{\left(z_0, z_1\right)\left(z_0, z_1\right)^*} = \sqrt{\|z_0\|^2 + \|z_1\|^2}$$

$$\text{and } \|z_0\|^2 = \left|\left(r_0, r_1\right)\right|^2 = r_0^2 + r_1^2, \; \|z_1\|^2 = \left|\left(r_2, r_3\right)\right|^2 = r_2^2 + r_3^2.$$

The norm (2.22) is always a nonnegative real number expressed as a standard Euclidean norm on $\mathbb{H}$. It is called the *modulus* of a quaternion q and is usually denoted with $|q|$.

The quaternions form a normed division algebra, meaning that the norm of the product of quaternions $q_0 = \left(z_0, z_1\right)$ and $q_1 = \left(z_2, z_3\right)$ equals the product of the norms $\|q_0\|$ and $\|q_1\|$; that is,

$$\|q_0 \cdot q_1\| = \|q_0\| \cdot \|q_1\| \tag{2.23}$$

---

[1] The *involution f* means a function that is its own inverse (e.g., $f(f(x)) = x$ for every x belonging to the domain of f. The involution is linear: $f(\alpha x_1 + \beta x_2) = \alpha f(x_1) + \beta f(x_2)$; $\alpha, \beta \in \mathbb{R}$ and the involution of a product is the product of involutions written in the reversed order: $f(x_1 \cdot x_2) = f(x_2) \cdot f(x_1)$ [4].

Proof: The evidence of (2.23) is immediate when applying the polar form of quaternions q_0 and q_1 given by (2.30). The reader is encouraged to prove it on his or her own.

Multiplicative Inverse

The *multiplicative inverse* of a quaternion $q \in \mathbb{H}$: $q = (z_0, z_1) = r_0 + r_1 \cdot e_1 + r_2 \cdot e_2 + r_3 \cdot e_3$ is

$$q^{-1} = \frac{q^*}{\|q\|^2} \tag{2.24}$$

Proof: If $q = (z_0, z_1)$, $z_0 = (r_0, r_1)$, $z_1 = (r_2, r_3)$, then

$$q^{-1} = (z_0, z_1)^{-1} = \left(\frac{z_0^*}{\|z_0\|^2 + \|z_1\|^2}, \frac{-z_1}{\|z_0\|^2 + \|z_1\|^2} \right)$$

where

$$\frac{z_0^*}{\|z_0\|^2 + \|z_1\|^2} = \left(\frac{r_0}{\|q\|^2}, \frac{-r_1}{\|q\|^2} \right), \quad -\frac{z_1}{\|z_0\|^2 + \|z_1\|^2} = \left(\frac{-r_2}{\|q\|^2}, \frac{-r_3}{\|q\|^2} \right)$$

and finally

$$q^{-1} = \frac{1}{\|q\|^2} (r_0, -r_1, -r_2, -r_3) = \frac{q^*}{\|q\|^2} \text{ , that proves (2.24).}$$

Summation

The sum of quaternions $q_0, q_1 \in \mathbb{H}$: $q_0 = a_0 + a_1 \cdot e_1 + a_2 \cdot e_2 + a_3 \cdot e_3$ and $q_1 = b_0 + b_1 \cdot e_1 + b_2 \cdot e_2 + b_3 \cdot e_3$ is also a quaternion given by

$$q_0 + q_1 = (a_0 + b_0) + (a_1 + b_1) \cdot e_1 + (a_2 + b_2) \cdot e_2 + (a_3 + b_3) \cdot e_3 \tag{2.25}$$

Proof: This result is evident and can be extended for any number of quaternions. According to (2.11), if $q_0 = (z_0, z_1)$ and $q_1 = (z_2, z_3)$, then their sum can be expressed as a complex ordered pair $(z_0 + z_2, z_1 + z_3)$ (see Table 2.1).

Multiplication

The product of quaternions $q_0, q_1 \in \mathbb{H}$: $q_0 = a_0 + a_1 \cdot e_1 + a_2 \cdot e_2 + a_3 \cdot e_3$ and $q_1 = b_0 + b_1 \cdot e_1 + b_2 \cdot e_2 + b_3 \cdot e_3$ gives the quaternion

$$
\begin{aligned}
q_0 \cdot q_1 = &\left(a_0 b_0 - a_1 b_1 - a_2 b_2 - a_3 b_3\right) + \left(a_0 b_1 + a_1 b_0 + a_2 b_3 - a_3 b_2\right) \cdot e_1 \\
&+ \left(a_0 b_2 + a_2 b_0 + a_3 b_1 - a_1 b_3\right) \cdot e_2 + \left(a_0 b_3 + a_3 b_0 + a_1 b_2 - a_2 b_1\right) \cdot e_3.
\end{aligned}
\tag{2.26}
$$

Proof: Let us prove the above result based on operations from Table 2.1. We have $q_0 \cdot q_1 = (z_0, z_1) \cdot (z_2, z_3)$, where $z_0 = (a_0, a_1)$, $z_1 = (a_2, a_3)$, $z_2 = (b_0, b_1)$, $z_3 = (b_2, b_3)$ and all $a_i, b_i \in \mathbb{R}$, $i = 0, 1, 2, 3$. Moreover, we can write

$$
q_0 \cdot q_1 = (z_0, z_1)(z_2, z_3) = \left(z_0 z_2 - z_3^* z_1, z_0 z_3 + z_2^* z_1\right)
\tag{2.27}
$$

Let us now expand the terms of (2.27):

$$
\begin{aligned}
z_0 z_2 - z_3^* z_1 &= (a_0, a_1)(b_0, b_1) - (b_2, b_3)^* (a_2, a_3) = (a_0, a_1)(b_0, b_1) - (b_2, -b_3)(a_2, a_3) \\
&= (a_0 b_0 - b_1 a_1, a_0 b_1 + b_0 a_1) - (b_2 a_2 + a_3 b_3, b_2 a_3 - a_2 b_3) \\
&= (a_0 b_0 - b_1 a_1 - b_2 a_2 - a_3 b_3, a_0 b_1 + b_0 a_1 - b_2 a_3 + a_2 b_3)
\end{aligned}
$$

$$
\begin{aligned}
z_0 z_3 + z_2^* z_1 &= (a_0, a_1)(b_2, b_3) + (b_0, b_1)^* (a_2, a_3) = (a_0, a_1)(b_2, b_3) + (b_0, -b_1)(a_2, a_3) \\
&= (a_0 b_2 - b_3 a_1, a_0 b_3 + b_2 a_1) + (b_0 a_2 + a_3 b_1, b_0 a_3 - a_2 b_1) \\
&= (a_0 b_2 - b_3 a_1 + b_0 a_2 + a_3 b_1, a_0 b_3 + b_2 a_1 + b_0 a_3 - a_2 b_1)
\end{aligned}
$$

Because the multiplication and summation in $\mathbb{R}$ is commutative, we get (2.26).

We should remember that the multiplication of quaternions is not commutative, (i.e., $q_0 \cdot q_1 \neq q_1 \cdot q_0$, $q_0, q_1 \in \mathbb{H}$). However, it is still *associative*. For example, for any three quaternions $q_0, q_1, q_2 \in \mathbb{H}$ we have

$$
(q_0 \cdot q_1) \cdot q_2 = q_0 \cdot (q_1 \cdot q_2)
\tag{2.28}
$$

The above result is a consequence of the commutativity of multiplication in $\mathbb{C}$ (2.9). We can easily prove (2.28) if we represent each quaternion as (2.11) and then use the multiplication property from Table 2.1. We leave this proof to a careful reader.

Moreover, it can be shown that the conjugation of the product of two quaternions $q_0, q_1 \in \mathbb{H}$ is the product of two conjugates written in the reversed order:

$$\left(q_0 \cdot q_1\right)^* = q_1^* q_0^* \tag{2.29}$$

This property can be generalized to more than two terms.

Proof: The verification of (2.29) is immediate. We can write the conjugate of the product (2.27) of quaternions $q_0 = (z_0, z_1)$ and $q_1 = (z_2, z_3)$ in the form

$$\left(q_0 \cdot q_1\right)^* = \left(z_0 z_2 - z_3^* z_1, z_0 z_3 + z_2^* z_1\right)^* = \left(z_0^* z_2^* - z_3 z_1^*, -z_0 z_3 - z_2^* z_1\right)$$

Then

$$q_1^* \cdot q_0^* = \left(z_2, z_3\right)^* \cdot \left(z_0, z_1\right)^* = \left(z_2^*, -z_3\right) \cdot \left(z_0^*, -z_1\right)$$
$$= \left(z_2^* z_0^* - z_1^* z_3, -z_2^* z_1 - z_0 z_3\right)$$

and due to the commutativity of multiplication in $\mathbb{C}$, we finally get (2.29).

2.1.2.2   The Polar Form of a Quaternion

The quaternion q given by (2.13) can equivalently be expressed in its polar form—which was introduced and proved by T. Bülow in [5]—that is

$$q = |q| \exp\left(e_1 \phi\right) \exp\left(e_3 \psi\right) \exp\left(e_2 \theta\right) \tag{2.30}$$

where $|q|$ is the modulus given by (2.22) and $(\phi, \theta, \psi) \in [-\pi, \pi) \times \left[-\frac{\pi}{2}, \frac{\pi}{2}\right) \times \left[-\frac{\pi}{4}, \frac{\pi}{4}\right]$ are Euler angles in 3-D space. This polar form will appear in Chapter 7 concerning the polar representation of quaternion analytic signals.

2.1.3   The Cayley-Dickson Algebra of Octonions

As has already been mentioned in (2.17), an octonion $o \in \mathbb{O}$ is a hypercomplex number defined by eight real numbers $r_i, i = 0, 1, \ldots, 7$ and seven imaginary units $e_i, i = 1, \ldots, 7$ in the form $o = r_0 + \sum_{i=1}^{7} r_i \cdot e_i$. The multiplication of imaginary units in the Cayley-Dickson algebra of octonions can easily be

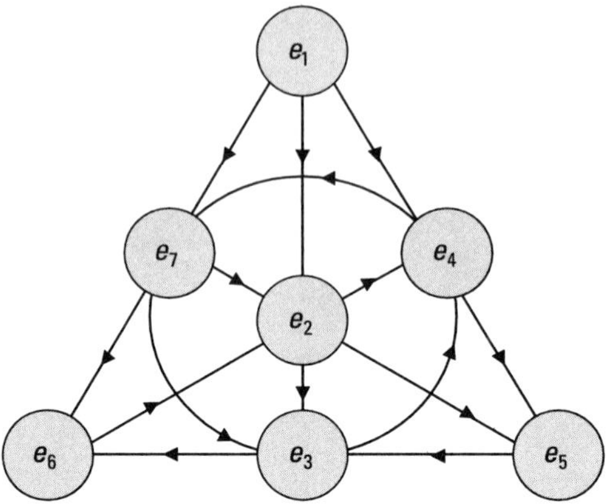

Figure 2.2 Multiplication of imaginary units in $\mathbb{O}$.

performed using the Table 2.3 or a diagram called the *Fano scheme*, shown in Figure 2.2. Similarly (as in Figure 2.1), the product of two imaginary units gets the "+" sign when we move in the direction of arrows. If the direction is opposite, the product is negative; for example: $e_1 \cdot e_4 = e_5$ and $e_4 \cdot e_7 = e_3$ while $e_6 \cdot e_3 = -e_5$ or $e_5 \cdot e_2 = -e_7$.

2.1.3.1 Chosen Properties of Octonions

Reversibility

The *reverse* to the octonion $o \in \mathbb{O}$: $o = (q_0, q_1) = r_0 + \sum_{i=1}^{7} r_i \cdot e_i$ where $q_0 = (z_0, z_1) = ((r_0, r_1), (r_2, r_3))$, $q_1 = (z_2, z_3) = ((r_4, r_5), (r_6, r_7))$ defined as

$$-o = -r_0 - \sum_{i=1}^{7} r_i \cdot e_i = (-q_0, -q_1) \tag{2.31}$$

is also an octonion.

Conjugation

The *conjugate* of an octonion $o \in \mathbb{O}$: $o = (q_0, q_1) = r_0 + \sum_{i=1}^{7} r_i \cdot e_i$ is also an octonion

$$o^* = (q_0, q_1)^* = r_0 - \sum_{i=1}^{7} r_i \cdot e_i$$
$$= r_0 - r_1 \cdot e_1 - r_2 \cdot e_2 - r_3 \cdot e_3 - r_4 \cdot e_4 - r_5 \cdot e_5 - r_6 \cdot e_6 - r_7 \cdot e_7 \tag{2.32}$$

Proof: Basing on (2.21) and general formula of Table 2.1 we can easily show that (2.32) is true.

The conjugation (2.32) is an involution meaning that the double conjugate of a octonion o returns the original: $\left(o^*\right)^* = o$.

Norm

The *norm* of an octonion $o \in \mathbb{O}$: $o = \left(q_0, q_1\right) = r_0 + \sum_{i=1}^{7} r_i \cdot e_i$ is

$$\|o\| = \sqrt{\|q_0\|^2 + \|q_1\|^2} = \sqrt{\sum_{i=0}^{7} r_i^2} = \sqrt{r_0^2 + r_1^2 + r_2^2 + r_3^2 + r_4^2 + r_5^2 + r_6^2 + r_7^2} \quad (2.33)$$

Proof: Please note that for $o = \left(q_0, q_1\right)$, we have

$$\|o\| = \left\|\left(q_0, q_1\right)\right\| = \sqrt{\left(q_0, q_1\right)^* \left(q_0, q_1\right)} = \sqrt{\|q_0\|^2 + \|q_1\|^2}$$

and using (2.22) for $q_0 = \left(z_0, z_1\right) = \left(\left(r_0, r_1\right), \left(r_2, r_3\right)\right)$ and $q_1 = \left(z_2, z_3\right) = \left(\left(r_4, r_5\right), \left(r_6, r_7\right)\right)$, we get (2.33).

The norm (2.32) is a nonnegative real number expressed as a standard Euclidean norm on $\mathbb{O}$. It is called the *modulus* of an octonion o denoted usually with $|o|$. It is shown that the norm of the product of octonions $o_0 = \left(q_0, q_1\right)$ and $o_1 = \left(q_2, q_3\right)$ equals the product of norms $\|o_0\|$ and $\|o_1\|$; for example,

$$\|o_0 \cdot o_1\| = \|o_0\| \cdot \|o_1\| \quad (2.34)$$

Proof: The norm of the product of o_0 and o_1 is

$$\|o_0 \cdot o_1\| = \left\|\left(q_0, q_1\right) \cdot \left(q_2, q_3\right)\right\| = \left\|\left(q_0 q_2 - q_3^* q_1, q_0 q_3 + q_2^* q_1\right)\right\|$$

$$= \sqrt{\left(q_0 q_2 - q_3^* q_1, q_0 q_3 + q_2^* q_1\right)^* \cdot \left(q_0 q_2 - q_3^* q_1, q_0 q_3 + q_2^* q_1\right)}$$

$$= \sqrt{\left(\left(q_0 q_2 - q_3^* q_1\right)^*, -q_0 q_3 - q_2^* q_1\right) \cdot \left(q_0 q_2 - q_3^* q_1, q_0 q_3 + q_2^* q_1\right)}$$

To simplify the notation, let us substitute in the above equation $x_0 = q_0 q_2 - q^*_3 q_1$, $x_1 = q_0 q_3 + q^*_2 q_1$. So it gets now the form

$$\|o_0 \cdot o_1\| = \sqrt{\left(x_0^*, -x_1\right)\left(x_0, x_1\right)} = \sqrt{\left(x_0^* x_0 + x_1^* x_1, x_0^* x_1 - x_0^* x_1\right)}$$

$$= \sqrt{\left(x_0^* x_0 + x_1^* x_1, 0\right)} = \sqrt{x_0^* x_0 + x_1^* x_1}$$

and further

$$\sqrt{x_0^* x_0 + x_1^* x_1} = \sqrt{\left(q_0 q_2 - q_3^* q_1\right)^*\left(q_0 q_2 - q_3^* q_1\right) + \left(q_0 q_3 + q_2^* q_1\right)^*\left(q_0 q_3 + q_2^* q_1\right)}$$

$$= \sqrt{\left(q_2^* q_0^* - q_1^* q_3\right)\left(q_0 q_2 - q_3^* q_1\right) + \left(q_3^* q_0^* + q_1^* q_2\right)\left(q_0 q_3 + q_2^* q_1\right)}$$

$$= \sqrt{q_2^* q_0^* q_0 q_2 + q_1^* q_3 q_3^* q_1 + q_3^* q_0^* q_0 q_3 + q_1^* q_2 q_2^* q_1}$$

$$= \sqrt{q_2^* \|q_0\|^2 q_2 + q_1^* \|q_3\|^2 q_1 + q_3^* \|q_0\|^2 q_3 + q_1^* \|q_2\|^2 q_1}$$

$$= \sqrt{\left(\|q_0\|^2 + \|q_1\|^2\right)\left(\|q_2\|^2 + \|q_3\|^2\right)} = \|o_0\| \cdot \|o_1\|$$

where we have applied the properties of quaternions given by (2.22) and (2.29). That finishes the proof of (2.34).

Multiplicative Inverse

The *multiplicative inverse* of an octonion $o \in \mathbb{O}$: $o = (q_0, q_1) = r_0 + \sum_{i=1}^{7} r_i \cdot e_i$ is

$$o^{-1} = \frac{o^*}{\|o\|^2} = \frac{r_0 - r_1 \cdot e_1 - r_2 \cdot e_2 - r_3 \cdot e_3 - r_4 \cdot e_4 - r_5 \cdot e_5 - r_6 \cdot e_6 - r_7 \cdot e_7}{r_0^2 + r_1^2 + r_2^2 + r_3^2 + r_4^2 + r_5^2 + r_6^2 + r_7^2} \tag{2.35}$$

Proof: If $o = (q_0, q_1)$, $q_0 = (z_0, z_1)$, $q_1 = (z_2, z_3)$, then

$$o^{-1} = \left(q_0, q_1\right)^{-1} = \left(\frac{q_0^*}{\|q_0\|^2 + \|q_1\|^2}, \frac{-q_1}{\|q_0\|^2 + \|q_1\|^2}\right) = \frac{\left(q_0, q_1\right)^*}{\|o\|^2}.$$

Summation

The sum of octonions $o_0, o_1 \in \mathbb{O}$: $o_0 = (q_0, q_1) = a_0 + \sum_{i=1}^{7} a_i \cdot e_i$, $o_1 = (q_2, q_3) = b_0 + \sum_{i=1}^{7} b_i \cdot e_i$ given by

$$o_0 + o_1 = \left(a_0 + b_0\right) + \sum_{i=1}^{7}\left(a_i + b_i\right) \cdot e_i \tag{2.36}$$

is also an octonion.

Proof: According to (2.14) the sum of $o_0 = (q_0, q_1)$ and $o_1 = (q_2, q_3)$ can be expressed as a complex ordered pair $(q_0 + q_2, q_1 + q_3)$ (see Table 2.1) and then, using (2.25), we come directly to (2.36). Of course, this property can be extended for any number of octonions.

Multiplication

The product of octonions $o_0, o_1 \in \mathbb{O}$: $o_0 = (q_0, q_1) = a_0 + \sum_{i=1}^{7} a_i \cdot e_i$, $o_1 = (q_2, q_3)$ $= b_0 + \sum_{i=1}^{7} b_i \cdot e_i$ expressed as

$$o_0 \cdot o_1 = (q_0, q_1) \cdot (q_2, q_3) = \left(q_0 q_2 - q_3^* q_1, q_0 q_3 + q_2^* q_1\right) \qquad (2.37)$$

is also an octonion. Since multiplication in $\mathbb{H}$ is *not commutative*, it is no more commutative in $\mathbb{O}$: $o_0 \cdot o_1 \neq o_1 \cdot o_0$.

Let us now study the subsequent properties of multiplication in the algebra of octonions. Looking at Table 2.3, we notice that, for any triple $(i,j,k) \in \{(1,2,3),(1,4,5),(1,7,6),(2,4,6), (2,5,7),(3,4,7),(3,6,5)\}$, we have: $e_i \cdot (e_j \cdot e_k) = (e_i \cdot e_j) \cdot e_k$ (*associativity*). However if $(i,j,k) \notin \{(1,2,3),(1,4,5),(1,7,6),(2,4,6), (2,5,7),(3,4,7),(3,6,5)\}$, we have $e_i \cdot (e_j \cdot e_k) = -(e_i \cdot e_j) \cdot e_k$ (*nonassociativity*).

Example: We can notice that $e_2 \cdot (e_4 \cdot e_7) = e_2 \cdot e_3 = e_1 = -(e_2 \cdot e_4) \cdot e_7 = -e_6 \cdot e_7$ while $e_3 \cdot (e_6 \cdot e_5) = e_3 \cdot e_3 = -1 = (e_3 \cdot e_6) \cdot e_5$.

In general, we say that multiplication in $\mathbb{O}$ is not associative; such associativity is known as *biassociativity* [6]. However, multiplication in $\mathbb{O}$ is *alternative*, that is, for any two elements $o_0, o_1 \in \mathbb{O}$:

$$o_0 \cdot (o_0 \cdot o_1) = (o_0 \cdot o_0) \cdot o_1, \quad (o_0 \cdot o_1) \cdot o_1 = o_0 \cdot (o_1 \cdot o_1) \qquad (2.38)$$

In particular, for imaginary units $e_i, i = 0, \ldots, 7$ from Table 2.2, we have

$$e_i \cdot (e_i \cdot e_j) = (e_i \cdot e_i) \cdot e_j = -e_j,$$

$$(e_i \cdot e_j) \cdot e_j = e_i \cdot (e_j \cdot e_j) = -e_i. \qquad (2.39)$$

Example: We have $e_2 \cdot (e_2 \cdot e_5) = e_2 \cdot e_7 = -e_5 = (e_2 \cdot e_2) \cdot e_5$ and $(e_2 \cdot e_5) \cdot e_5 = e_7 \cdot e_5 = -e_2 = e_2 \cdot (e_5 \cdot e_5)$

2.2 Selected Clifford Algebras

In this section, we will describe basic definitions and properties of Clifford algebras of biquaternions and bioctonions. Clifford algebras are named after

William Kingdon Clifford, the nineteenth century English mathematician and physicist, who, in 1873, defined the algebra of *biquaternions* [7] called *Clifford biquaternions*. It should be noted that this algebra differs from the algebra of complexified quaternions (original name: biquaternions) defined by Hamilton in [8] as quaternions with complex coefficients; for example, hypercomplex numbers of the form: $z_0 + z_1 \cdot e_1 + z_2 \cdot e_2 + z_3 \cdot e_3$ where $z_i \in \mathbb{C}$. The detailed description of the properties of Clifford algebras is found in [9, 10].

The basis of any Clifford algebra of order 2^n comprises both single imaginary units $1, e_1, e_2, \ldots$ and their products:

$$\left\{ e_{i_1} e_{i_2} \ldots e_{i_k} : 1 \le i_1 \le \ldots \le i_k \le n, \, 0 \le k \le n \right\} \tag{2.40}$$

It is possible to define Clifford algebras satisfying $e_i^2 = 1$ or $e_i^2 = -1$. Let us denote with p (the number of elements of the basis satisfying $e_i^2 = 1$) and with q (the number of elements with $e_i^2 = -1$ and $p + q = n$). Any Clifford algebra is usually denoted with $Cl_{p,q}(\mathbb{R})$. Therefore, the exemplary Clifford algebras are:

- $Cl_{0,1}(\mathbb{R})$—the second-order algebra of complex numbers $(Cl_{0,1}(\mathbb{R}) \equiv \mathbb{C})$
- $Cl_{1,0}(\mathbb{R})$—the second-order algebra of double numbers
- $Cl_{0,2}(\mathbb{R})$—the fourth-order algebra of quaternions $(Cl_{0,2}(\mathbb{R}) \equiv \mathbb{H})$
- $Cl_{1,1}(\mathbb{R})$—the fourth-order algebra of coquaternions
- $Cl_{0,3}(\mathbb{R})$—the eighth-order algebra of Clifford biquaternions
- $Cl_{0,4}(\mathbb{R})$—the sixteenth-order algebra of Clifford bioctonions, etc.

Let us mention that octonions described in Section 2.1.3 are not of the Clifford algebra because their basis $\{1, e_1, \ldots, e_7\}$ does not contain products of imaginary units (2.40).

Let us describe general properties of Clifford algebras. It should be noted that they are *noncommutative*:

$$e_i \cdot e_j = -e_j \cdot e_i \text{ for } i \neq j \tag{2.41}$$

but are *associative*

$$e_i \cdot \left(e_j \cdot e_k \right) = \left(e_i \cdot e_j \right) \cdot e_k \tag{2.42}$$

2.2.1 The Clifford Algebra of Biquaternions

Basic properties of Clifford biquaternions are described in detail in [10]. The basis of algebra of Clifford biquaternions $Cl_{0,3}(\mathbb{R})$ is:

$$\left\{1, e_1, e_2, e_1 e_2, e_3, e_1 e_3, e_2 e_3, e_1 e_2 e_3 = \omega_q\right\} \tag{2.43}$$

In the following sections, we will use the simplified notation: $e_i e_j = e_{ij}$ for $i \neq j$, $i < j < 3$ and $e_1 e_2 e_3 = e_{123} = \omega_q$ called the *pseudoscalar*. It differs from other basis elements with the sign of its square, since $e_{123}^2 = \omega_q^2 = +1$. Other basis elements (2.43) satisfy:

$$\begin{aligned} e_i^2 &= -1 \ \text{ for } \ i = 1,2,3 \\ e_{ij}^2 &= -1 \ \text{ for } \ i,j = 1,2,3 \text{ and } i < j \end{aligned} \tag{2.44}$$

Using the basis (2.43), we define a *Clifford biquaternion* $q_{Cl} \in Cl_{0,3}(\mathbb{R})$ in a general form:

$$q_{Cl} = r_0 + r_1 \cdot e_1 + r_2 \cdot e_2 + r_3 \cdot e_{12} + r_4 \cdot e_3 + r_5 \cdot e_{13} + r_6 \cdot e_{23} + r_7 \cdot \omega_q \tag{2.45}$$

where all $r_i \in \mathbb{R}$, $i = 0,1,\ldots,7$. Its conjugate is

$$q_{Cl}^* = r_0 - r_1 \cdot e_1 - r_2 \cdot e_2 - r_3 \cdot e_{12} - r_4 \cdot e_3 - r_5 \cdot e_{13} - r_6 \cdot e_{23} - r_7 \cdot \omega_q \tag{2.46}$$

and the *norm* is given by

$$\left\| q_{Cl} \right\| = \sqrt{q_{Cl} \cdot q_{Cl}^*} = \sqrt{r_0^2 + r_1^2 + r_2^2 + r_3^2 + r_4^2 + r_5^2 + r_6^2 - r_7^2} \tag{2.47}$$

Please note the minus sign appearing in (2.47) as a consequence of $\omega_q^2 = +1$. It means that the norm of a Clifford biquaternion is not the standard Euclidean norm expressed as a square root of a *sum* of squares of vector elements. That is why we use the name *seminorm*, adapted from the standard mathematical terminology.

2.2.1.1 Chosen Properties of $Cl_{0,3}(\mathbb{R})$

Let us study some chosen properties of Clifford algebra of biquaternions. Firstly from (2.41), we know that the algebra $Cl_{0,3}(\mathbb{R})$ is in general noncommutative; for example,

$$e_{ij} = -e_{ji} \ \text{ for } i \neq j \tag{2.48}$$

$$e_{ij} \cdot e_{ik} = -e_{ik} \cdot e_{ij} \ \text{ for } i \neq j \neq k, \ i < j, \ i < k \tag{2.49}$$

$$e_{ij} \cdot e_{jk} = -e_{jk} \cdot e_{ij} \ \text{ for } i \neq j \neq k, \ i < j, \ j < k \tag{2.50}$$

$$e_{ij} \cdot e_{kj} = -e_{kj} \cdot e_{ij} \quad \text{for } i \neq j \neq k,\, i < j,\, k < j \tag{2.51}$$

Proof: Keeping in mind the property (2.42) we can write:

$$e_{ij} \cdot e_{ik} = \left(e_{ij} \cdot e_i\right) \cdot e_k = -\left(e_i \cdot e_{ij}\right) \cdot e_k = e_{jk},$$

$$e_{ik} \cdot e_{ij} = \left(e_{ik} \cdot e_i\right) \cdot e_j = -\left(e_i \cdot e_{ik}\right) \cdot e_j = e_{kj} = -e_{jk}$$

which proves (2.49). For (2.50), we have:

$$e_{ij}e_{jk} = \left(e_i \cdot e_j\right)e_{jk} = e_i \cdot \left(e_j \cdot e_{jk}\right) = -e_{ik}$$

$$e_{jk} \cdot e_{ij} = \left(-e_{kj}\right) \cdot e_{ij} = \left(-e_k \cdot e_j\right) \cdot e_{ij} = e_k \cdot \left(e_j \cdot e_{ji}\right) = -e_{ki} = e_{ik}$$

Concerning (2.51), we have

$$e_{ij}e_{kj} = e_{ij}\left(-e_{jk}\right) = -\left(e_{ij}e_j\right)e_k = e_{ik}$$

$$e_{kj}e_{ij} = e_{kj}\left(-e_{ji}\right) = -\left(e_{kj}e_j\right)e_i = e_{ki} = -e_{ik}$$

Example: The rules (2.49) and (2.50) apply to $e_{12} \cdot e_{13}$, $e_{12} \cdot e_{23}$, $e_{13} \cdot e_{23}$. We have

$$e_{12} \cdot e_{13} = \left(e_{12} \cdot e_1\right) \cdot e_3 = -\left(e_1 \cdot e_{12}\right) \cdot e_3 = e_{23}$$

$$e_{13} \cdot e_{12} = \left(e_{13} \cdot e_1\right) \cdot e_2 = -\left(e_1 \cdot e_{13}\right) \cdot e_2 = e_{32} = -e_{23}$$

$$e_{12} \cdot e_{23} = \left(e_1 \cdot e_2\right) \cdot e_{23} = e_1 \cdot \left(e_2 \cdot e_{23}\right) = -e_{13}$$

$$e_{23} \cdot e_{12} = e_{23} \cdot \left(e_1 \cdot e_2\right) = -e_{23} \cdot \left(e_2 \cdot e_1\right) = \left(e_{32} \cdot e_2\right) \cdot e_1 = -e_{31} = e_{13}$$

$$e_{13} \cdot e_{23} = e_{13} \cdot \left(e_2 \cdot e_3\right) = e_{13} \cdot \left(-e_3 \cdot e_2\right) = -\left(e_{13} \cdot e_3\right) \cdot e_2 = e_{12}$$

$$e_{23} \cdot e_{13} = e_{23} \cdot \left(e_1 \cdot e_3\right) = e_{23} \cdot \left(-e_3 \cdot e_1\right) = -\left(e_{23} \cdot e_3\right) \cdot e_1 = e_{21} = -e_{12}$$

However, in some cases, the elements of the basis (2.43) commute:

$$e_i \cdot e_{jk} = e_{jk} \cdot e_i \quad \text{for } i \neq j \neq k, j < k \tag{2.52}$$

since $e_i \cdot e_{jk} = e_{ij} \cdot e_k = -e_{ji} \cdot e_k = -e_j \cdot e_{ik} = e_j \cdot e_{ki} = e_{jk} \cdot e_i$. It can be easily proved that it is true for $e_1 \cdot e_{23}$, $e_2 \cdot e_{13}$, and $e_3 \cdot e_{12}$.

Finally, all multiplications of imaginary units by ω_q are commutative; for example,

$$e_i \cdot \omega_q = \omega_q \cdot e_i \quad \text{for } i = 1,2,3$$
$$e_{ij} \cdot \omega_q = \omega_q \cdot e_{ij} \quad \text{for } i,j \in \{1,2,3\}, i \neq j, i < j \tag{2.53}$$

Table 2.4 shows the multiplication rules in $Cl_{0,3}(\mathbb{R})$. The gray-colored cells represent products of imaginary units that commute.

Bearing in mind the multiplication properties in the Clifford algebra of biquaternions, we notice that any biquaternion defined by (2.45) can be expressed as a hypercomplex sum of two quaternions $q_0, q_1 \in \mathbb{H}$; for example,

$$q_{Cl} = q_0 + q_1 \cdot \omega_q \tag{2.54}$$

where $q_0 = r_0 + r_1 \cdot e_1 + r_2 \cdot e_2 + r_4 \cdot e_3$ and $q_1 = r_7 - r_6 \cdot e_1 + r_5 \cdot e_2 - r_3 \cdot e_3$. That is why the name *biquaternion* is justified.

Table 2.4
Multiplication Rules in $Cl_{0,3}(\mathbb{R})$

$\times$	1	e_1	e_2	e_{12}	e_3	e_{13}	e_{23}	ω_q
1	1	e_1	e_2	e_{12}	e_3	e_{13}	e_{23}	ω_q
e_1	e_1	-1	e_{12}	$-e_2$	e_{13}	$-e_3$	ω_q	$-e_{23}$
e_2	e_2	$-e_{12}$	-1	e_1	e_{23}	$-\omega_q$	$-e_3$	e_{13}
e_{12}	$e_1 e_2$	e_2	$-e_1$	-1	ω_q	e_{23}	$-e_{13}$	$-e_3$
e_3	e_3	$-e_{13}$	$-e_{23}$	ω_q	-1	e_1	e_2	$-e_{12}$
e_{13}	e_{13}	e_3	$-\omega_q$	$-e_{23}$	$-e_1$	-1	e_{12}	e_2
e_{23}	e_{23}	ω_q	e_3	e_{13}	$-e_2$	$-e_{12}$	-1	$-e_1$
ω	ω_q	$-e_{23}$	e_{13}	$-e_3$	$-e_{12}$	e_2	$-e_1$	1

2.2.2 The Clifford Algebra of Bioctonions

Let us present some basic definitions and properties of the Clifford algebra of bioctonions, $Cl_{0,4}(\mathbb{R})$. Its basis (2.40) includes sixteen elements:

$$\left\{ 1, e_1, e_2, e_{12}, e_3, e_{13}, e_{23}, e_{123}, e_4, e_{14}, e_{24}, e_{34}, e_{124}, e_{134}, e_{234}, e_{1234} = \omega_o \right\} \tag{2.55}$$

with the pseudoscalar ω_o satisfying $(e_1 e_2 e_3 e_4)^2 = e_{1234}^2 = \omega_o^2 = -1$. For other basis elements we have

$$\begin{aligned}
e_i^2 &= -1 \quad \text{for } i = 1,2,3,4 \\
e_{ij}^2 &= -1 \quad \text{for } i,j = 1,2,3,4 \text{ and } i < j \\
e_{ijk}^2 &= +1 \quad \text{for } i,j,k = 1,2,3,4 \text{ and } i < j < k
\end{aligned} \tag{2.56}$$

In Table 2.5, we show all multiplication rules in the algebra of Clifford bioctonions. Using the basis (2.55), we define a *Clifford bioctonion* $o_{Cl} \in Cl_{0,4}(\mathbb{R})$ in a general form:

$$\begin{aligned}
o_{Cl} = {}& r_0 + r_1 \cdot e_1 + r_2 \cdot e_2 + r_3 \cdot e_{12} + r_4 \cdot e_3 + r_5 \cdot e_{13} + r_6 \cdot e_{23} + r_7 \cdot e_{123} + r_8 \cdot e_4 \\
& + r_9 \cdot e_{14} + r_{10} \cdot e_{24} + r_{11} \cdot e_{124} + r_{12} \cdot e_{34} + r_{13} \cdot e_{134} + r_{14} \cdot e_{234} + r_{15} \cdot \omega_o
\end{aligned} \tag{2.57}$$

where all $r_i \in \mathbb{R}$, $i = 0, 1, \ldots, 15$. Its conjugate is

$$\begin{aligned}
o_{Cl}^* = {}& r_0 - r_1 \cdot e_1 - r_2 \cdot e_2 - r_3 \cdot e_{12} - r_4 \cdot e_3 - r_5 \cdot e_{13} - r_6 \cdot e_{23} - r_7 \cdot e_{123} - r_8 \cdot e_4 \\
& - r_9 \cdot e_{14} - r_{10} \cdot e_{24} - r_{11} \cdot e_{124} - r_{12} \cdot e_{34} - r_{13} \cdot e_{134} - r_{14} \cdot e_{234} - r_{15} \cdot \omega_o
\end{aligned} \tag{2.58}$$

and the *norm* is expressed as

$$\begin{aligned}
\left\| o_{Cl} \right\| &= \sqrt{o_{Cl} \cdot o_{Cl}^*} \\
&= \sqrt{r_0^2 + r_1^2 + r_2^2 + r_3^2 + r_4^2 + r_5^2 + r_6^2 - r_7^2 + r_8^2 + r_9^2 + r_{10}^2 - r_{11}^2 + r_{12}^2 - r_{13}^2 - r_{14}^2 + r_{15}^2}
\end{aligned} \tag{2.59}$$

Let us notice four minus signs appearing in (2.59) as a consequence of the rules $e_{ijk}^2 = +1$ for $i,j,k \in \{1,2,3,4\}$ and $i < j < k$.

The Clifford bioctonion (2.57) can be expressed as a hypercomplex sum of two biquaternions $q_{Cl}^0, q_{Cl}^1 \in Cl_{0,3}(\mathbb{R})$ given by (2.54) as follows

Table 2.5
Multiplication Rules in the Algebra of Clifford Bioctonions

$\times$	1	e_1	e_2	e_{12}	e_3	e_{13}	e_{23}	e_{123}	e_4	e_{14}	e_{24}	e_{124}	e_{34}	e_{134}	e_{234}	e_{1234}
1	1	e_1	e_2	e_{12}	e_3	e_{13}	e_{23}	e_{123}	e_4	e_{14}	e_{24}	e_{124}	e_{34}	e_{134}	e_{234}	ω
e_1	e_1	-1	e_{12}	$-e_2$	e_{13}	$-e_3$	e_{123}	$-e_{23}$	e_{14}	$-e_4$	e_{124}	$-e_{24}$	e_{134}	$-e_{34}$	ω	$-e_{234}$
e_2	e_2	$-e_{12}$	-1	e_1	e_{23}	$-e_{123}$	$-e_3$	e_{13}	e_{24}	$-e_{124}$	$-e_4$	e_{14}	e_{234}	$-\omega$	$-e_{34}$	e_{134}
e_{12}	e_{12}	e_2	$-e_1$	-1	e_{123}	e_{23}	$-e_{13}$	$-e_3$	e_{124}	e_{24}	$-e_{14}$	$-e_4$	ω	$-e_{234}$	$-e_{134}$	$-e_{34}$
e_3	e_3	$-e_{13}$	$-e_{23}$	e_{123}	-1	e_1	e_2	$-e_{12}$	e_{34}	$-e_{134}$	$-e_{234}$	ω	$-e_4$	e_{14}	e_{24}	e_{124}
e_{13}	e_{13}	e_3	$-e_{123}$	$-e_{23}$	$-e_1$	-1	e_{12}	e_2	e_{134}	e_{34}	$-\omega$	$-e_{234}$	$-e_{14}$	$-e_4$	e_{124}	$-e_{24}$
e_{23}	e_{23}	e_{123}	e_3	e_{13}	$-e_2$	$-e_{12}$	-1	$-e_1$	e_{234}	$-\omega$	e_{34}	e_{134}	$-e_{24}$	$-e_{124}$	$-e_4$	e_{14}
e_{123}	e_{123}	$-e_{23}$	e_{13}	$-e_3$	$-e_{12}$	e_2	$-e_1$	1	ω	$-e_{234}$	e_{134}	$-e_{34}$	$-e_{124}$	e_{24}	$-e_{14}$	e_4
e_4	e_4	$-e_{14}$	$-e_{24}$	e_{124}	$-e_{34}$	e_{134}	e_{234}	ω	-1	e_1	e_2	$-e_{12}$	e_3	$-e_{13}$	$-e_{23}$	$-e_{123}$
e_{14}	e_{14}	e_4	$-e_{124}$	$-e_{24}$	$-e_{134}$	$-e_{34}$	$-\omega$	e_{234}	$-e_1$	-1	e_{12}	e_2	e_{13}	$-e_3$	$-e_{123}$	e_{23}
e_{24}	e_{24}	e_{124}	e_4	e_{14}	$-e_{234}$	$-\omega$	$-e_{34}$	$-e_{134}$	$-e_2$	$-e_{12}$	-1	$-e_1$	e_{23}	e_{123}	e_3	$-e_{13}$
e_{124}	e_{124}	$-e_{24}$	e_{14}	$-e_4$	$-\omega$	e_{234}	$-e_{134}$	e_{34}	$-e_{12}$	e_2	$-e_1$	1	e_{123}	$-e_{23}$	e_{13}	e_3
e_{34}	e_{34}	e_{134}	e_{234}	ω	e_4	e_{14}	e_{24}	e_{124}	$-e_3$	$-e_{13}$	$-e_{23}$	$-e_{123}$	-1	$-e_1$	$-e_2$	$-e_{12}$
e_{134}	e_{134}	$-e_{34}$	ω	$-e_{234}$	e_{14}	$-e_4$	e_{124}	$-e_{24}$	$-e_{13}$	e_3	$-e_{123}$	e_{23}	$-e_1$	1	$-e_{12}$	e_2
e_{234}	e_{234}	$-\omega$	$-e_{34}$	e_{134}	e_{24}	$-e_{124}$	$-e_4$	e_{14}	$-e_{23}$	$-e_{123}$	e_3	$-e_{13}$	$-e_2$	e_{12}	1	$-e_1$
e_{1234}	ω	e_{234}	$-e_{134}$	$-e_{34}$	e_{124}	e_{24}	$-e_{14}$	e_4	$-e_{123}$	$-e_{23}$	e_{13}	e_3	$-e_{12}$	$-e_2$	e_1	-1

$$o_{Cl} = q^0_{Cl} + q^1_{Cl} \cdot \omega_o = \left(q_0 + q_1 \cdot \omega_q \right) + \left(q_2 + q_3 \cdot \omega_q \right) \cdot \omega_o \qquad (2.60)$$

where $q^0_{Cl} = q_0 + q_1 \cdot \omega_q$, $q^1_{Cl} = q_2 + q_3 \cdot \omega_q$ and $q_0 = r_0 + r_1 \cdot e_1 + r_2 \cdot e_2 + r_4 \cdot e_3$

$$q_1 = r_7 - r_6 \cdot e_1 + r_5 \cdot e_2 - r_3 \cdot e_3$$
$$q_2 = r_{15} - r_{14} \cdot e_1 + r_{13} \cdot e_2 + r_{11} \cdot e_3$$
$$q_3 = r_8 + r_9 \cdot e_1 + r_{10} \cdot e_2 + r_{12} \cdot e_3$$

2.2.2.1 Chosen Properties of $Cl_{0,4}(\mathbb{R})$

In this section, we present the properties of multiplication of elements of the basis of the Clifford algebra of bioctonions. All relations are shown without any proof. A reader interested in this subject can follow the derivation presented in [11].

Relations of Commutation in $Cl_{0,4}(\mathbb{R})$

$$e_i \cdot e_{jk} = e_{jk} \cdot e_i \ \text{ for } \ i \neq j \neq k, j < k \qquad (2.61)$$

$$e_{ij} \cdot e_{kl} = e_{kl} \cdot e_{ij} \ \text{ for } i \neq j \neq k \neq l, i < j, k < l \qquad (2.62)$$

$$e_i \cdot e_{ijk} = e_{ijk} \cdot e_i \ \text{ for } i \neq j \neq k, i < j < k \qquad (2.63)$$

$$e_j \cdot e_{ijk} = e_{ijk} \cdot e_j \ \text{ for } i \neq j \neq k, i < j < k \qquad (2.64)$$

$$e_k \cdot e_{ijk} = e_{ijk} \cdot e_k \ \text{ for } i \neq j \neq k, i < j < k \qquad (2.65)$$

$$e_{ij} \cdot e_{ijk} = e_{ijk} \cdot e_{ij} \ \text{ for } i \neq j \neq k, i < j < k \qquad (2.66)$$

$$e_{jk} \cdot e_{ijk} = e_{ijk} \cdot e_{jk} \ \text{ for } i \neq j \neq k, i < j < k \qquad (2.67)$$

Relations of Noncommutation in $Cl_{0,4}(\mathbb{R})$

$$e_i \cdot e_{ij} = -e_{ij} \cdot e_i \ \text{ for } i \neq j, i < j \qquad (2.68)$$

$$e_i \cdot e_{jkl} = -e_{jkl} \cdot e_i \ \text{ for } i \neq j \neq k \neq l, j < k < l \qquad (2.69)$$

Let us remark that products of elements e_{ijk}, $i \neq j \neq k$, $i < j < k$ *do not commute*. We do not present any evidence here because, by using all of the above relations, it is easy to develop forms of mutual products of $e_{123}, e_{124}, e_{134}, e_{234}$.

Further, it should be noted that—in some cases—multiplication of elements of the basis (2.55) by ω_o is commutative. Sometimes it is not. Let us study the simplest case of $\omega_o \cdot e_i$, $i = 1,2,3,4$. We have

$$\omega_o \cdot e_i = \begin{cases} -e_i \cdot \omega_o, & i = 1,2 \\ e_i \cdot \omega_o, & i = 3,4 \end{cases} \tag{2.70}$$

The consequences of (2.70) are changes in signs in multiplications $\omega_o \cdot e_{ij}$, $i,j \in \{1,2,3,4\}$, $i \neq j$, $i < j$ and $\omega_o \cdot e_{ijk}$, $i,j,k \in \{1,2,3,4\}$, $i \neq j \neq k$, $i < j < k$. We observe that only $\omega_o \cdot e_{12}$, and $\omega_o \cdot e_{34}$, *commute*. Lastly, the multiplication of ω_o by e_{ijk}, $i \neq j \neq k$, $i < j < k$ satisfies the following:

$$\omega_o \cdot e_{123} = e_{123} \cdot \omega_o, \qquad \omega_o \cdot e_{124} = e_{124} \cdot \omega_o$$
$$\omega_o \cdot e_{134} = -e_{134} \cdot \omega_o, \qquad \omega_o \cdot e_{234} = -e_{234} \cdot \omega_o \tag{2.71}$$

2.3 Comparison of Algebras

In Sections 2.1 and 2.2, we briefly presented some algebras used for the purpose of the theory of complex and hypercomplex analytic signals. These are the Cayley-Dickson algebras of complex numbers, quaternions, and octonions, and the Clifford algebras of biquaternions and bioctonions. Of course, the family of different hypercomplex algebras is much wider, as shown in Figure 2.3. The symbol $\subset$ means "is included." It should be noted that the properties of different algebras change when we move from bottom to top.

2.4 Applications of Hypercomplex Algebras in Signal Processing

Nowadays, it is observed that hypercomplex algebras find more and more applications in various domains. The most commonly used are quaternions, which is due to their structure and nice properties. The definition of a quaternion given by (2.13) shows that it is suitable to apply in color image processing. Sangwine et al. were the first to interpret its vector part in terms of three components: R (red), G (green), and B (blue) of a color image [12–15]. This allowed for the development of some advanced methods of color image

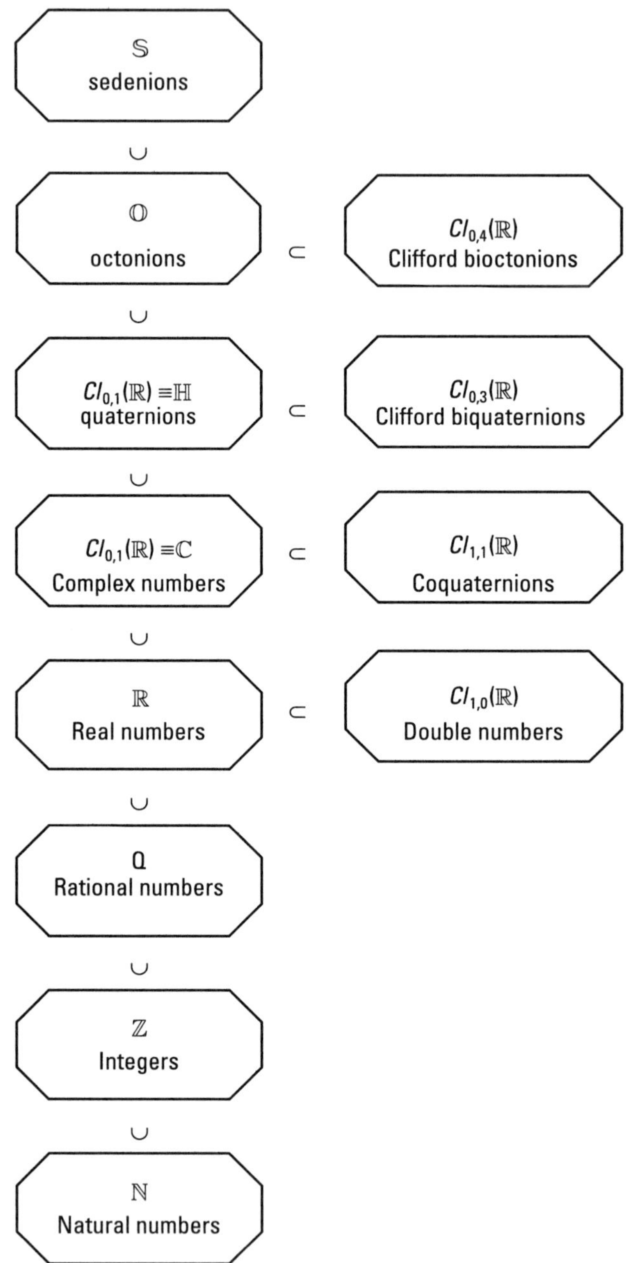

Figure 2.3　Relations between different algebras.

analysis, decomposition and compression [16], pattern recognition [17], and edge detection [12–14, 17–19], as well as to elaborate novel hypercomplex correlation techniques [20], filtering methods [14, 21, 22], and watermarking procedures [23–26].

The Fourier transformation methods based on quaternion algebra have also been proposed [12, 14, 27–31]. The Quaternion Fourier transformation will be described in Chapter 4. It should be noted that in the last few years, quaternions have also proliferated into the domain of wavelet transformation [32–36]. Quaternions also provide a very convenient notation for representing the rotations of a rigid body in 3-D space. This is due to the simple representation of rotations and a compact form of dynamics equations. Consequently, they are effectively applied in computer graphics [37–39], computer vision and robotics [40, 41], astrophysics [42], theoretical physics [43–45], and even in geodesy [46].

Concerning the octonions, we noted in Sections 2.1.3 and 2.3 that they have a more complicated form and worse properties than quaternions. Perhaps that is why, until recently, their practical applications were limited to mathematical physics, especially to fields such as electrodynamics, electromagnetics, mechanics, and gravity theory [47, 48], as well to quantum physics [49]. Finally, the sedenions are rarely applied in theoretical physics [48] and quantum physics [51, 52]. We see that the hypercomplex Cayley-Dickson algebras of the order 2^N, $N \geq 3$, are still waiting for new applications in various fields.

The construction and properties of Clifford algebras made them convenient to apply in various fields [53–56]. Their potential has been especially appreciated by researchers working in mathematical physics [39, 54, 55], as well as in robotics and computer vision [56, 59], and also in neural networks [60]. Recently, biquaternions and their different forms (reduced, commutative reduced) have also found applications in digital signal processing, especially in the design of hypercomplex filter banks [61, 62]. In color image processing, they have been used in template matching and color-sensitive edge detection [63], as well as in image recognition [64]. In multidimensional signal processing, the Clifford algebras have found applications in frequency analysis. The Clifford Fourier transform, presented in Chapter 4, is commonly used in 3-D signal processing.

2.4.1 Complex and Hypercomplex Algebras in Multidimensional Signal Theory

In this book, we are especially interested in applications of hypercomplex algebras in signal processing. Complex and hypercomplex signals are n-dimensional

(n-D) generalizations of corresponding complex and hypercomplex numbers belonging to Cayley-Dickson and Clifford algebras.

The n-D *complex-valued* (abbreviated as *complex*) signal ψ_c defined over the n-dimensional space $\boldsymbol{x} = (x_1, x_2, \ldots, x_n)$, $\boldsymbol{x} \in \mathbb{R}^n$, is given by

$$\psi_c(\boldsymbol{x}) = \operatorname{Re}\psi_c(\boldsymbol{x}) + \operatorname{Im}\psi_c(\boldsymbol{x}) \cdot e_1 \tag{2.72}$$

where $\operatorname{Re}\psi_c(\boldsymbol{x})$ is its *real part* and $\operatorname{Im}\psi_c(\boldsymbol{x})$ is its *imaginary* part. The values of ψ_c are complex numbers (described in Section 2.1 concerning Cayley-Dickson algebras). The conjugate of (2.72) is

$$\psi_c^*(\boldsymbol{x}) = \operatorname{Re}\psi_c(\boldsymbol{x}) - \operatorname{Im}\psi_c(\boldsymbol{x}) \cdot e_1 \tag{2.73}$$

and its norm

$$\left\|\psi_c(\boldsymbol{x})\right\| = \sqrt{\operatorname{Re}^2\psi_c(\boldsymbol{x}) + \operatorname{Im}^2\psi_c(\boldsymbol{x})} \tag{2.74}$$

It is evident that all operations on complex signals are similar to operations on complex numbers. In Chapter 5, a special class of complex signals—the n-D *complex signals with single-orthant spectra*—will be described.

The general form of an n-D *hypercomplex* (hypercomplex-valued) signal defined over the n-dimensional space $\boldsymbol{x} = (x_1, x_2, \ldots, x_n)$, $\boldsymbol{x} \in \mathbb{R}^n$ is

$$\psi(\boldsymbol{x}) = r_0(\boldsymbol{x}) + \sum_{i=1}^{m} r_i(\boldsymbol{x}) \cdot e_i \tag{2.75}$$

with $r_0(\boldsymbol{x})$ as its real component and m imaginary components, all being n-D functions, m depends on the order 2^N of a given Cayley-Dickson algebra $(m = 2^N - 1)$. For $m = 3$, we get the general form of an n-D *quaternion signal* (four components):

$$\psi_q(\boldsymbol{x}) = r_0(\boldsymbol{x}) + r_1(\boldsymbol{x}) \cdot e_1 + r_2(\boldsymbol{x}) \cdot e_2 + r_3(\boldsymbol{x}) \cdot e_3 \tag{2.76}$$

For $m = 7$, the n-D *octonion signal* (eight components) is given by

$$\psi_o(\boldsymbol{x}) = r_0(\boldsymbol{x}) + \sum_{i=1}^{7} r_i(\boldsymbol{x}) \cdot e_i \tag{2.77}$$

Of course, in this way we can get next higher-order n-D hypercomplex signals, but it is out of the scope of this book. A careful reader will notice that the dimension of the signal space should not be mixed with the order of the algebra (equal $2^2 = 4$ for quaternions and $2^3 = 8$ for octonions). For example, if we deal with 2-D RGB images represented by a quaternion, the signal space has the order 2, while the order of the quaternion algebra is 4. In Chapter 5 we will discuss quaternion (of order 4) analytic signals defined in 2-D, and in Chapter 6 we will introduce *rank-2 signals* as 3-D signals of a quaternion (order 4) structure.

In analogy to quaternions (2.11), any n-D quaternion (quaternion-valued) signal (2.76) is a complex sum of two complex signals defined by (2.72)

$$\psi_q(x) = \psi_{c_0}(x) + \psi_{c_1}(x) \cdot e_2 \tag{2.78}$$

The same applies to n-D octonion (octonion-valued) signals represented by two n-D quaternion signals (and in consequence by four n-D complex signals) in the form

$$\psi_o(x) = \psi_{q_0}(x) + \psi_{q_1}(x) \cdot e_4 \tag{2.79}$$

The norm of the n-D Cayley-Dickson hypercomplex signal is defined analogously to the norm of a corresponding hypercomplex number (Table 2.1) and equals

$$\|\psi(x)\| = \sqrt{\sum_{i=0}^{m} r_i^2(x)} \tag{2.80}$$

Note that the square of the norm (2.80) defines the so called *local amplitude* of the n-D signal (see Chapter 7).

Let us now have a look at the definitions of Clifford biquaternion and bioctonion signals. They derive directly from (2.45) and (2.57). We define the n-D *Clifford biquaternion signal* as

$$\begin{aligned}
q_{Cl}(x) = r_0(x) + r_1(x) \cdot e_1 + r_2(x) \cdot e_2 + r_3(x) \cdot e_{12} + r_4(x) \cdot e_3 \\
+ r_5(x) \cdot e_{13} + r_6(x) \cdot e_{23} + r_7(x) \cdot \omega_q
\end{aligned} \tag{2.81}$$

Its conjugate is

$$\begin{aligned}
q_{Cl}^*(x) = r_0(x) - r_1(x) \cdot e_1 - r_2(x) \cdot e_2 - r_3(x) \cdot e_{12} \\
- r_4(x) \cdot e_3 - r_5(x) \cdot e_{13} - r_6(x) \cdot e_{23} - r_7(x) \cdot \omega_q
\end{aligned} \tag{2.82}$$

Let us note that the *norm* of the signal (2.81) is not the standard Euclidean norm due to the minus sign appearing as a consequence of multiplication rules in $Cl_{0,3}(\mathbb{R})$. We have

$$\left\| q_{Cl}(x) \right\| = \sqrt{q_{Cl}(x) \cdot q_{Cl}^*(x)}$$
$$= \sqrt{r_0^2(x) + r_1^2(x) + r_2^2(x) + r_3^2(x) + r_4^2(x) + r_5^2(x) + r_6^2(x) - r_7^2(x)}$$

$$(2.83)$$

sometimes called a *seminorm* of (2.81).

The *n*-D *Clifford bioctonion signal* is a generalized Clifford bioctonion number given by (2.56)

$$o_{Cl}(x) = r_0(x) + r_1(x) \cdot e_1 + r_2(x) \cdot e_2 + r_3(x) \cdot e_{12} + r_4(x) \cdot e_3 + r_5(x) \cdot e_{13}$$
$$+ r_6(x) \cdot e_{23} + r_7(x) \cdot e_{123} + r_8(x) \cdot e_4 + r_9(x) \cdot e_{14} + r_{10}(x) \cdot e_{24}$$
$$+ r_{11}(x) \cdot e_{124} + r_{12}(x) \cdot e_{34} + r_{13}(x) \cdot e_{134} + r_{14}(x) \cdot e_{234} + r_{15}(x) \cdot \omega_o$$

$$(2.84)$$

The conjugate of (2.84) is

$$o_{Cl}^*(x) = r_0(x) - r_1(x) \cdot e_1 - r_2(x) \cdot e_2 - r_3(x) \cdot e_{12} - r_4(x) \cdot e_3 - r_5(x) \cdot e_{13}$$
$$- r_6(x) \cdot e_{23} - r_7(x) \cdot e_{123} - r_8(x) \cdot e_4 - r_9(x) \cdot e_{14} - r_{10}(x) \cdot e_{24}$$
$$- r_{11}(x) \cdot e_{124} - r_{12}(x) \cdot e_{34} - r_{13}(x) \cdot e_{134} - r_{14}(x) \cdot e_{234} - r_{15}(x) \cdot \omega_o$$

$$(2.85)$$

and in its *norm* (seminorm), we once again notice minus signs as in (2.58) as follows

$$\left\| o_{Cl}(x) \right\| = \sqrt{o_{Cl}(x) \cdot o_{Cl}^*(x)}$$
$$= \sqrt{r_0^2(x) + \ldots + r_6^2(x) - r_7^2(x) + \ldots + r_{10}^2(x) - r_{11}^2(x) + r_{12}^2(x) - r_{13}^2(x) - r_{14}^2(x) + r_{15}^2(x)}$$

$$(2.86)$$

2.5 Summary

In this chapter, we presented mathematical bases of the theory of chosen algebras of complex and hypercomplex numbers. We put our attention on Cayley-Dickson algebras of complex numbers, quaternions, and octonions, and compared their definitions and properties with Clifford biquaternions and bioctonions.

In Section 2.1.1, we described the Cayley-Dickson construction, a very elegant way of defining hypercomplex algebras of order 2^N known as Cayley-Dickson algebras. In Sections 2.1.2 and 2.1.3, we studied the properties of quaternions and octonions enriched with proofs of some results. We noticed that by increasing the order of the algebra, we lose some properties. For example, the multiplication in $\mathbb{C}$ is commutative, while in higher order Cayley-Dickson algebras is, in general, not commutative. Consequently, the order of multiplication in $\mathbb{H}$ and in $\mathbb{O}$. cannot be changed. Bearing in mind the similarity of definitions, we were interested in selected Clifford algebras known as *Clifford biquaternions* and *Clifford bioctonions*; these were briefly presented in Section 2.2.

The scheme from Figure 2.3 (Section 2.3) clearly showed mutual relations between different algebras, including those that are not presented in this book. Section 2.4 described different fields of the latest practical applications of complex and hypercomplex algebras. In the following chapters, the reader will become familiar with their applications in the domain of multidimensional signal processing, especially in multidimensional analytic signal theory.

References

[1]   Conway, J., and D. Smith, *On Quaternions and Octonions: Their Geometry, Arithmetic, and Symmetry*, Natick, MA: A. K. Peters Ltd., 2003.

[2]   Hamilton, W. R., "On quaternions," *Proc. Royal Irish Academy*, Vol. 3, 1847, pp. 1–16.

[3]   Cawagas, R. E., "On the Structure and Zero Divisors of the Cayley-Dickson Sedenion Algebra," *Discussiones Mathematicae, General Algebra and Applications*, Vol. 24, 2004, pp. 251–265.

[4]   Ell, T. A., and S. J. Sangwine, "Quaternion involutions and anti-involutions," *Computers and Mathematics with Applications*, Vol. 53, 2007, pp. 137–143.

[5]   Bülow, T., "Hypercomplex Spectral Signal Representations for the Processing and Analysis of Images," Ph.D. thesis, University of Kiel, Germany, 1999.

[6]   Saltzmann, H., et al., *Compact Projective Planes: With an Introduction to Octonion Geometry*, Berlin, Germany: Walter de Gruyter & Co., 1995.

[7]   Clifford, W. K., "Preliminary sketch of biquaternions," *Proc. London Math. Soc.*, 1873, pp. 381–396.

[8]   Hamilton, W. R., *Lectures on Quaternions*, Dublin, Ireland: Hodges and Smith, 1853. Available online at Cornell University Library: http://historical.library.cornell.edu /math/.

[9]   Snygg, J., *Clifford Algebra : A Computational Tool for Physicists*, New York: Oxford University Press, 1997.

[10]   Franchini, S., Vassallo, G., and F. Sorbell, "A brief introduction to Clifford algebra," Universita degli studi di Palermo, Dipartimento di Ingegneria Informatica, Technical Report No. 2, February 2010.

[11]   Snopek, K. M., *Studies of complex and Hypercomplex Multidimensional Analytic Signals*, Prace Naukowe Elektronika, Vol. 190, Oficyna Wydawnicza Politechniki Warszawskiej, Warszawa, 2013.

[12]   Sangwine, S. J., "Fourier transforms of colour images using quaternion or hypercomplex numbers," *Electron. Lett.*, Vol. 32, No. 21, Oct. 1996, pp. 1979–1980.

[13]   Sangwine, S. J., "Colour in image processing," *Electronic and Communication Engineering Journal*, Vol. 12, No. 5, 2000, pp. 211–219.

[14]   Sangwine, S. J., and T. A. Ell, "Colour image filters based on hypercomplex convolution," *IEEE Proc. Vision, Image and Signal Processing*, Vol. 49, 2000, pp. 89–93.

[15]   Sangwine, S. J., C. J. Evans, and T. A. Ell, "Colour-sensitive edge detection using hypercomplex filters," *Proc. 10th European Signal Processing Conf. EUSIPCO*, Vol. 1., Tampere, Finland, 2000, pp. 107–110.

[16]   Le Bihan, N., and S. J. Sangwine, "Color Image Decomposition Using Quaternion Singular Value Decomposition," *Intern. Conf. on Visual Information Engineering*, July 7–9, 2003, pp. 113–116.

[17]   Pei, S.-C., J.-J. Ding, and J. Chang, "Color pattern recognition by quaternion correlation," *IEEE Int. Conf. Image Process.*, Thessaloniki, Greece, October 7–10, 2010, pp. 894–897.

[18]   Wang, J., and L. Liu, "Specific Color-pair Edge Detection using Quaternion Convolution," *3rd International Congress on Image and Signal Processing (CISP'2010)*, Yantai, China, October 16–18, 2010, pp. 1138–1141.

[19]   Gao, C., J. Zhou, F. Lang, Q. Pu, and C. Liu C., "Novel Approach to Edge Detection of Color Image Based on Quaternion Fractional Directional Differentiation," *Advances in Automation and Robotics*, Vol. 1, pp. 163–170, 2012.

[20]   Moxey, C. E, S. J. Sangwine, and T. A. Ell, "Hypercomplex Correlation Techniques for Vector Images," *IEEE Trans. Sig. Proc.*, Vol. 31, No. 7, July 2003, pp. 1941–1953.

[21]   Denis, P., P. Carre, and C. Fernandez-Maloigne, "Spatial and spectral quaternionic approaches for colour images," *Computer Vision and Image Understanding*, Vol. 107, Elsevier, 2007, pp. 74–87.

[22] Took, C. C., and D. P. Mandic, "The Quaternion LMS Algorithm for Adaptive Filtering of Hypercomplex Processes," *IEEE Trans. Signal Processing*, Vol. 57, No. 4, April 2009, pp. 1316–1327.

[23] Bas, P., N. Le Bihan, and J.-M. Chassery, "Color Image Watermarking Using Quaternion Fourier Transform," *Proc. ICASSP*, Hong Kong, 2003.

[24] Ma, X., Y. Xu, and X. Yang, "Color image watermarking using local quaternion Fourier spectra analysis," *IEEE Int. Conf. Multimedia and Expo*, Hannover, June 23–26, 2008, pp. 233–236.

[25] Li, C., B. Li, L. Xiao, Y. Hu, and L. Tian, "A Watermarking Method Based on Hypercomplex Fourier Transform and Visual Attention," *J. Information & Computational Science*, Vol. 9, No. 15, 2012, pp. 4485–4492.

[26] Wang, X., C. Wang, H. Yang, and P. Niu, "A robust blind color image watermarking in quaternion Fourier transform domain," *J. Systems and Software*, Vol. 86, No. 2, February 2013, pp. 255–277.

[27] Alexiadis, D. S., and G. D Sergiadis, "Estimation of Motions in Color Image Sequences Using Hypercomplex Fourier Transforms," *IEEE Trans. Image Processing*, Vol. 18, No. 1, January 2009, pp. 168–187.

[28] Sangwine, S. J., and T. A. Ell, "Hypercomplex Fourier transforms of color images," *IEEE Int. Conf. Image Process.*, Vol. 1, Thessaloniki, Greece, October 7–10, 2001, pp. 137–140.

[29] Ell, T. A., and S. J. Sangwine, "Hypercomplex Fourier Transforms of Color Images," *IEEE Trans. Image Processing*, Vol. 16, No.1, January 2007, pp. 22–35.

[30] Said, S., N. Le Bihan, and S. J. Sangwine, "Fast Complexified Quaternion Fourier Transform," *IEEE Trans. Signal Proc.*, Vol. 56, No. 4, April 2008, pp. 1522–1531.

[31] Khalil, M. I., "Applying Quaternion Fourier Transforms for Enhancing Color Images," *Int. J. of Image, Graphics and Signal Processing*, MECS, Vol. 4, No. 2, 2012, pp. 9–15.

[32] Bayro-Corrochano, E., "The theory and use of the quaternion wavelet transform," *J. Math. Imaging and Vision*, Vol. 24, 2006, pp. 19–35.

[33] Zhou, J., J. Xu, and X. Yang, "Quaternion wavelet phase based stereo matching for uncalibrated images," *Pattern Recognition Letters*, Vol. 28, 2007, pp. 1509–1522.

[34] Chan, W., H. Choi, and R. Baraniuk, "Coherent multiscale image processing using dual-tree quaternion wavelets," *IEEE Trans. Image Process.*, Vol. 17, No. 7, July 2008, pp. 1069–1082.

[35] Bahri, M., "Quaternion Algebra-Valued Wavelet Transform," *Applied Math. Sciences*, Vol. 5, No. 71, 2011, pp. 3531–3540.

[36] Gai, S., G. Yang, and S. Zhang, "Multiscale texture classification using reduced quaternion wavelet transform, " *Int. J. Electronics and Communication*, Vol. 67, No. 3, March 2013, pp. 233–241.

[37] Lengyel, E., *Mathematics for 3D Game Programming and Computer Graphics*, Hingham, MA: Charles River Media, 2001.

[38] Mukundan, R., "Quaternions: From Classical Mechanics to Computer Graphics, and Beyond," *Proc. 7th ATCM Conf*, Melaka, Malaysia, December 17-21, 2002, pp. 97–106.

[39] Farrell, J., *Mathematics for Game Developers*, Boston: Premier Press, Inc., 2004.

[40] Yuan, J. S. C., "Closed Loop Manipulator Control with Quaternion Feedback," *IEEE Trans. Robotics and Automation*, Vol. 4, No. 4, 1988, pp. 434–439.

[41] Funda, J., and R. P. Paul, "A Comparison of Transforms and Quaternions in Robotics," *Proceedings, IEEE Int. Conf. Robotics and Automation*, Vol. 2, Philadelphia, PA, September 24–29, 1988, pp. 886–891.

[42] Andreis, D., and E. S. Canuto, "Orbit dynamics and kinematics with full quaternions," *Proc. of the American Control Conference*, Boston, MA, June 30–July 2, 2004, pp. 3660–3665.

[43] Girard, P. R., *Quaternions, Clifford Algebras and Relativistic Physics*, Berlin, Germany: Birkhäuser, 2008.

[44] Chrisitianto, V., and F. Smarandache, "A Derivation of Maxwell Equations in Quaternion Space, " *Progress in Physics*, Vol. 2, April 2010, pp. 23–27.

[45] Kwaśniewski, A. K., "Glimpses of the Octonions and Quaternions History and Today's Applications in Quantum Physics," *Adv. Appl. Clifford Algebra*, Vol. 22, 2012, pp. 87–105.

[46] Meister, L., "Mathematical Modelling in Geodesy Based on Quaternion Algebra," *Phys.Chem. Earth (A)*, Vol. 25, No. 9–11, 2000, pp. 661–665.

[47] Kaplan, A., "Quaternions and Octonions in Mechanics," *Revista de la Unión Matemática Argentina*, Vol. 49, No. 2, 2008, pp. 45–53.

[48] Chanyal, B. C., P. S. Bisht, and O. P. S. Negi, "Generalized Octonion Electrodynamics," *Int. J., Theor. Physics*, Vol. 49, 2010, pp. 1333–1343.

[49] Köplinger, J., "Dirac equation on hyperbolic octonions," *Appl. Math. Comp.*, Vol. 182, No. 1, November 2006, pp. 443–446.

[50] Borges, M. F., M. D. Roque, and J. A. Marao, "Sedenions of Cayley-Dickson and the Cauchy-Riemann Like Relations," *Int. J. Pure and Applied Mathematics*, Vol. 68, No. 2, 2011, pp. 165–188.

[51] Köplinger, J., "Signature of gravity in conic sedenions," *Appl. Math. Comp.*, Vol. 188, 2007, pp. 942–947.

[52] Köplinger, J., "Gravity and electromagnetism on conic sedenions," *Appl. Math. Comp.*, Vol. 188, 2007, pp. 948–953.

[53] Dorst, L., and S. Mann, "Geometric Algebra: A Computational Framework for Geometrical Applications," *IEEE Computer Graphics and Applications*, Vol. 22, No.3, May–June 2002, pp. 24–31.

[54] Abłamowicz, R. (Ed.), *Clifford Algebras: Applications to Mathematics, Physics, and Engineering*, Berlin, Germany: Birkäuser, 2004.

[55] Abłamowicz, R., and G. Sobczyk, *Lectures on Clifford (Geometric) Algebras and Applications*, Berlin: Birkäuser, 2004.

[56] Hitzer, E. M. S., "Applications of Clifford's Geometric Algebra," *Adv. Appl. Clifford Algebras*, Vol. 23, No. 2, 2013, pp. 377–404.

[57] Panicaud, B., "Clifford Algebra ($\mathbb{C}$) for Applications to Field Theories," *Int. J. Theoretical Physics*, Vol. 50, No. 10, 2011, pp. 3186–3204.

[58] Alexeyeva, L. A., "Biquaternions algebra and its applications by solving of some theoretical physics equations," *Int. J. Clifford Analysis, Clifford Algebras and their Applications*, Vol. 7, No. 1, 2012.

[59] Sommer, G. (ed.), *Geometric Computing with Clifford Algebras: Theoretical Foundations and Applications in Computer Vision and Robotics*, Berlin: Springer-Verlag, 2001.

[60] Buchholz, S., and G. Sommer, *On Clifford neurons and Clifford multilayer perceptrons in Neural Networks*, Vol. 21, 2008, pp. 925–935.

[61] Dimitrov, V. S., T. V. Cooklev, and B. D. Donevsky, "On the Multiplication of Reduced Biquaternions and Applications," *Inf. Process. Letters*, Vol. 43, No. 3, Sept. 1992, pp. 161–164.

[62] Alfsmann, D., and H. G. Gockler, "Design of Hypercomplex Allpass-Based Paraunitary Filter Banks applying Reduced Biquaternions," *The International Conference on Computer as a Tool, EUROCON'2005*, Belgrade, Serbia&Montenegro, November 21–24, 2005, pp. 92–95.

[63] Pei, S.-C., J.-H. Chang, and J.-J. Ding, "Commutative Reduced Biquaternions and Their Fourier Transform for Signal and Image Processing Applications," *IEEE Trans. Sig. Proc.*, Vol. 52, No. 7, July 2004, pp. 2012–2031.

[64] Rundblad-Labunets, E., and V. Labunets, "Spatial-Color Clifford Algebras for Invariant Image Recognition," in *Geometric Computing with Clifford Algebras*, G. Sommer (ed.), Berlin: Springer-Verlag, 2001, pp. 155–185.

3

Orthants of the *n*-Dimensional Cartesian Space and Single-Orthant Operators

3.1 The Notion of an Orthant

In 1992, Stefan Hahn presented a paper with definitions and a description of *n*-D analytic signals with single-orthant spectra [1]. The following description is devoted to readers unfamiliar with the notion of orthants. We deal with two domains: the signal-domain and the frequency-domain. The *n*-D signal is a function of the *n*-tuple $\boldsymbol{x} = (x_1, x_2, \ldots, x_n)$. Its Fourier spectrum is a function of the *n*-tuple $\boldsymbol{f} = (f_1, f_2, \ldots, f_n)$ in the frequency-domain Cartesian rectangular coordinate system. Let us recall that there are two equivalent definitions of *n*-D analytic signals:

- The analytic signal is defined by the inverse *n*-D Fourier transform of a single-orthant spectrum.
- The analytic signal is defined by the convolution of a real signal $u(\boldsymbol{x})$ with a complex delta distribution.

Detailed descriptions of both methods are presented in Chapter 5. For convenience, let us illustrate both methods for 1-D signals (usually functions of time). The Fourier transform of a real time signal $u(t)$ defined by the integral

$$U(f) = \int_{-\infty}^{\infty} u(t)\exp[-j2\pi ft]\,dt = \mathrm{Re}(f) - j\mathrm{Im}(f)$$

is called the complex Fourier spectrum of $u(t)$ with support in both negative and positive frequencies. The 1-D analytic signal is defined by the inverse Fourier transform of a one-sided spectrum at positive frequencies. This one-sided spectrum is given by a product of the unit step $2 \times \mathbf{1}(f) = 1 + \mathrm{sgn}\,(f)$ (the use of a double unit step is a matter of convention) and $U(f)$: $\Gamma(f) = [1 + \mathrm{sgn}(f)]U(f)$.

The inverse Fourier transform yields a complex function called analytic signal $\psi(t) = \int_{-\infty}^{\infty} \left[1 + \mathrm{sgn}(f)\right]U(f)\exp[j2\pi ft]\,df = u(t) + jv(t)$.

The imaginary part $v(t)$ is called the Hilbert transform of $u(t)$. Because a product in the frequency domain corresponds to a convolution in the time domain, the analytic signal can be written in the form $\psi(t) = \left[\delta(t) + j(1/\pi t)\right] * u(t)$, we applied the Fourier pairs $\delta(t) \overset{F}{\Leftrightarrow} 1$ and $1/\pi t \overset{F}{\Leftrightarrow} \mathrm{sgn}(f)$.

In the above descriptions, the 1-D frequency domain *space* has the form of a line being a union of two half-lines separated by the point of the origin. The 2-D frequency domain has the form of a plane being a union of four quadrants separated by the coordinate axes $f_1 = 0$ and $f_2 = 0$. Then, the 3-D frequency-space is a union of eight octants separated by the coordinate axes $f_1 = 0$, $f_2 = 0$ and $f_3 = 0$ (see Figure 3.1). In general, the n-D frequency-space is a union of 2^n orthants separated by n coordinate axes. Of course, the same names apply for the signal-space.

Table 3.1
Orthants of n-D Cartesian Space

n	Name of Orthants
1	2 half-lines
2	4 quadrants
3	8 octants
n	2^n orthants

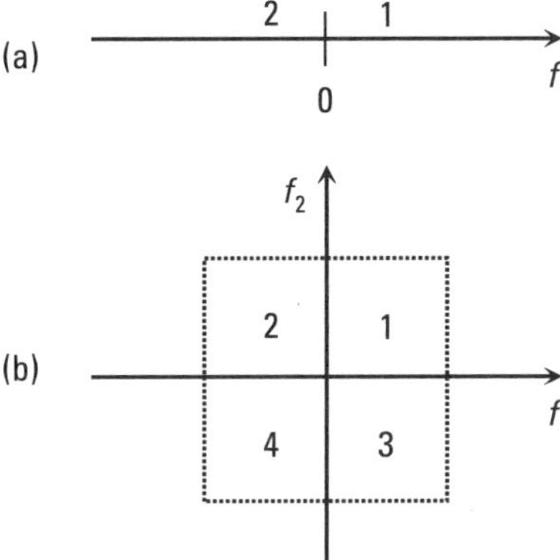

Figure 3.1   Labelling of orthants. (a) Two half-lines labeled 1 and 2. (b) Four quadrants labeled 1, 2 3 and 4. Note that in the half-plane $f_1 > 0$ the quadrants are labeled with odd numbers.

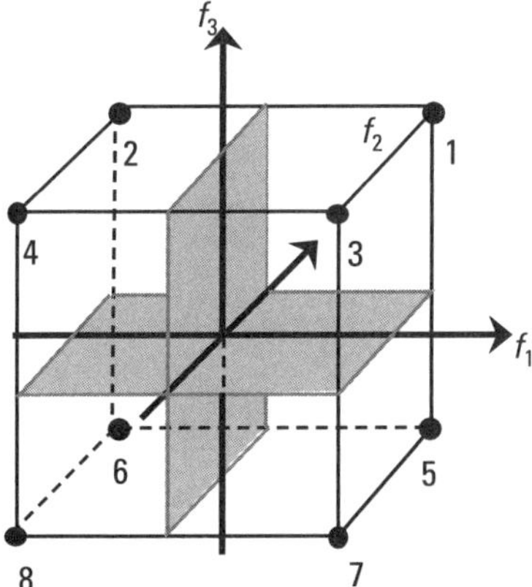

Figure 3.2   Labelling of octants. The octants in the half-space $f_1 > 0$ are labeled with odd numbers.

3.2   Single-Orthant Operators

The single-orthant spectrum of the analytic signal is defined by the relation:

$$\text{single-orthant spectrum} = \text{single-orthant operator} \times \text{Fourier spectrum}$$

$$(3.1)$$

The single-orthant operator can be defined using the unit step distribution. The 1-D unit step can be defined using the signum distribution

$$\mathbf{1}\left(f_1\right) = 0.5 + 0.5\,\mathrm{sgn}\left(f_1\right) \tag{3.2}$$

The signum distribution can be conveniently defined using the Mikusiński approach (Reference [10] in Chapter 1) and tanh approximating function

$$s_1 = \mathrm{sgn}\,f_1 = \lim_{a \to \infty}\left[\tanh\left(a\,f_1\right)\right] \tag{3.3}$$

where

$$\tanh\left(a\,f_1\right) = \frac{e^{a f_1} - e^{-a f_1}}{e^{a f_1} + e^{-a f_1}}. \tag{3.4}$$

In the limit a $\to \infty$, we get

$$\mathrm{sgn}\left(f_1\right) = \left\{\begin{array}{ll} 1, & f_1 > 0 \\ 0, & f_1 = 0 \\ -1, & f_1 < 0 \end{array}\right. \tag{3.5}$$

Note that omitting the zero for $f_1 = 0$ is false. Equation (3.3) is illustrated in Figure 3.3.

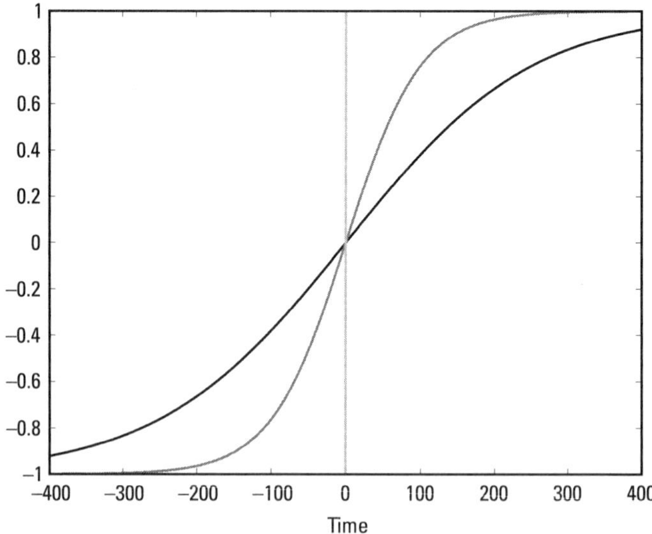

Figure 3.3.   Illustration of (3.3), for a = 0.004, 0.1 and10. The last value yields a nearly perfect signum function.

For 1-D signals, (3.1) is

$$U_{hl}\left(f_1\right) = \left(1 + s_1\right)U\left(f_1\right) = 2 \cdot \mathbf{1}\left(f_1\right)U\left(f_1\right) \tag{3.6}$$

where $U(f_1)$ is the Fourier transform of the real signal $u(x_1)$ (usually $x_1 = t$) and the subscript hl denotes a half-line. The use of the factor 2 is a matter of convention. This factor indicates that the energy of the analytic signal is a sum of energies of the real signal and of its Hilbert transform, which are equal.

For 2-D signals, the single-quadrant operator for 1st quadrant $f_1 > 0$, $f_2 > 0$, is

$$2^2 \cdot \mathbf{1}\left(f_1\right) \cdot \mathbf{1}\left(f_2\right) = 1 + s_1 + s_2 + s_1 s_2 \tag{3.7}$$

where $s_i = \mathrm{sgn} f_i$, as in (3.3). For 3-D signals, the single-quadrant operator for the first octant $f_1 > 0, f_2 > 0, f_3 > 0$, is

$$2^3 \cdot \mathbf{1}\left(f_1\right) \cdot \mathbf{1}\left(f_2\right) \cdot \mathbf{1}\left(f_3\right) = 1 + s_1 + s_2 + s_1 s_2 + s_3 + s_1 s_3 + s_2 s_3 + s_1 s_2 s_3 \tag{3.8}$$

In general, the n-D single-orthant operator is a product of n 1-D operators and is a separable function (distribution).

3.3 Decomposition of Real Functions into Even and Odd Terms

Each n-D function defined in $\mathbb{R}^n$ (Cartesian rectangular coordinates) may be decomposed into a union of terms regarding the evenness and oddness. In 1-D, a real signal is in general a union of two terms

$$u(x) = u_e + u_o \tag{3.9}$$

The even and odd terms are

$$u_e(x) = \frac{u(x) + u(-x)}{2} \qquad u_o(x) = \frac{u(x) - u(-x)}{2} \tag{3.10}$$

Similarly, the 2-D signal is a union of four terms: even-even (*ee*), even-odd (*eo*), odd-even (*oe*) and odd-odd (*oo*) as follows

$$u\left(x_1, x_2\right) = u_{ee}\left(x_1, x_2\right) + u_{eo}\left(x_1, x_2\right) + u_{oe}\left(x_1, x_2\right) + u_{oo}\left(x_1, x_2\right) \tag{3.11}$$

where

$$u_{ee}\left(x_1,x_2\right)=\frac{u\left(x_1,x_2\right)+u\left(-x_1,x_2\right)+u\left(x_1,-x_2\right)+u\left(-x_1,-x_2\right)}{4} \tag{3.12}$$

$$u_{eo}\left(x_1,x_2\right)=\frac{u\left(x_1,x_2\right)+u\left(-x_1,x_2\right)-u\left(x_1,-x_2\right)-u\left(-x_1,-x_2\right)}{4} \tag{3.13}$$

$$u_{oe}\left(x_1,x_2\right)=\frac{u\left(x_1,x_2\right)-u\left(-x_1,x_2\right)+u\left(x_1,-x_2\right)-u\left(-x_1,-x_2\right)}{4} \tag{3.14}$$

$$u_{oo}\left(x_1,x_2\right)=\frac{u\left(x_1,x_2\right)-u\left(-x_1,x_2\right)-u\left(x_1,-x_2\right)+u\left(-x_1,-x_2\right)}{4} \tag{3.15}$$

The above decomposition (3.11) is relative and changes with translation and rotation of the Cartesian coordinates. If the signal is an even-even function, its Fourier transform is a real function. Otherwise, it is a complex (hypercomplex) function (see Chapter 4). Figure 3.4 shows the decomposition of the single-quadrant cube function $\Pi_a(x_1-(a/2),x_2-(a/2))$ into even-even, even-odd, odd-even and odd-odd parts. We observe the factor ¼ appearing with a sign according to the sign of u in a given quadrant. Next, Figure 3.5 presents the plots of the parts u_{ee} and u_{oo} of the 2-D Gaussian nonseparable signal

$$u\left(x_1,x_2\right)=\frac{1}{2\pi\sigma_1\sigma_2\sqrt{1-\rho^2}}$$

$$\cdot\exp\left\{\frac{-1}{2\left(1-\rho^2\right)}\left(\frac{\left(x_1-\mu_1\right)^2}{\sigma_1^2}+\frac{\left(x_2-\mu_2\right)^2}{\sigma_2^2}-2\rho\frac{\left(x_1-\mu_1\right)\left(x_2-\mu_2\right)}{\sigma_1\sigma_2}\right)\right\} \tag{3.16}$$

where $\mu_1=\mu_2=0$, $\sigma_1=1$, $\sigma_2=2$, $\rho=0.5$. Please note that if $\rho=0$, we obtain the separable 2-D Gaussian signal presented in detail in [1–3].

Continuing, the 3-D signal is a union of eight terms indexed *eee*, *eeo*, *eoe*, *eoo*, *oee*, *oeo*, *ooe* and *ooo*. The order is given by a sequence of binary numbers inserting $e \to 0$ and $o \to 1$. Equation (3.9) takes the form of a sum of eight terms divided by 8.

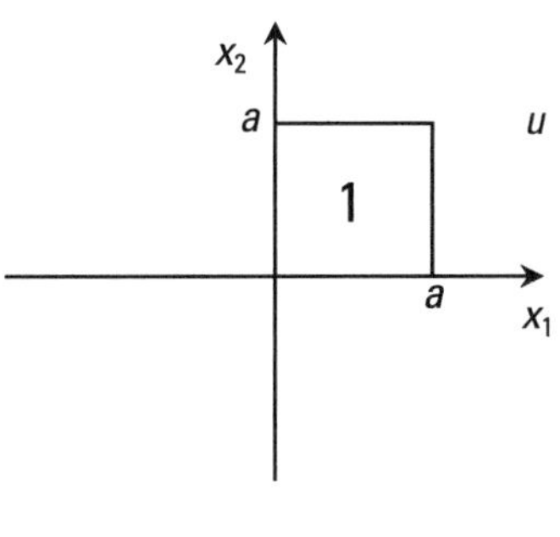

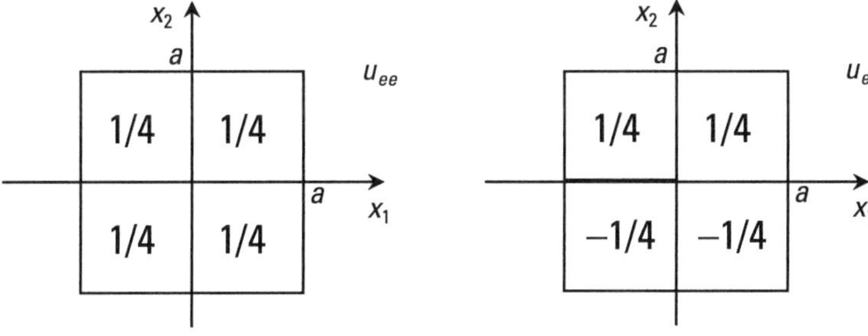

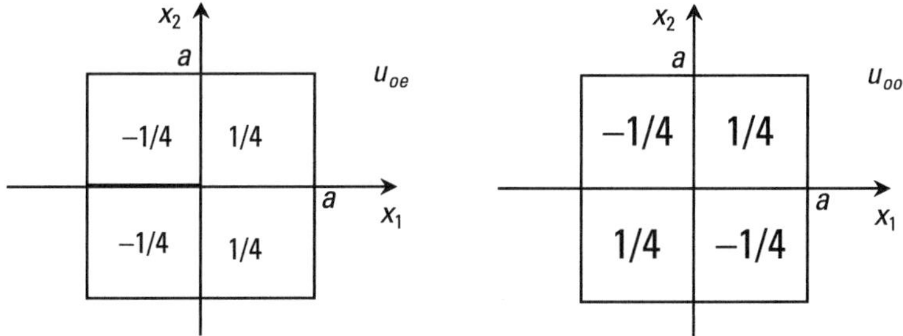

Figure 3.4   The decomposition of the single-quadrant cube signal u into the parts u_{ee}, u_{eo}, u_{oe}, and u_{oo}.

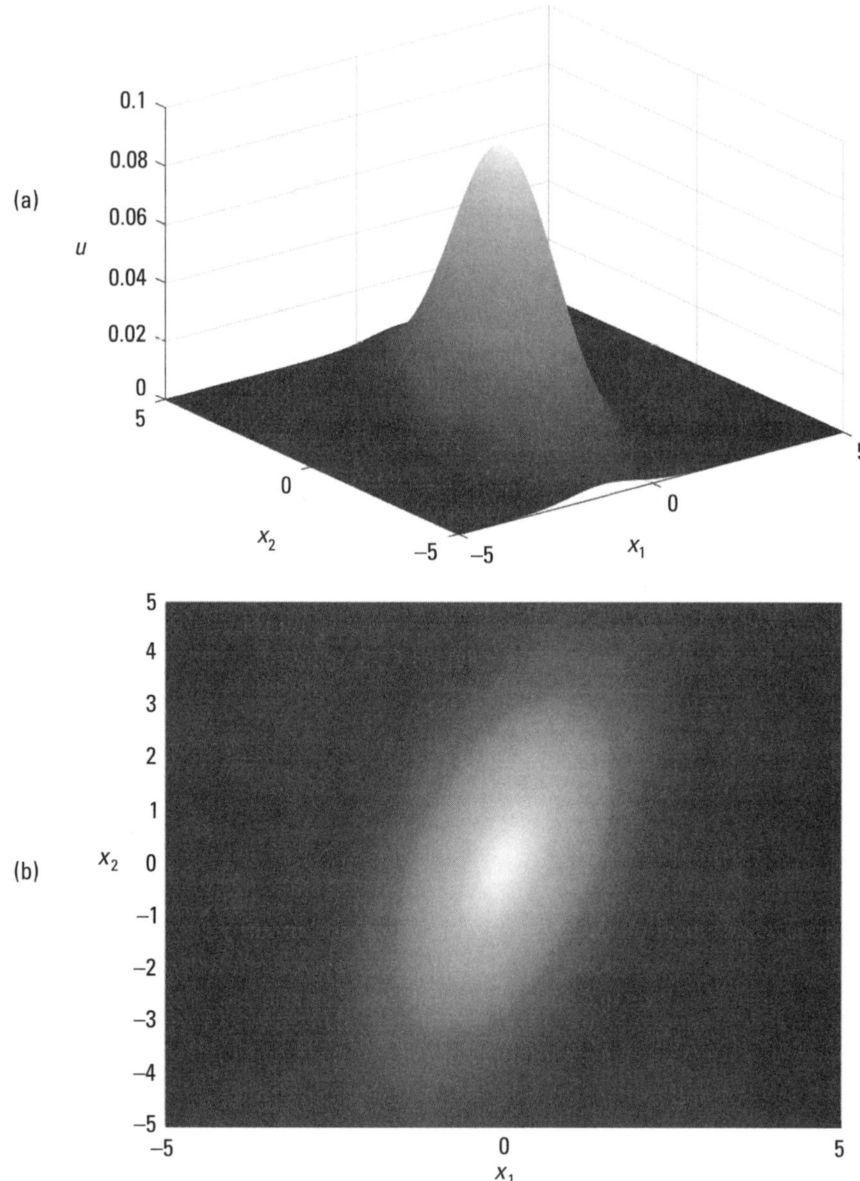

Figure 3.5 The decomposition of the nonseparable 2-D Gaussian signal u into the parts u_{ee} and u_{oo}. (a) 3-D display of u. (b) 2-D display of u. (c) The part $u_{ee}(x_1, x_2)$. (d) $-u_{oe}(x_1, x_2)$.

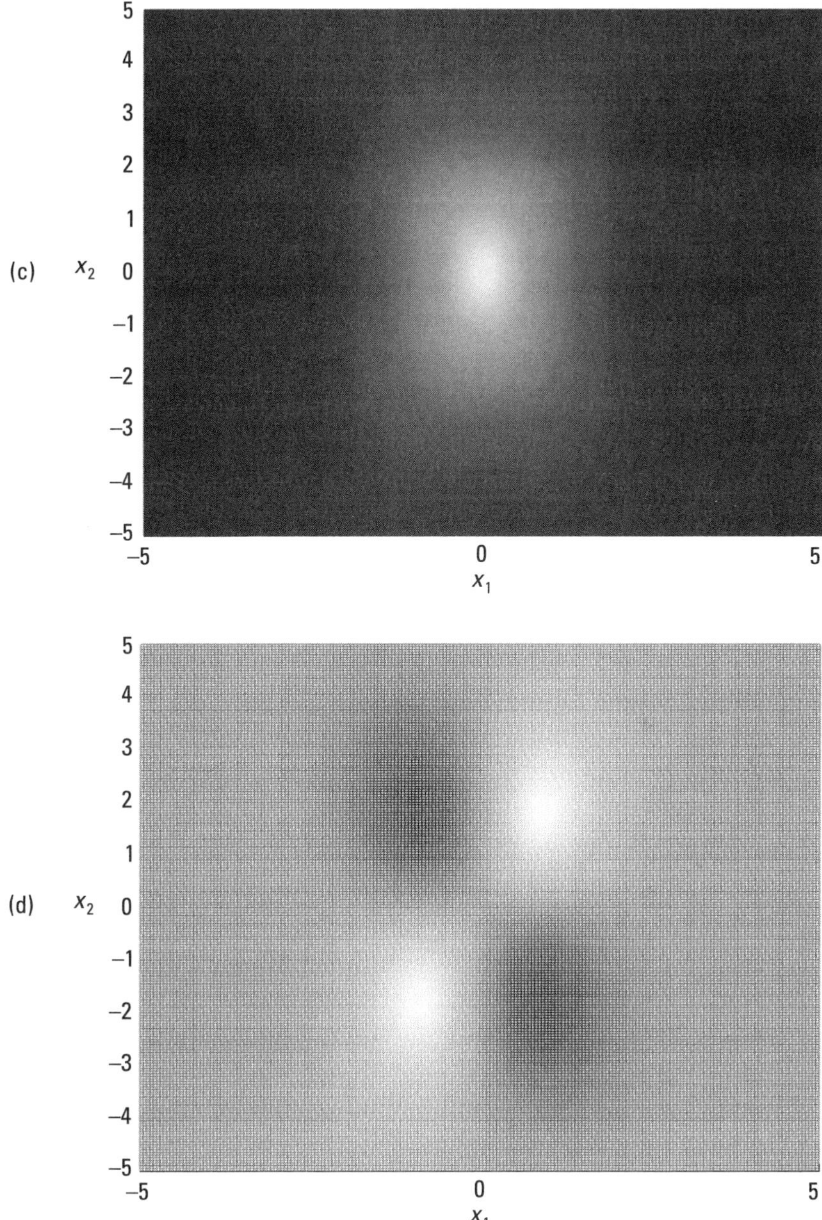

Figure 3.5   Continued

References

[1]   Hahn, S. L., "Multidimensional Complex Signals with Single-Orthant Spectra," *Proc. IEEE*, Vol. 80, No.8, August 1992, pp. 1287–1300.

[2]   Hahn, S. L., *Hilbert Transforms in Signal Processing*, Norwood, MA: Artech House, Inc., 1996.

[3]   Hahn, S., "Amplitudes, Phases and Complex Frequencies of 2-D Gaussian Signals," *Bull. Polish Academy Sci., Tech. Sci.*, Vol. 40, No. 3, 1992, pp. 289–311.

4

Fourier Transformation in Analysis of *n*-Dimensional Signals

This chapter presents two different approaches to the frequency analysis of n-dimensional signals. The first approach is the complex Fourier transformation—a well-known and commonly used tool in frequency analysis. The second approach applies the hypercomplex Fourier transformation. The 1-D complex Fourier transformations have been the subject of a multitude of scientific papers and books dealing with time signals [1–3]. In image processing, the 2-D complex Fourier transformation has also found many interesting applications. Our intention is not to once again recall their properties. We will instead focus on their relation to the Cayley-Dickson Fourier transformations defined basing on the hypercomplex algebras described in Section 2.1.

Section 4.1 is devoted to basics of the theory of complex and hypercomplex Fourier transformations of 1-D, 2-D, and 3-D signals. We will show that, in the complex case, the full information about a frequency content of a n-D real signal is in a half-space $f_1 > 0$. While using the hypercomplex approach, it is limited to a single frequency orthant of $\mathbb{R}^n$. We will also decompose the Fourier spectra into even-odd components and study their mutual relations.

Section 4.2 presents the theoretical basics of Cayley-Dickson transformations, including the well-known quaternion Fourier transform (QFT) and the more recently defined octonion Fourier transform (OFT). In Section 4.3, we present formulas relating QFT and OFT to 2-D and 3-D complex Fourier transforms proving equivalence of both approaches. The theory is illustrated with examples of spectra of different test signals.

Section 4.4 describes some applications of the presented theory.

4.1   Complex *n*-D Fourier Transformation

Consider an *n*-D signal $u(\boldsymbol{x})$, where $\boldsymbol{x} = (x_1, x_1, \ldots, x_n)$ is a *n*-D signal-domain variable. The *n* D complex Fourier transformation (FT) defines the *n*-D frequency function $U(\boldsymbol{f}), \boldsymbol{f} = (f_1, f_1, \ldots, f_n)$, called the *frequency spectrum*, or more concisely, the "spectrum of *u*." Generally, $U(\boldsymbol{f})$ is a complex-valued function defined by the integral

$$U(\boldsymbol{f}) = \int_{\mathbb{R}^n} u(\boldsymbol{x}) \prod_{i=1}^{n} \exp\left(-j2\pi f_i x_i\right) d^n \boldsymbol{x} \qquad (4.1)$$

Let us remark that for *n* = 1, we get the 1-D spectrum $U(f)$ of a signal $u(t)$, where *t* denotes the time variable:

$$U(f) = \int_{\mathbb{R}} u(t) \exp(-j2\pi ft) dt \qquad (4.2)$$

Example 4.1 Spectrum of the 1-D Gaussian Signal

The 1-D Gaussian signal is defined as

$$u(t) = \frac{1}{\sigma\sqrt{2\pi}} \exp\left[-\frac{(t-\mu)^2}{2\sigma^2}\right] \qquad (4.3)$$

In probability theory, *u* is a probability density function of a normal (Gaussian) distribution, μ is the expected value, and σ^2 is the variance. In signal theory, μ represents a time-shift of a signal and σ determines the slope of the signal plot. Figure 4.1(a) shows two Gaussian signals for $\mu = 0$, $\sigma = 1$ and Figure 4.1(b) shows a Gaussian signal for $\mu = 2$, $\sigma = 2$. The Gaussian signal (4.3) has a Gaussian-shaped spectrum $U(f)$ given by

$$U(f) = \exp\left(-8\pi^2\sigma^2 f^2\right)\exp(-j2\pi f\mu) \qquad (4.4)$$

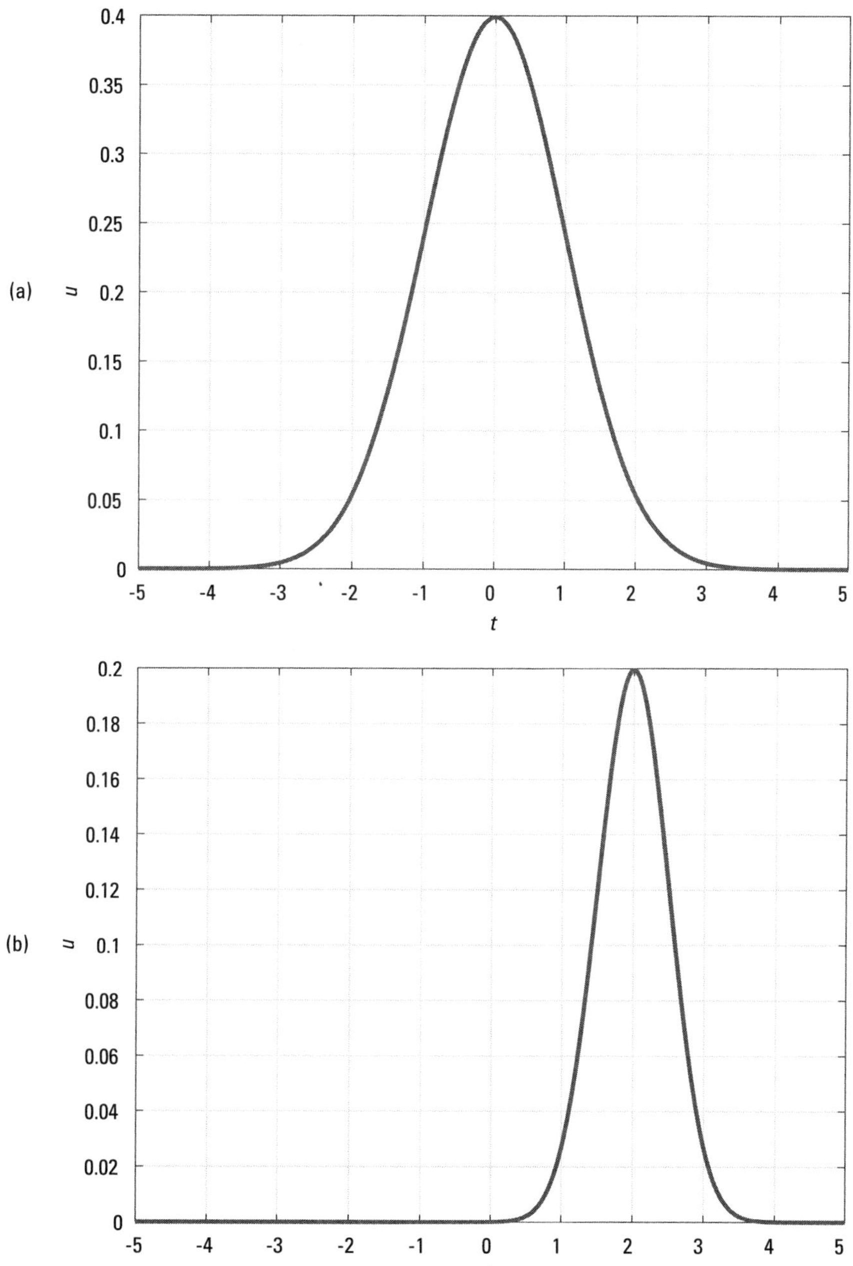

Figure 4.1 The 1-D Gaussian signal: (a) $\mu = 0$, $\sigma = 1$, (b) $\mu = 2$, $\sigma = 2$.

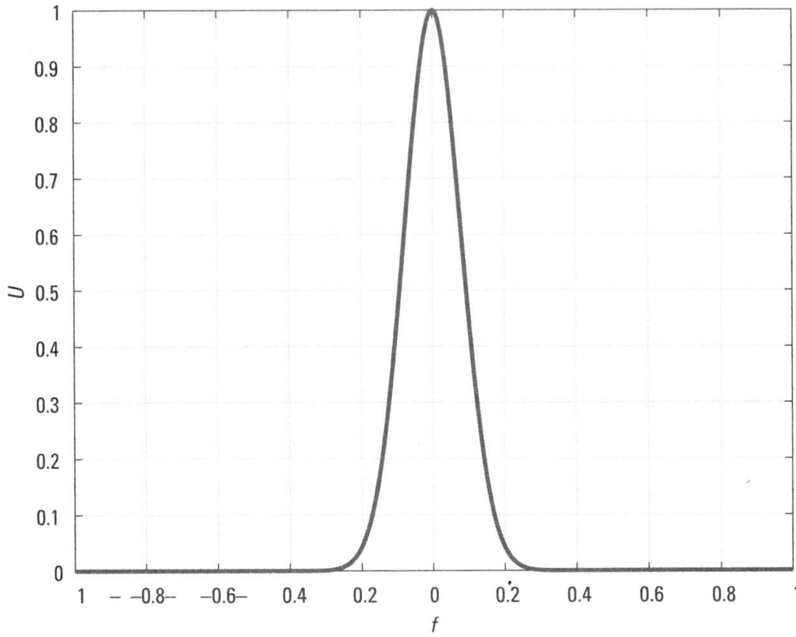

Figure 4.2 The spectrum of the 1-D Gaussian signal: $\mu = 0$, $\sigma = 1$.

where we used the time-shift property of the 1-D Fourier transformation (see Appendix A). Figure 4.2 presents the spectrum of $u(t) = 1/\sqrt{2\pi}\,\exp\!\left(-t^2/2\right)$ ($\mu = 0$, $\sigma = 1$ in (4.3)) given by $U(f) = \exp\!\left(-8\pi^2 f^2\right)$ We see that the frequency support of the spectrum is much smaller than its time support (see Figure 4.1(a)). It is a consequence of the scaling property of the 1-D FT (see Appendix A). On the other hand, the spectrum amplitude $U(0)$ is bigger than the signal's maximum value.

If $n = 2$, (4.1) defines the 2-D Fourier spectrum $U(f_1,f_2)$ of a 2-D signal $u(x_1,x_2)$:

$$U\left(f_1,f_2\right) = \int_{\mathbb{R}^2} u\left(x_1,x_2\right)\exp\!\left[-j2\pi\left(f_1 x_1 + f_2 x_2\right)\right]dx_1\,dx_2 \qquad (4.5)$$

Example 4.2 Spectrum of the 2-D Gaussian Signal

In Chapter 3, we presented the definition of the 2-D nonseparable Gaussian signal (see (3.16)) and showed its exemplary plot. For $\mu_1 = \mu_2 = 0$, the signal is defined as

$$u(x_1,x_2) = \frac{1}{2\pi\sigma_1\sigma_2\sqrt{1-\rho^2}} \cdot \exp\left\{\frac{-1}{2(1-\rho^2)}\left(\frac{x_1^2}{\sigma_1^2} + \frac{x_2^2}{\sigma_2^2} - \frac{2\rho x_1 x_2}{\sigma_1\sigma_2}\right)\right\} \qquad (4.6)$$

and its spectrum is given by

$$U(f_1,f_2) = \exp\left[-2\pi^2\left(\sigma_1^2 f_1^2 + \sigma_2^2 f_2^2 + 2\rho\sigma_1\sigma_2 f_1 f_2\right)\right] \qquad (4.7)$$

Figure 4.3 shows mesh and surface plots of this spectrum for $\sigma_1 = 1$, $\sigma_2 = 2$, and $\rho = 0.5$. Let us compare the frequency supports of the spectrum to the signal–domain supports from Figure 3.3. We see that the wider the signal along the *x*-axis, the narrower its spectrum along the corresponding *f*-axis (the consequence of the scaling property of FT; see Appendix C).

The transformation from the frequency domain into the signal domain defines the *n*-D *inverse Fourier transform* given by

$$u(\boldsymbol{x}) = \int_{\mathbb{R}^n} U(\boldsymbol{f})\prod_{i=1}^{n}\exp\left(j2\pi f_i x_i\right)d^n\boldsymbol{f} \qquad (4.8)$$

We see that (4.8) differs from (4.1) only in the change of sign inside the exponential. This property is called the *duality of the Fourier transformation*.[1] For $n = 1$, we get a very known formula of the 1-D inverse FT:

$$u(t) = \int_{\mathbb{R}} U(f)\exp(j2\pi ft)df \qquad (4.9)$$

and for $n = 2$:

$$u(x_1,x_2) = \int_{\mathbb{R}^2} U(f_1,f_2)\exp\left[j2\pi(f_1 x_1 + f_2 x_2)\right]df_1\,df_2 \qquad (4.10)$$

In Chapter 3, it was shown that *n*-D signals can be represented as sums of components of different parity. The number of these components is always 2^n. It is evident for $n = 1$, since $u(t) = u_e(t) + u_o(t)$ (i.e., we have $2 = 2^1$ components

---

[1] If we deal with a *n*-D spectrum as a function of $\omega = (\omega_1, \omega_2, \ldots, \omega_n)$, $\omega_i = 2\pi f_i$ the direct and inverse FT are defined as $U(\omega) = \int_{\mathbb{R}^n} u(x)\prod_{i=1}^{n}\exp(-j\omega_i x_i)d^n x$, $u(x) = \frac{1}{(2\pi)^n}\int_{\mathbb{R}^n} U(\omega)\prod_{i=1}^{n}\exp(j\omega_i x_i)d^n\omega$ respectively. The scaling factor $1/(2\pi)^n$ is needed in order for the inverse transform to reconstruct the original signal with the correct amplitude. In (4.8), there is no scaling factor.

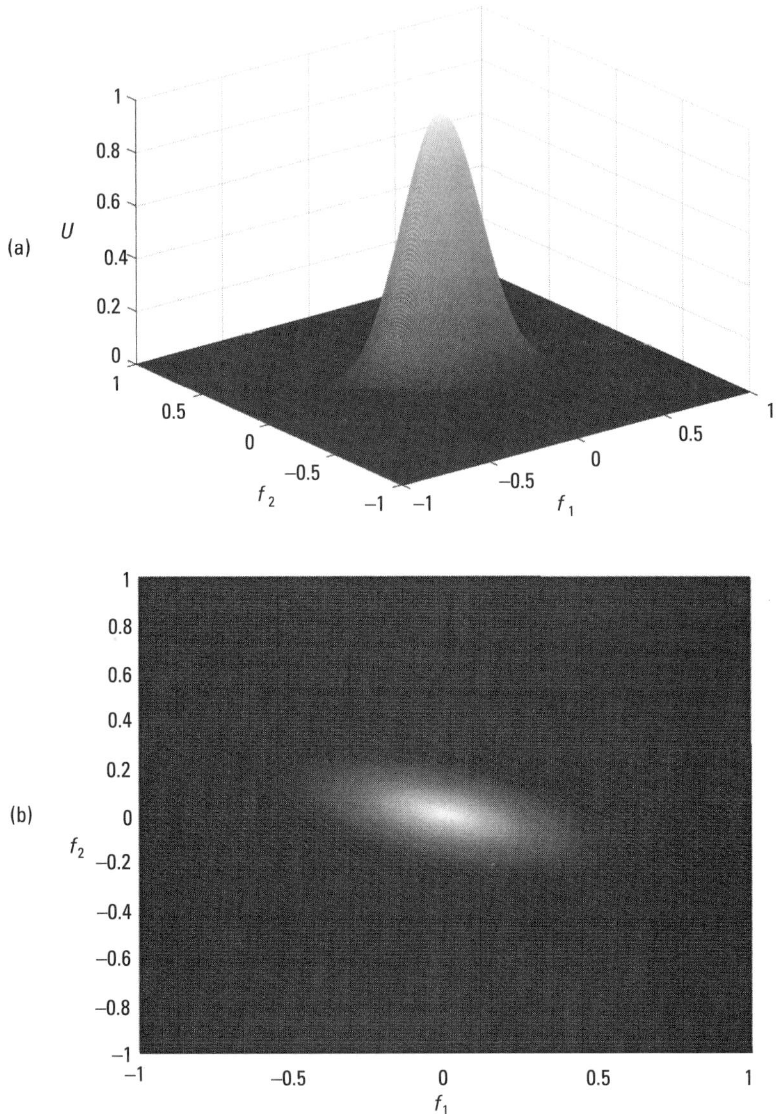

Figure 4.3   The spectrum of the 2-D Gaussian signal $\mu_1 = \mu_2 = 0$, $\sigma_1 = 1$, $\sigma_2 = 2$, $\rho = 0.5$. (a) Mesh, (b) surface plots.

even (subscript *e*) and odd (subscript *o*) regarding *t*). For $n = 2$, the signal $u(x_1, x_2)$ is a sum of $4 = 2^2$ components: even-even $u_{ee}(x_1, x_2)$, even-odd $u_{eo}(x_1, x_2)$, odd-even $u_{oe}(x_1, x_2)$, and odd-odd $u_{oo}(x_1, x_2)$ where the first letter in the subscript indicates parity regarding x_1 and the second one indicates parity with regard to x_2. Let us describe in detail cases $n = 1, 2, 3$.

4.1.1   Spectrum of a 1-D Real Signal in Terms of its Even and Odd Components

Let us recall that a signal $u(t)$ is a sum of its even and odd components: $u(t) = u_e(t) + u_o(t)$. From (4.2) we get

$$U(f) = \int_{\mathbb{R}} u_e(t)\cos(2\pi ft)\,dt - j\int_{\mathbb{R}} u_o(t)\sin(2\pi ft)\,dt \qquad (4.11)$$

We see that $U(f)$ is, in general, a complex-valued function

$$U(f) = U_e(f) - j \cdot U_o(f) = |U(f)|\exp\left[j\arg U(f)\right] \qquad (4.12)$$

with the real part $U_e(f)$ given by the 1-D *cosine Fourier transform* of the even component $u_e(t)$ (called the *real part of the spectrum* of u or the *real spectrum*); that is

$$U_e(f) = \int_{\mathbb{R}} u_e(t)\cos(2\pi ft)\,dt \qquad (4.13)$$

and the imaginary part $U_o(f)$ being the 1-D *sine Fourier transform* of the odd component $u_o(t)$ (known as the *imaginary part of the spectrum* or the *imaginary spectrum*):

$$U_o(f) = \int_{\mathbb{R}} u_o(t)\sin(2\pi ft)\,dt \qquad (4.14)$$

Let us remark that some authors apply a minus sign in (4.14) and express the complex-valued spectrum as $U(f) = U_e(f) + j \cdot U_o(f)$.

The absolute value of $U(f)$ is called the *amplitude spectrum* and its argument $\varphi(f) = \arg U(f)$ is called the *phase spectrum*. If u is a real signal, the real spectrum $U_e(f)$ is an even function, while the imaginary spectrum $U_o(f)$ is an odd one, that is

$$U_e(-f) = U_e(f),$$
$$U_o(-f) = -U_o(f). \tag{4.15}$$

As a consequence

$$U(-f) = U_e(-f) - jU_o(-f) = U_e(f) + jU_o(f) = U^*(f) \tag{4.16}$$

meaning that the 1-D complex Fourier transform of a 1-D real signal is *Hermitian symmetric*. So, the *full information about the frequency content of a 1-D real-valued signal is included in a half-axis $f > 0$* (in a 1-D single frequency orthant).

The amplitude spectrum $|U(f)|$ of a 1-D real signal $u(t)$ given by

$$|U(f)| = \sqrt{U_e^2(f) + U_o^2(f)} \tag{4.17}$$

is also an even function of f, because $|U(-f)| = \sqrt{U_e^2(-f) + U_o^2(-f)} = |U(f)|$.
However, the *phase spectrum* $\varphi(f) = \arg U(f)$ defined by relations

$$\cos\varphi(f) = \frac{U_e(f)}{|U(f)|}, \quad \sin\varphi(f) = \frac{-U_o(f)}{|U(f)|} \tag{4.18}$$

is an odd function of f. Hence, for real signals we have

$$|U(f)| = |U(-f)|, \quad \varphi(f) = -\varphi(-f). \tag{4.19}$$

Example 4.3 Amplitude and Phase Spectra of a One-Sided Exponential Signal
The one-sided exponential signal is given by

$$u(t) = \exp(-\alpha t) \cdot \mathbf{1}(t), \quad \alpha > 0 \tag{4.20}$$

and its Fourier spectrum is a complex-valued function of f:

$$U(f) = \frac{1}{\alpha + j2\pi f} \tag{4.21}$$

(see Appendix B). Then, the amplitude spectrum has the form

$$|U(f)| = \frac{1}{\sqrt{\alpha^2 + 4\pi^2 f^2}} \tag{4.22}$$

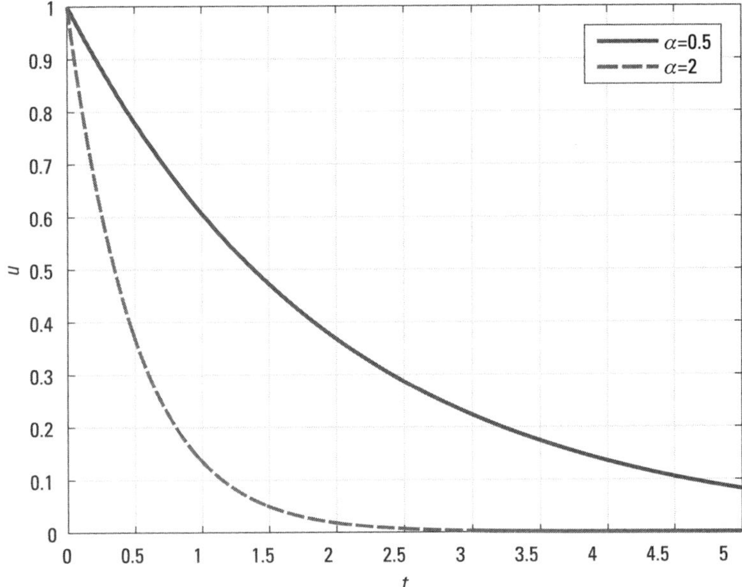

Figure 4.4　The one-sided exponential signal (4.20) for $\alpha = 0.5$, $\alpha = 2$.

and, finally, the phase spectrum

$$\varphi(f) = -\arg(\alpha + j2\pi f) = -\mathrm{atan2}(2\pi f, \alpha) = -\arctan\left(\frac{2\pi f}{\alpha}\right) \quad (4.23)$$

In Figure 4.4, we observe two plots of the one-sided exponential signal (4.20) for $\alpha = 0.5$ and $\alpha = 2$. We see that the slope of the curve depends on the value of the parameter α. The *smaller* the value of α, the *wider* the signal time-domain support. Figure 4.5(a) shows amplitude, and Figure 4.5(b) shows phase spectra of this signal. Once again, we notice the consequences of the scaling property of the Fourier transform (i.e., for smaller values of α, the frequency support of the amplitude spectrum is *narrower* and the slope of the phase characteristic of the signal is larger); see Appendix A.

4.1.2　Spectrum of a 2-D Real Signal in Terms of its Even and Odd Components

The 2-D real signal $u(x_1, x_2)$ is a sum of four components: $u(x_1, x_2) = u_{ee}(x_1, x_2) + u_{eo}(x_1, x_2) + u_{oe}(x_1, x_2) + u_{oo}(x_1, x_2)$. Its spectrum (4.5) can easily be developed as a complex-valued function

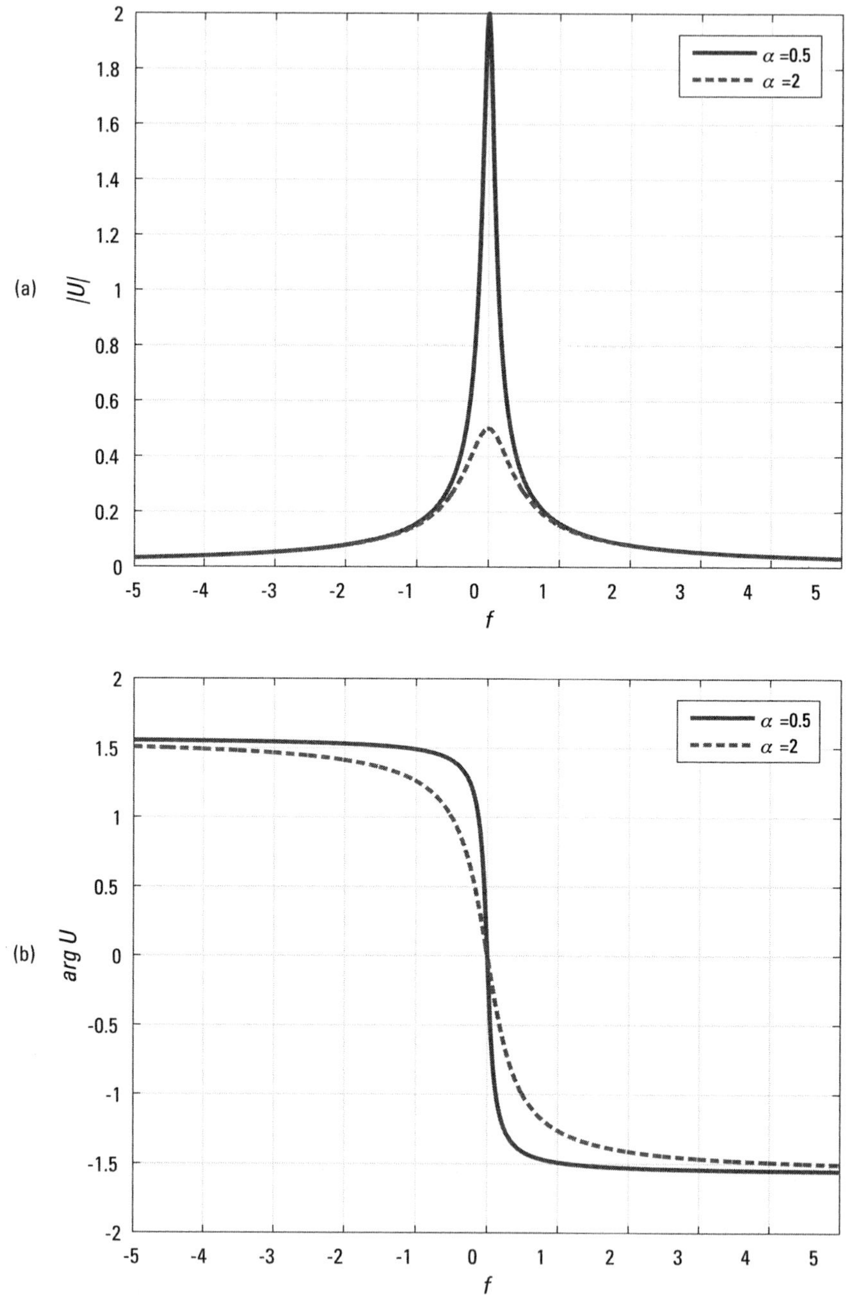

Figure 4.5 The (a) amplitude and (b) phase spectra of the one-sided exponential signal (4.20) for $\alpha = 0.5$, $\alpha = 2$.

$$U(f_1, f_2) = U_{ee} - U_{oo} - j(U_{eo} + U_{oe}) \qquad (4.24)$$

where

$$U_{ee}(f_1, f_2) = \int_{\mathbb{R}^2} u_{ee} \cos(2\pi f_1 x_1) \cos(2\pi f_2 x_2)\, dx_1\, dx_2 \qquad (4.25)$$

$$U_{oo}(f_1, f_2) = \int_{\mathbb{R}^2} u_{oo} \sin(2\pi f_1 x_1) \sin(2\pi f_2 x_2)\, dx_1\, dx_2 \qquad (4.26)$$

$$U_{eo}(f_1, f_2) = \int_{\mathbb{R}^2} u_{eo} \cos(2\pi f_1 x_1) \sin(2\pi f_2 x_2)\, dx_1\, dx_2 \qquad (4.27)$$

$$U_{oe}(f_1, f_2) = \int_{\mathbb{R}^2} u_{oe} \sin(2\pi f_1 x_1) \cos(2\pi f_2 x_2)\, dx_1\, dx_2 \qquad (4.28)$$

The function $U_{ee}(f_1, f_2)$ is known as the 2-D *cosine Fourier transform* of $u_{ee}(x_1, x_2)$, while $U_{oo}(f_1, f_2)$ is the 2-D *sine Fourier transform* of $u_{oo}(x_1, x_2)$. The relations (4.27)–(4.28) define so called *mixed* (*sine-cosine*) FTs of even-odd and odd-even components of the signal $u(x_1, x_2)$. We have the following symmetry properties

$$
\begin{aligned}
U_{ee}(f_1, f_2) &= U_{ee}(-f_1, f_2) = U_{ee}(f_1, -f_2) = U_{ee}(-f_1, -f_2), \\[4pt]
U_{oo}(f_1, f_2) &= -U_{oo}(-f_1, f_2) = -U_{oo}(f_1, -f_2) = U_{oo}(-f_1, -f_2), \\[4pt]
U_{eo}(f_1, f_2) &= U_{eo}(-f_1, f_2) = -U_{eo}(f_1, -f_2) = -U_{eo}(-f_1, -f_2), \\[4pt]
U_{oe}(f_1, f_2) &= -U_{oe}(-f_1, f_2) = U_{oe}(f_1, -f_2) = -U_{oe}(-f_1, -f_2).
\end{aligned}
\qquad (4.29)
$$

Consequently, we can write

$$U(-f_1, f_2) = U_{ee}(f_1, f_2) + U_{oo}(f_1, f_2) - j\left[U_{eo}(f_1, f_2) - U_{oe}(f_1, f_2)\right] \qquad (4.30)$$

$$U(-f_1, -f_2) = U_{ee}(f_1, f_2) - U_{oo}(f_1, f_2) + j\left[U_{eo}(f_1, f_2) + U_{oe}(f_1, f_2)\right] \qquad (4.31)$$

$$U(f_1, -f_2) = U_{ee}(f_1, f_2) + U_{oo}(f_1, f_2) - j\left[-U_{eo}(f_1, f_2) + U_{oe}(f_1, f_2)\right] \qquad (4.32)$$

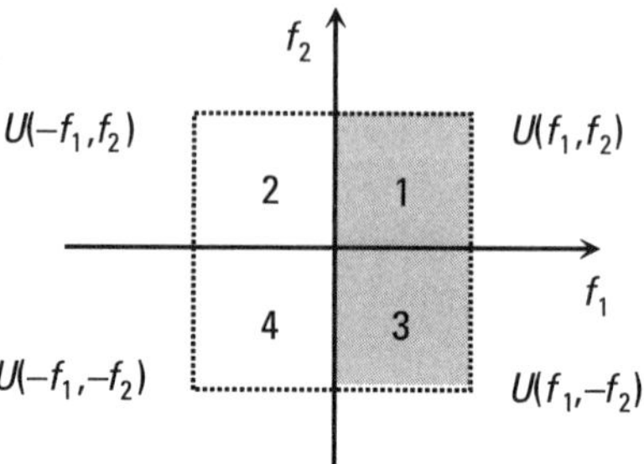

Figure 4.6   Half-plane support of the 2-D Fourier spectrum of a 2-D real signal.

Then, we notice that

$$U\left(-f_1,f_2\right)=U^*\left(f_1,-f_2\right) \tag{4.33}$$

$$U\left(-f_1,-f_2\right)=U^*\left(f_1,f_2\right) \tag{4.34}$$

The above relations are known as the *Hermitian symmetry* properties of the 2-D Fourier transform $U(f_1,f_2)$ of a 2-D real signal $u(x_1,x_2)$ (see Appendix C). We see that the *full information about the frequency content of a 2-D real signal is contained in a half-plane* $(f_1 > 0,f_2)$ (Figure 4.6).

The *amplitude spectrum* of a 2-D real signal has a general form

$$A\left(f_1,f_2\right)=\left|U\left(f_1,f_2\right)\right|=\sqrt{\left(U_{ee}-U_{oo}\right)^2+\left(U_{eo}+U_{oe}\right)^2} \tag{4.35}$$

and its *phase spectrum* is given by

$$\varphi\left(f_1,f_2\right)=\arg U\left(f_1,f_2\right)=\operatorname{atan2}\left(-U_{eo}-U_{oe},U_{ee}-U_{oo}\right) \tag{4.36}$$

Example 4.4 Spectrum of the 2-D Pyramid Signal

The 2-D pyramid signal (Figure 4.7(a)) is given by a function of the form

$$u\left(x_1,x_2\right)=\begin{cases}\dfrac{h}{a}\left(a-\left|x_1\right|-\left|x_2\right|\right), & \left(x_1,x_2\right)\in S\\[2mm]0, & \left(x_1,x_2\right)\notin S\end{cases} \tag{4.37}$$

where h is the height and S is the support of the pyramid, as shown in Figure 4.7(b) for $h = 1$, $a = 1$. The spectrum of (4.37) is given by

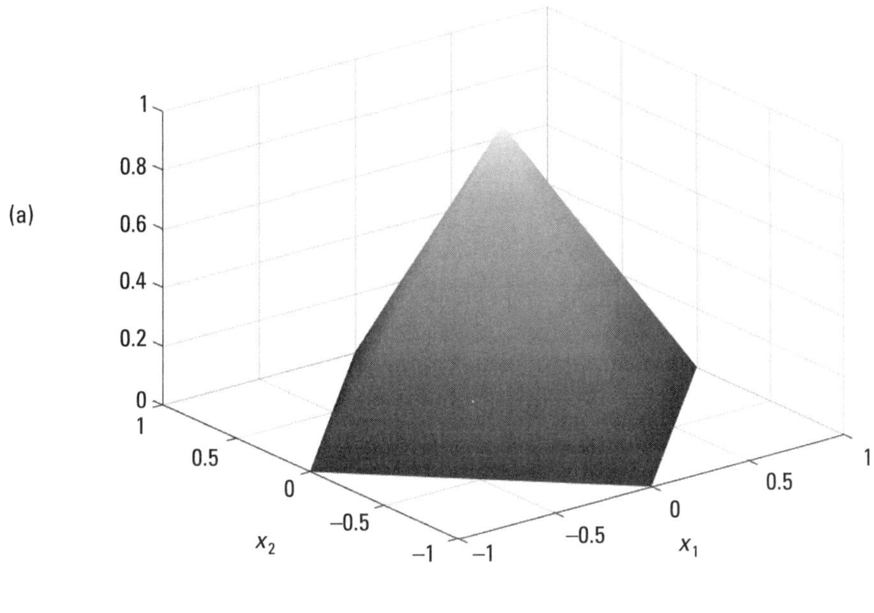

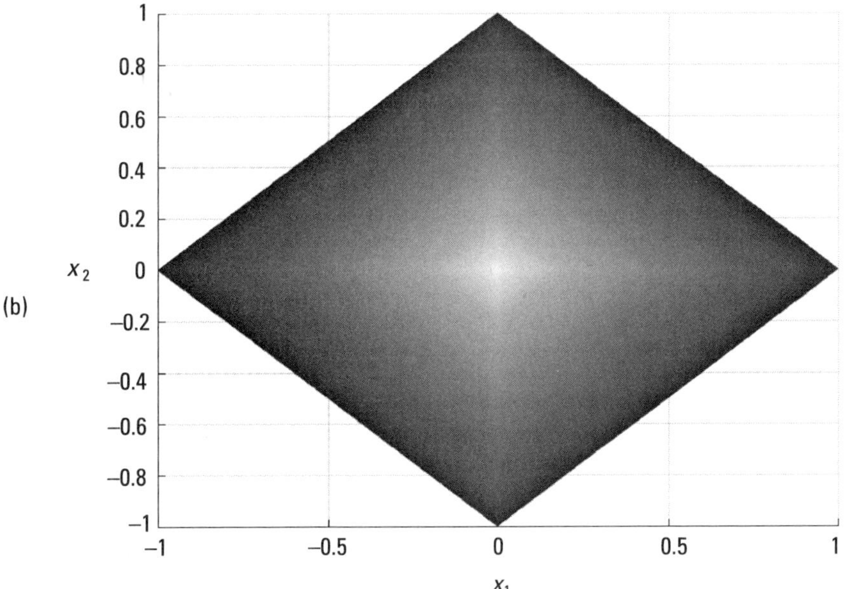

Figure 4.7   (a) A 2-D pyramid signal (4.37), $h = 1$, $a = 1$, and (b) its surface plot. (c) Spectrum $U(f_1, f_2)$. (d) Amplitude spectrum. (e) Phase spectrum.

(c)

(d)

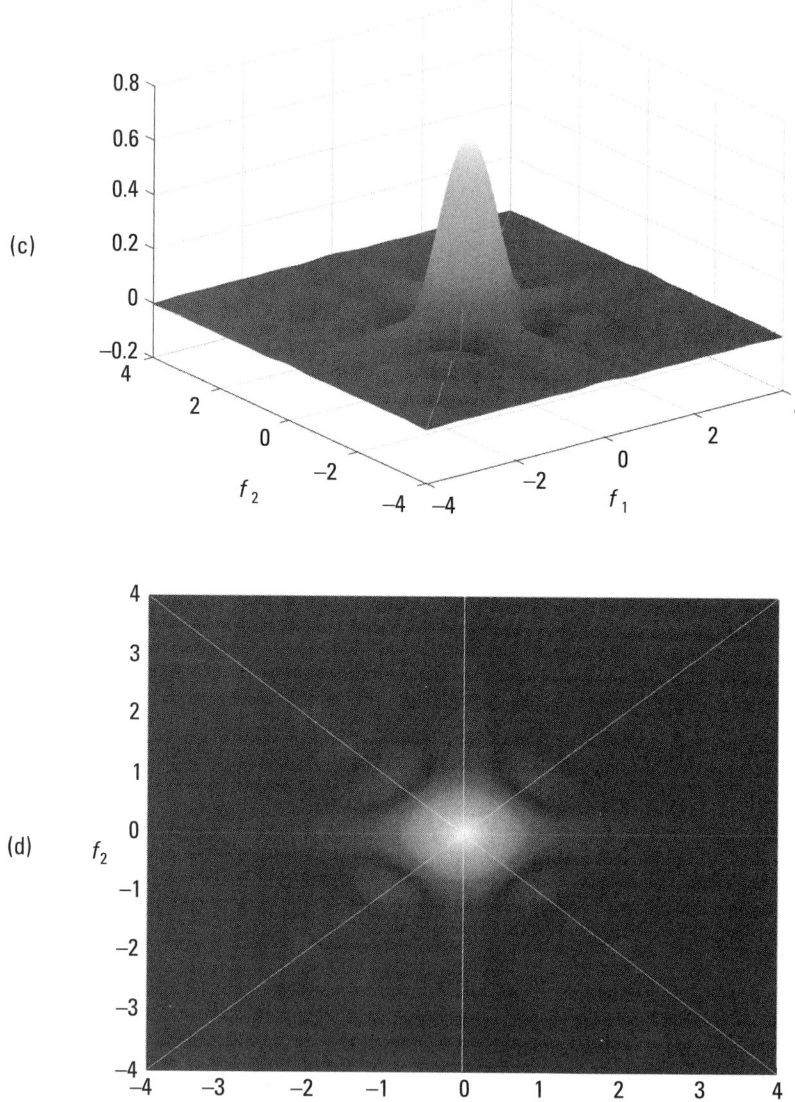

Figure 4.7   Continued

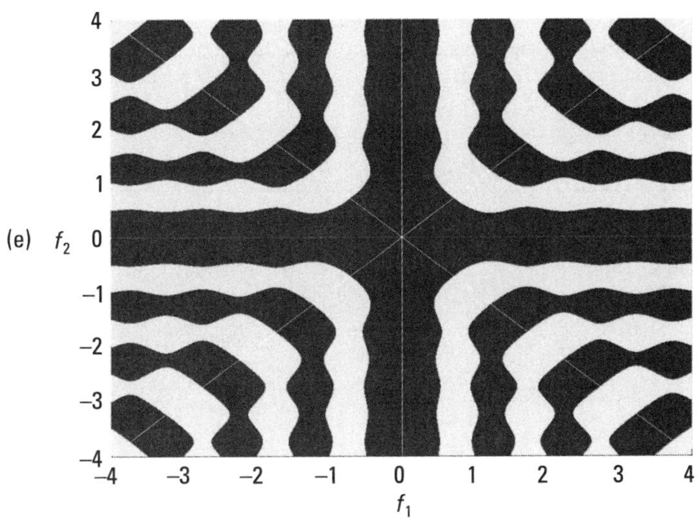

Figure 4.7 Continued

$$U\left(f_1,f_2\right)=\frac{4h}{4\pi^2\left(f_2^2-f_1^2\right)}\left[\frac{\sin\left(2\pi f_1 a\right)}{2\pi f_1 a}-\frac{\sin\left(2\pi f_2 a\right)}{2\pi f_2 a}\right] \tag{4.38}$$

The plot of (4.38), the surface plots of the amplitude spectrum (4.35) and the phase spectrum (4.36) are presented in Figure 4.7(c–e).

4.1.3 Spectrum of a 3-D Real Signal in Terms of its Even and Odd Components

The 3-D real signal $u(x_1,x_2,x_3)$ is a sum of eight terms: $u(x_1,x_2,x_3) = x_{eee} + x_{eeo} + x_{eoe} + x_{eoo} + x_{oee} + x_{oeo} + x_{ooe} + x_{ooo}$ (all being functions of (x_1,x_2,x_3)). It can easily be shown that the 3-D complex FT (4.1) is

$$U\left(f_1,f_2,f_3\right)=U_{eee}-U_{ooe}-U_{oeo}-U_{eoo}+j\left(-U_{oee}-U_{eoe}-U_{eeo}+U_{ooo}\right) \tag{4.39}$$

where

$$U_{eee}\left(f_1,f_2,f_3\right)=\int_{\mathbb{R}^3}u_{eee}\cos\left(2\pi f_1 x_1\right)\cos\left(2\pi f_2 x_2\right)\cos\left(2\pi f_3 x_3\right)d^3x \tag{4.40}$$

$$U_{ooe}\left(f_1,f_2,f_3\right)=\int_{\mathbb{R}^3}u_{ooe}\sin\left(2\pi f_1 x_1\right)\sin\left(2\pi f_2 x_2\right)\cos\left(2\pi f_3 x_3\right)d^3x \tag{4.41}$$

Table 4.1

Symmetry Properties of the Even-Odd Components of $U(f_1, f_2, f_3)$

No.	Signs of f_i			U_{eee}	U_{ooe}	U_{oeo}	U_{eoo}	U_{oee}	U_{eoe}	U_{eeo}	U_{ooo}
	f_1	f_2	f_3								
1	+	+	+	+	−	−	−	−	−	−	+
2	−	+	+	+	+	+	−	+	−	−	−
3	+	−	+	+	+	−	+	−	+	−	+
4	−	−	+	+	−	+	+	+	+	−	+
5	+	+	−	+	−	+	+	−	−	+	+
6	−	+	−	+	+	−	+	+	−	+	+
7	+	−	−	+	+	+	−	−	+	+	+
8	−	−	−	+	−	−	−	+	+	+	−

$$U_{oeo}\left(f_1, f_2, f_3\right) = \int_{\mathbb{R}^3} u_{oeo} \sin\left(2\pi f_1 x_1\right)\cos\left(2\pi f_2 x_2\right)\sin\left(2\pi f_3 x_3\right)d^3\mathbf{x} \qquad (4.42)$$

$$U_{eoo}\left(f_1, f_2, f_3\right) = \int_{\mathbb{R}^3} u_{eoo} \cos\left(2\pi f_1 x_1\right)\sin\left(2\pi f_2 x_2\right)\sin\left(2\pi f_3 x_3\right)d^3\mathbf{x} \qquad (4.43)$$

$$U_{oee}\left(f_1, f_2, f_3\right) = \int_{\mathbb{R}^3} u_{oee} \sin\left(2\pi f_1 x_1\right)\cos\left(2\pi f_2 x_2\right)\cos\left(2\pi f_3 x_3\right)d^3\mathbf{x} \qquad (4.44)$$

$$U_{eoe}\left(f_1, f_2, f_3\right) = \int_{\mathbb{R}^3} u_{eoe} \cos\left(2\pi f_1 x_1\right)\sin\left(2\pi f_2 x_2\right)\cos\left(2\pi f_3 x_3\right)d^3\mathbf{x} \qquad (4.45)$$

$$U_{eeo}\left(f_1, f_2, f_3\right) = \int_{\mathbb{R}^3} u_{eeo} \cos\left(2\pi f_1 x_1\right)\cos\left(2\pi f_2 x_2\right)\sin\left(2\pi f_3 x_3\right)d^3\mathbf{x} \qquad (4.46)$$

$$U_{ooo}\left(f_1, f_2, f_3\right) = \int_{\mathbb{R}^3} u_{ooo} \sin\left(2\pi f_1 x_1\right)\sin\left(2\pi f_2 x_2\right)\sin\left(2\pi f_3 x_3\right)d^3\mathbf{x} \qquad (4.47)$$

The functions (4.40)–(4.47) satisfy the symmetry properties as shown in Table 4.1. So, basing on (4.39) we can easily show that the 3-D Fourier spectrum is *Hermitian symmetric*, that is

$$U\left(f_1, f_2, f_3\right) = U^*\left(-f_1, -f_2, -f_3\right),$$

$$U\left(-f_1, f_2, f_3\right) = U^*\left(f_1, -f_2, -f_3\right),$$

$$U\left(f_1, -f_2, f_3\right) = U^*\left(-f_1, f_2, -f_3\right), \tag{4.48}$$

$$U\left(-f_1, -f_2, f_3\right) = U^*\left(f_1, f_2, -f_3\right).$$

We see that the *full information about the frequency content of a 3-D real signal is contained in a half-space* $\left(f_1 > 0, f_2, f_3\right)$ (Figure 4.8).

4.2 Cayley-Dickson Fourier Transformation

4.2.1 General Formulas

The *n-D Cayley-Dickson Fourier transformation* is defined on the basis of the Cayley-Dickson algebra described in Chapter 2, Section 2.1. Its kernel differs from the kernel of the *n*-D complex FT given by (4.1), as will be explained below.

The general form of the *n-D Cayley-Dickson Fourier transform* $U_{CD}(f)$, $f = \left(f_1, f_1, \ldots, f_n\right)$ of a *n*-D real signal $u(x)$, $x = \left(x_1, x_1, \ldots, x_n\right)$ is

$$U_{CD}(f) = \int_{\mathbb{R}^n} u(x) \prod_{i=1}^{n} \exp\left(-e_k 2\pi f_i x_i\right) d^n x \tag{4.49}$$

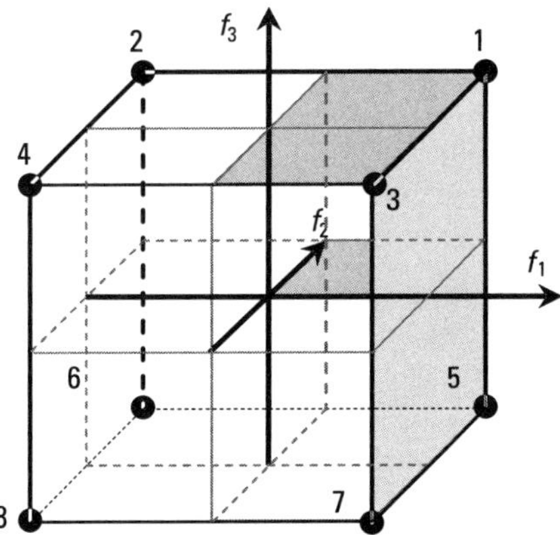

Figure 4.8   The half-space support of the Fourier spectrum of a 3-D real signal.

where the subscript $k = 2^{i-1}$. For $n = 1$, (4.49) is exactly the same as (4.2). For $n = 2$, the definition (4.49) is equivalent to the *Quaternion Fourier transform* $U_q(f_1,f_2)$ of a 2-D real signal $u(x_1,x_2)$ defined by T. Bülow [4] in the form

$$U_q\left(f_1,f_2\right)= \int_{\mathbb{R}^2} u\left(x_1,x_2\right)e^{-e_1 2\pi f_1 x_1}e^{-e_2 2\pi f_2 x_2}\, dx_1\, dx_2 \qquad (4.50)$$

Introducing in (4.50) the following simplified notation for $i = 1,2$:

$$c_i = \cos\left(2\pi f_i x_i\right),\ s_i = \sin\left(2\pi f_i x_i\right) \qquad (4.51)$$

we expand the kernel as follows

$$\exp\left(-e_1 2\pi f_1 x_1\right)\exp\left(-e_2 2\pi f_2 x_2\right)=\left(c_1 - s_1\cdot e_1\right)\left(c_2 - s_2\cdot e_2\right)$$
$$= c_1 c_2 - s_1 c_2\cdot e_1 - c_1 s_2\cdot e_2 + s_1 s_2\cdot e_3 \qquad (4.52)$$

We see that the kernel (4.52) has a form of a quaternion-valued function (we applied the multiplication rule of multiplication in $\mathbb{H}$: $e_1\cdot e_2 = e_3$).

Let us consider now the case $n = 3$. The Cayley-Dickson spectrum of a 3-D real signal $u(x_1,x_2,x_3)$ is expressed as

$$U_{CD}\left(f_1,f_2,f_3\right)= \int_{\mathbb{R}^3} u\left(x_1,x_2,x_3\right)e^{-e_1 2\pi f_1 x_1}e^{-e_2 2\pi f_2 x_2}e^{-e_4 2\pi f_3 x_3}\, dx_1 dx_2 dx_3 \qquad (4.53)$$

In the above definition, the order of multiplication of exponentials forming the kernel of the transformation is from left to right. We will show below that this product is an octonion-valued function and this fact inspired the authors of [5] to call (4.53) the *octonion Fourier transform* of a 3-D real signal $u(x_1,x_2,x_3)$. The choice of imaginary units $\{e_1,e_2,e_4\}$ in exponents is arbitrary and has been inspired by the definition (4.50) of the QFT, where the order is $\{e_1,e_2\}$. Of course, other sets of imaginary units in the kernel of the Cayley-Dickson FT could be used to get the octonion kernel function, such as $\{e_1,e_7,e_2\}$ or $\{e_2,e_3,e_6\}$ (the reader can easily find other triples basing on Table 2.3 or Figure 2.2 of Chapter 2). The kernel of (4.53) is expanded as

$$\left(e^{-e_1 2\pi f_1 x_1}e^{-e_2 2\pi f_2 x_2}\right)e^{-e_4 2\pi f_3 x_3} =\left[\left(c_1 - s_1\cdot e_1\right)\left(c_2 - s_2\cdot e_2\right)\right]\left(c_3 - s_3\cdot e_4\right)$$
$$= c_1 c_2 c_3 - s_1 c_2 c_3\cdot e_1 - c_1 s_2 c_3\cdot e_2 + s_1 s_2 c_3\cdot e_3 - c_1 c_2 s_3\cdot e_4 \qquad (4.54)$$
$$+ s_1 c_2 s_3\cdot e_5 + c_1 s_2 s_3\cdot e_6 - s_1 s_2 s_3\cdot e_7$$

where we adopted the notation (4.51) and applied multiplication rules of the algebra of octonions $\mathbb{O}$ (see Chapter 2, Table 2.3).

The *inverse n-D Cayley-Dickson FT* is defined as:

$$u(x) = \int_{\mathbb{R}^n} U_{CD}(f) \prod_{i=0}^{n-1} \exp\left(e_k 2\pi f_{n-i} x_{n-i}\right) d^n f \qquad (4.55)$$

where $k = 2^{n-i}$. For $n = 2$, we get the definition of the *inverse quaternion Fourier transform*:

$$u(x_1, x_2) = \int_{\mathbb{R}^2} U_q(f_1, f_2) \exp\left(e_2 2\pi f_2 x_2\right) \exp\left(e_1 2\pi f_1 x_1\right) df_1 df_2 \qquad (4.56)$$

If $n = 3$, the formula (4.55) defines the *inverse octonion Fourier transform*, [5].

$$u(x_1, x_2, x_3)$$
$$= \int_{\mathbb{R}^3} U_{CD}(f_1, f_2, f_3) \exp\left(e_4 2\pi f_3 x_3\right) \exp\left(e_2 2\pi f_2 x_2\right) \exp\left(e_1 2\pi f_1 x_1\right) df_1 df_2 df_3$$

$$(4.57)$$

Similarly as in Sections 4.1.1–4.1.3, we will now express the Cayley-Dickson FT of a real signal as a sum of even-odd terms with regard to its frequency variables. Let us consider first the case of $n = 2$.

4.2.2 Quaternion Fourier Spectrum in Terms of its Even and Odd Components

We know that the 2-D real signal $u(x_1, x_2)$ is a sum of its four components: even-even, even-odd, odd-even and odd-odd, respectively: $u(x_1, x_2) = u_{ee}(x_1, x_2) + u_{eo}(x_1, x_2) + u_{oe}(x_1, x_2) + u_{oo}(x_1, x_2)$. We introduce this form into (4.50) and get

$$U_q(f_1, f_2) = U_{ee} - U_{oe} \cdot e_1 - U_{eo} \cdot e_2 + U_{oo} \cdot e_3 \qquad (4.58)$$

where all terms are defined as in (4.25)–(4.28). We remember that U_{ee}, U_{oe}, U_{eo} and U_{oo} satisfy symmetry relations shown in (4.29). Therefore we obtain

$$U_q(-f_1, f_2) = U_{ee} + U_{oe} \cdot e_1 - U_{eo} \cdot e_2 - U_{oo} \cdot e_3 \qquad (4.59)$$

$$U_q\left(f_1,-f_2\right)=U_{ee}-U_{oe}\cdot e_1+U_{eo}\cdot e_2-U_{oo}\cdot e_3 \qquad (4.60)$$

$$U_q\left(-f_1,-f_2\right)=U_{ee}+U_{oe}\cdot e_1+U_{eo}\cdot e_2+U_{oo}\cdot e_3 \qquad (4.61)$$

Let us show that the Quaternion Fourier spectrum in successive quadrants of the frequency plane is expressed as a hypercomplex function of the spectrum in the first quadrant $\left(f_1 > 0, f_2 > 0\right)$, that is

$$U_q\left(-f_1,f_2\right)=-e_2\cdot U_q\left(f_1,f_2\right)\cdot e_2 \qquad (4.62)$$

$$U_q\left(f_1,-f_2\right)=-e_1\cdot U_q\left(f_1,f_2\right)\cdot e_1 \qquad (4.63)$$

$$U_q\left(-f_1,-f_2\right)=-e_3\cdot U_q\left(f_1,f_2\right)\cdot e_3 \qquad (4.64)$$

called *involutions* [6, 7].[2]

Proof of (4.62): Using the multiplication rules in the algebra $\mathbb{H}$

$$\left(e_i\cdot e_j\right)\cdot e_i = \begin{cases} e_j & \text{for } i \neq j \\ -e_i & \text{for } i = j \end{cases} \qquad (4.65)$$

let us develop the right side of (4.62):

$$-e_2\cdot U_q\left(f_1,f_2\right)\cdot e_2 = -e_2\cdot\left(U_{ee}-U_{oe}\cdot e_1-U_{eo}\cdot e_2+U_{oo}\cdot e_3\right)\cdot e_2$$

$$= U_{ee}+U_{oe}\cdot e_1-U_{eo}\cdot e_2-U_{oo}\cdot e_3 = U_q\left(-f_1,f_2\right)$$

getting the formula (4.59) and proving the symmetry relation (4.62). In the exact same manner (keeping the order of multiplication from left to right) we can prove formulas (4.63)–(4.64). We leave this to the reader.

As a consequence of (4.62)–(4.64), we see that *using the quaternion approach, the whole information about the real signal $u(x_1,x_2)$ is contained in a single quadrant of the frequency plane*, usually chosen as $\left(f_1 > 0, f_2 > 0\right)$ (Figure 4.9).

---

[2] The *involution* f means a function that is its own inverse (e.g., $f(f(x)) = x$ for every x belonging to the domain of f). The involution is linear: $f(\alpha x_1 + \beta x_2) = \alpha f(x_1) + \beta f(x_2)$; $\alpha,\beta \in \mathbb{R}$ and the involution of a product is the product of involutions written in the reversed order: $f(x_1 \cdot x_2) = f(x_2) \cdot f(x_1)$ [7].

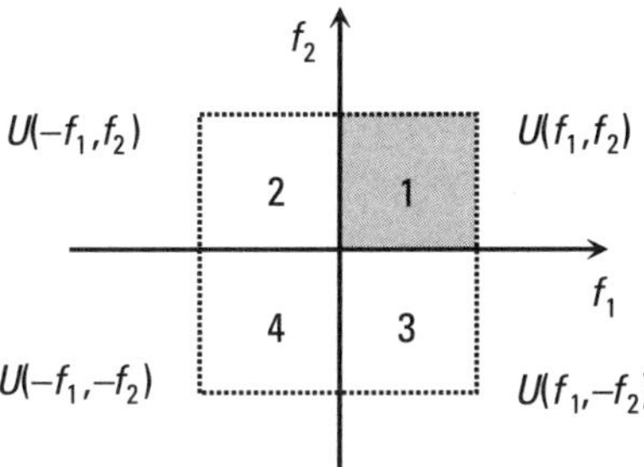

Figure 4.9 First-quadrant support of the 2-D quaternion Fourier spectrum of a 2-D real signal.

4.2.3 Octonion Fourier Spectrum in Terms of its Even and Odd Components

As a 3-D real signal $u(x_1,x_2,x_3)$ is a sum of eight terms: $u(x_1,x_2,x_3) = u_{eee} + u_{eeo} + u_{eoe} + u_{eoo} + u_{oee} + u_{oeo} + u_{ooe} + u_{ooo}$ (all being functions of (x_1,x_2,x_3)), we introduce it to (4.53) as follows

$$U_{CD}(f_1,f_2,f_3)$$
$$= U_{eee} - U_{oee} \cdot e_1 - U_{eoe} \cdot e_2 + U_{ooe} \cdot e_3 - U_{eeo} \cdot e_4 + U_{oeo} \cdot e_5 + U_{eoo} \cdot e_6 - U_{ooo} \cdot e_7$$

$$(4.66)$$

and express the Octonion Fourier transform of a 3-D real signal as an *octonion* sum of the same terms as in (4.39). In Table 4.2, we show the signs of all terms of the octonion FT in all octants of the 3-D frequency space.

Let us show that the Octonion Fourier spectrum in successive octants of the frequency space is expressed as a hypercomplex function of the spectrum in the first octant $(f_1 > 0, f_2 > 0, f_3 > 0)$. To simplify the notation, we denote with α_i the involution:

$$\alpha_i(U_{CD}) = (e_i \cdot U_{CD}) \cdot e_i \qquad (4.67)$$

It means that, for every imaginary unit from the octonion basis $\{1, e_1, e_2, \ldots, e_7\}$, we can write $\alpha_i(e_j) = (e_i \cdot e_j) \cdot e_i$. If $i = j$, $\alpha_i(e_i) = (e_i \cdot e_i) \cdot e_i = -1 \cdot e_i = -e_i$. In the case of $i \neq j$, $\alpha_i(e_j) = (e_i \cdot e_j) \cdot e_i = (-e_j \cdot e_i) \cdot e_i$ and then, using the *alternativity* of octonions (see Chapter 2, (2.38)), we get $\alpha_i(e_j) = (-e_j \cdot e_i) \cdot e_i = -(e_j \cdot e_i) \cdot e_i = -e_j(e_i \cdot e_i) = -e_j \cdot (-1) = e_j$. Shortly:

$$\alpha_i(e_j) = \begin{cases} e_j & \text{for } i \neq j \\ -e_i & \text{for } i = j \end{cases} \qquad (4.68)$$

Table 4.2
Symmetry Properties of the Even-Odd Components of $U_{CD}(f_1, f_2, f_3)$

No	Signs of f_i			U_{CD}	$U_{eee} \cdot 1$	$U_{oee} \cdot e_1$	$U_{eoe} \cdot e_2$	$U_{ooe} \cdot e_3$	$U_{eeo} \cdot e_4$	$U_{oeo} \cdot e_5$	$U_{eoo} \cdot e_6$	$U_{ooo} \cdot e_7$
	f_1	f_2	f_3									
1	+	+	+	$U_{CD}(f_1, f_2, f_3)$	+	−	−	+	−	+	+	−
2	−	+	+	$U_{CD}(-f_1, f_2, f_3)$	+	+	−	−	−	−	+	+
3	+	−	+	$U_{CD}(f_1, -f_2, f_3)$	+	−	+	−	−	+	−	+
4	−	−	+	$U_{CD}(-f_1, -f_2, f_3)$	+	+	+	+	−	−	−	−
5	+	+	−	$U_{CD}(f_1, f_2, -f_3)$	+	−	−	+	+	−	−	+
6	−	+	−	$U_{CD}(-f_1, f_2, -f_3)$	+	+	−	−	+	+	−	−
7	+	−	−	$U_{CD}(f_1, -f_2, -f_3)$	+	−	+	−	+	−	+	−
8	−	−	−	$U_{CD}(-f_1, -f_2, -f_3)$	+	+	+	+	+	+	+	+

Using (4.67) and (4.68), we get the following symmetry relations:

$$U_{CD}\left(-f_1,f_2,f_3\right)=\alpha_1\left(\alpha_3\left(\alpha_5\left(\alpha_7\left(U_{CD}\left(f_1,f_2,f_3\right)\right)\right)\right)\right)$$
$$=\alpha_7\left(\alpha_5\left(\alpha_3\left(\alpha_1\left(U_{CD}\left(f_1,f_2,f_3\right)\right)\right)\right)\right)$$
$$=-\alpha_2\left(\alpha_4\left(\alpha_6\left(U_{CD}\left(f_1,f_2,f_3\right)\right)\right)\right)$$
$$=-\alpha_6\left(\alpha_4\left(\alpha_2\left(U_{CD}\left(f_1,f_2,f_3\right)\right)\right)\right)$$

$$(4.69)$$

$$U_{CD}\left(f_1,-f_2,f_3\right)=\alpha_2\left(\alpha_3\left(\alpha_6\left(\alpha_7\left(U_{CD}\left(f_1,f_2,f_3\right)\right)\right)\right)\right)$$
$$=\alpha_7\left(\alpha_6\left(\alpha_3\left(\alpha_2\left(U_{CD}\left(f_1,f_2,f_3\right)\right)\right)\right)\right)$$
$$=-\alpha_1\left(\alpha_4\left(\alpha_5\left(U_{CD}\left(f_1,f_2,f_3\right)\right)\right)\right)$$
$$=-\alpha_5\left(\alpha_4\left(\alpha_1\left(U_{CD}\left(f_1,f_2,f_3\right)\right)\right)\right)$$

$$(4.70)$$

$$U_{CD}\left(-f_1,-f_2,f_3\right)=\alpha_1\left(\alpha_2\left(\alpha_5\left(\alpha_6\left(U_{CD}\left(f_1,f_2,f_3\right)\right)\right)\right)\right)$$
$$=\alpha_6\left(\alpha_5\left(\alpha_2\left(\alpha_1\left(U_{CD}\left(f_1,f_2,f_3\right)\right)\right)\right)\right)$$
$$=-\alpha_3\left(\alpha_4\left(\alpha_7\left(U_{CD}\left(f_1,f_2,f_3\right)\right)\right)\right)$$
$$=-\alpha_7\left(\alpha_4\left(\alpha_3\left(U_{CD}\left(f_1,f_2,f_3\right)\right)\right)\right)$$

$$(4.71)$$

$$U_{CD}\left(f_1,f_2,-f_3\right)=\alpha_3\left(\alpha_4\left(\alpha_5\left(\alpha_6\left(U_{CD}\left(f_1,f_2,f_3\right)\right)\right)\right)\right)$$
$$=\alpha_6\left(\alpha_5\left(\alpha_4\left(\alpha_3\left(U_{CD}\left(f_1,f_2,f_3\right)\right)\right)\right)\right)$$
$$=-\alpha_1\left(\alpha_2\left(\alpha_7\left(U_{CD}\left(f_1,f_2,f_3\right)\right)\right)\right)$$
$$=-\alpha_7\left(\alpha_2\left(\alpha_1\left(U_{CD}\left(f_1,f_2,f_3\right)\right)\right)\right)$$

$$(4.72)$$

$$U_{CD}\left(-f_1,f_2,-f_3\right) = \alpha_1\left(\alpha_3\left(\alpha_4\left(\alpha_6\left(U_{CD}\left(f_1,f_2,f_3\right)\right)\right)\right)\right)$$
$$= \alpha_6\left(\alpha_4\left(\alpha_3\left(\alpha_1\left(U_{CD}\left(f_1,f_2,f_3\right)\right)\right)\right)\right)$$
$$= -\alpha_2\left(\alpha_5\left(\alpha_7\left(U_{CD}\left(f_1,f_2,f_3\right)\right)\right)\right)$$
$$= -\alpha_7\left(\alpha_5\left(\alpha_2\left(U_{CD}\left(f_1,f_2,f_3\right)\right)\right)\right) \tag{4.73}$$

$$U_{CD}\left(f_1,-f_2,-f_3\right) = \alpha_2\left(\alpha_3\left(\alpha_4\left(\alpha_5\left(U_{CD}\left(f_1,f_2,f_3\right)\right)\right)\right)\right)$$
$$= \alpha_5\left(\alpha_4\left(\alpha_3\left(\alpha_2\left(U_{CD}\left(f_1,f_2,f_3\right)\right)\right)\right)\right)$$
$$= -\alpha_1\left(\alpha_6\left(\alpha_7\left(U_{CD}\left(f_1,f_2,f_3\right)\right)\right)\right)$$
$$= -\alpha_7\left(\alpha_6\left(\alpha_1\left(U_{CD}\left(f_1,f_2,f_3\right)\right)\right)\right) \tag{4.74}$$

$$U_{CD}\left(-f_1,-f_2,-f_3\right) = \alpha_1\left(\alpha_2\left(\alpha_4\left(\alpha_7\left(U_{CD}\left(f_1,f_2,f_3\right)\right)\right)\right)\right)$$
$$= \alpha_7\left(\alpha_4\left(\alpha_2\left(\alpha_1\left(U_{CD}\left(f_1,f_2,f_3\right)\right)\right)\right)\right)$$
$$= -\alpha_3\left(\alpha_5\left(\alpha_6\left(U_{CD}\left(f_1,f_2,f_3\right)\right)\right)\right)$$
$$= -\alpha_6\left(\alpha_5\left(\alpha_3\left(U_{CD}\left(f_1,f_2,f_3\right)\right)\right)\right) \tag{4.75}$$

We see that using the octonion approach, the *full information about the frequency content of a 3-D real signal is included in a single octant* $\left(f_1 > 0, f_2 > 0, f_3 > 0\right)$ *of the 3-D frequency space* (Figure 4.10).

Let us now prove the relation (4.69).

Proof: Based on Table 4.2, we see that $U_{CD}\left(-f_1,f_2,f_3\right)$ differs from $U_{CD}\left(f_1,f_2,f_3\right)$ in signs of components corresponding to e_1,e_3,e_5,e_7. We notice that, according to (4.68), each involution α_i changes the sign of $U_{eee} \cdot 1$ and the sign of the term corresponding to e_i. In Table 4.3 the effects of superposition of four involutions $\alpha_1 \circ \alpha_3 \circ \alpha_5 \circ \alpha_7$ are shown. We observe that the bottom row is exactly the same as the second from Table 4.2. In an analogous way, we can prove other relations (4.69)–(4.75).

Table 4.3
Superposition of Involutions $\alpha_1 \circ \alpha_3 \circ \alpha_5 \circ \alpha_7$

Involution		$U_{eee}\cdot 1$	$U_{oee}\cdot e_1$	$U_{eoe}\cdot e_2$	$U_{ooe}\cdot e_3$	$U_{eeo}\cdot e_4$	$U_{oeo}\cdot e_5$	$U_{eoo}\cdot e_6$	$U_{ooo}\cdot e_7$
	$U_{CD}(f_1,f_2,f_3) = U$	+	−	−	+	−	+	+	−
α_1	$(e_1 \cdot U) \cdot e_1 = V$	−	+	−	+	−	+	+	−
α_3	$(e_3 \cdot V) \cdot e_3 = W$	+	+	−	−	−	+	+	−
α_5	$(e_5 \cdot W) \cdot e_5 = X$	−	+	−	−	−	−	+	−
α_7	$(e_7 \cdot X) \cdot e_5 = U_{CD}(-f_1,f_2,f_3)$	+	+	−	−	−	−	+	+

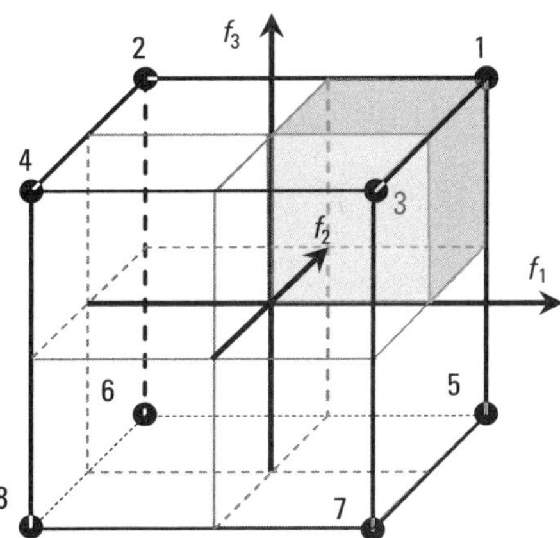

Figure 4.10 The single-octant support of the octonion Fourier spectrum of a 3-D real signal.

4.3 Relations Between Complex and Hypercomplex Fourier Transforms

In this Section, we will show the equivalence of complex and hypercomplex approaches in 2-D and 3-D signal analysis. By having a complex Fourier spectrum of a 2-D/3-D real signal, we can easily derive analytic formulas of quaternion and octonion Fourier spectra. The reader can follow up proofs of both formulas and also notice usefulness of even-odd decomposition of complex Fourier spectra.

4.3.1 Relation Between QFT and 2-D FT

Let us consider a 2-D real signal $u(x_1,x_2)$ and its 2-D complex Fourier transform $U(f_1,f_2)$ given by (4.5). Its Quaternion FT is given by (4.50). It can be shown that the QFT is related to $U(f_1,f_2)$ by the relation

$$U_q(f_1,f_2) = U(f_1,f_2)\frac{1-e_3}{2} + U(f_1,-f_2)\frac{1+e_3}{2} \qquad (4.76)$$

It should be noted that the order of terms in (4.76) plays no role in case if the spectrum $U(f_1,f_2)$ is a real function. If $U(f_1,f_2)$ is a complex-valued function, the order cannot be changed (bearing in mind the properties of multiplication in Cayley-Dickson algebras (the absence of commutativity is described in Chapter 2)).

Proof of (4.76): Starting with (4.50), we can write

$$U(f_1,f_2) = \iint_{\mathbb{R}^2} u(x_1,x_2)\exp(-e_1 2\pi f_1 x_1)\exp(-e_1 2\pi f_2 x_2)\,dx_1\,dx_2$$
$$U(f_1,-f_2) = \iint_{\mathbb{R}^2} u(x_1,x_2)\exp(-e_1 2\pi f_1 x_1)\exp(e_1 2\pi f_2 x_2)\,dx_1\,dx_2 \qquad (4.77)$$

We calculate two sums

$$\frac{U(f_1,f_2)+U(f_1,-f_2)}{2} = \int_{\mathbb{R}^2} u(x_1,x_2)\exp(-e_1 2\pi f_1 x_1)\cos(2\pi f_2 x_2)\,dx_1 dx_2 \qquad (4.78)$$

$$\frac{U(f_1,f_2)-U(f_1,-f_2)}{2} = \int_{\mathbb{R}^2} u(x_1,x_2)\exp(-e_1 2\pi f_1 x_1)\left[-e_1\sin(2\pi f_2 x_2)\right]\,dx_1 dx_2 \qquad (4.79)$$

Then, we multiply (4.79) from the right side by $(-e_3)$ and get

$$\frac{U(f_1,f_2)-U(f_1,-f_2)}{2}(-e_3)$$
$$= \int_{\mathbb{R}^2} u(x_1,x_2)\exp(-e_1 2\pi f_1 x_1)\left[-e_2\sin(2\pi f_2 x_2)\right]\,dx_1 dx_2 \qquad (4.80)$$

Adding (4.78) and (4.80), we obtain

$$\frac{U\left(f_1,f_2\right)+U\left(f_1,-f_2\right)}{2}+\frac{U\left(f_1,f_2\right)-U\left(f_1,-f_2\right)}{2}\left(-e_3\right)$$
$$=\int_{\mathbb{R}^2} u\left(x_1,x_2\right)\exp\left(-e_1 2\pi f_1 x_1\right)\exp\left(-e_2 2\pi f_2 x_2\right)dx_1\,dx_2$$

(4.81)

which proves (4.76).

It should be noted that (4.76) can easily be verified using the even-odd decomposition (4.24) and (4.58). We have

$$U\left(f_1,f_2\right)\cdot\frac{1-e_3}{2}+U\left(f_1,-f_2\right)\cdot\frac{1+e_3}{2}$$
$$=\left[U_{ee}-U_{oo}-\left(U_{eo}+U_{oe}\right)\cdot e_1\right]\cdot\frac{1-e_3}{2}$$
$$+\left[U_{ee}+U_{oo}-\left(U_{oe}-U_{eo}\right)\cdot e_1\right]\cdot\frac{1+e_3}{2}$$
$$=U_{ee}-U_{oe}\cdot e_1-U_{eo}\cdot e_2+U_{oo}\cdot e_3$$

and obtain the result identical to (4.58).

Let us remark that in Chapter 7, we will come back to complex and hypercomplex analysis of 2-D and 3-D real signals. Namely, we will analyze the polar components of 2-D complex and quaternion analytic signals and show that the number of amplitude-phase components is the same in both cases. Moreover, the polar components are related by formulas that can be simplified in case of separable signals.

4.3.2 Relation Between OFT and 3-D FT

The formula relating the 3-D OFT (4.53) and the 3-D FT (4.1) has been derived for the first time in [8]; after some modifications, it was presented with a proof in its final version in [9]. It is as follows:

$$U_{CD}\left(f_1,f_2,f_3\right)=\frac{1}{4}\left[U\left(f_1,f_2,f_3\right)+U\left(f_1,-f_2,f_3\right)\right]\cdot\left(1-e_5\right)$$
$$+\frac{1}{4}\left[U\left(f_1,f_2,-f_3\right)+U\left(f_1,-f_2,-f_3\right)\right]\cdot\left(1+e_5\right)$$
$$+\frac{1}{4}e_3\cdot\left[U\left(-f_1,f_2,f_3\right)-U\left(-f_1,-f_2,f_3\right)\right]\cdot\left(1+e_5\right)$$
$$+\frac{1}{4}e_3\cdot\left[U\left(-f_1,f_2,-f_3\right)-U\left(-f_1,-f_2,-f_3\right)\right]\cdot\left(1-e_5\right)$$

(4.82)

We do not present here the proof of (4.82). However, we can verify it basing on the decomposition of the 3-D complex FT into even/odd terms (4.39)

$$U\left(f_1, f_2, f_3\right) = U_{eee} - U_{ooe} - U_{oeo} - U_{eoo} + (-U_{oee} - U_{eoe} - U_{eeo} + U_{ooo}) \cdot e_1$$

and the symmetry properties of even-odd components of $U\left(f_1, f_2, f_3\right)$ shown in Table 4.1. Using multiplication rules in $\mathbb{O}$ (see Chapter 2, Table 2.3), we get for each term of (4.82):

$$\tfrac{1}{4} U_{eee}\left[1 - e_5 + 1 + e_5 + e_3 \cdot \left(1 + e_5\right) + e_3 \cdot \left(1 - e_5\right)\right] = U_{eee} \qquad (4.83)$$

$$\tfrac{1}{4} U_{ooe}\left[e_3 \cdot 2 \cdot \left(1 + e_5\right) + e_3 \cdot 2 \cdot \left(1 - e_5\right)\right] = U_{ooe} \cdot e_3 \qquad (4.84)$$

$$\tfrac{1}{4} U_{oeo}\left[e_3 \cdot (-2) \cdot \left(1 - e_5\right) + e_3 \cdot 2 \cdot \left(1 + e_5\right)\right] = U_{oeo} \cdot e_5 \qquad (4.85)$$

$$\tfrac{1}{4} U_{eoo}\left[e_3 \cdot (-2) \cdot \left(1 + e_5\right) + e_3 \cdot 2 \cdot \left(1 - e_5\right)\right] = U_{eoo} \cdot e_6 \qquad (4.86)$$

$$\tfrac{1}{4} U_{oee}\left[e_1 \cdot (-2) \cdot \left(1 - e_5\right) + e_1 \cdot (-2) \cdot \left(1 + e_5\right)\right] = -U_{oee} \cdot e_1 \qquad (4.87)$$

$$\tfrac{1}{4} U_{eoe}\left[e_3 \cdot \left(-2 \cdot e_1\right) \cdot \left(1 + e_5\right) + e_3 \cdot \left(-2 \cdot e_1\right) \cdot \left(1 - e_5\right)\right] = -U_{eoe} \cdot e_2 \qquad (4.88)$$

$$\tfrac{1}{4} U_{eeo}\left[\left(-2 \cdot e_1\right) \cdot \left(1 - e_5\right) + 2 \cdot e_1 \cdot \left(1 + e_5\right)\right] = -U_{eeo} \cdot e_4 \qquad (4.89)$$

$$\tfrac{1}{4} U_{ooo}\left[e_3 \cdot \left(-2 \cdot e_1\right) \cdot \left(1 + e_5\right) + e_3 \cdot 2 \cdot e_1 \cdot \left(1 - e_5\right)\right] = -U_{ooo} \cdot e_7 \qquad (4.90)$$

and finally we get the formula (4.66)

$$U_{CD}\left(f_1, f_2, f_3\right)$$
$$= U_{eee} - U_{oee} \cdot e_1 - U_{eoe} \cdot e_2 + U_{ooe} \cdot e_3 - U_{eeo} \cdot e_4 + U_{oeo} \cdot e_5 + U_{eoo} \cdot e_6 - U_{ooo} \cdot e_7$$

The above general formula (4.66) expresses the Octonion Fourier spectrum as an octonion sum of even-odd terms of the 3-D complex Fourier spectrum $U(f_1, f_2, f_3)$. Let us recall that "e" in the subscript means the even parity of $U(f_1, f_2, f_3)$ regarding a corresponding variable f_i and "o" is the odd parity. We see that the components are the same in both approaches (complex and hypercomplex), which once again proves their equivalence.

4.4 Survey of Applications of Complex and Hypercomplex Fourier Transformations

For many years and even recently, the most important role in signal processing has been dedicated to the 1-D and 2-D complex Fourier transformation. We could enumerate *ad infinitum* some different fields of their applications, such as spectrum analysis, signal filtration, signal sampling, modulation, time-frequency analysis, wavelet processing, stochastic signal processing, image analysis and processing, audio and image watermarking, and data compression. In [1–3, 10–13], one can find a lot of interesting applications of the complex Fourier transformation.

In this book, however, we would like to put the reader's attention on a continuously growing role of the hypercomplex Fourier analysis. In recent years, the Quaternion Fourier transform has become an alternative to the 2-D complex FT in color image processing [4, 14–20]. In [16], Sangwine and Ell defined the discrete QFT, and in [21] they decomposed it into a pair of 2-D complex FTs. Consequently, the fast Fourier transform (FFT) techniques can be applied in quaternion color image processing [22]. Various watermarking techniques based on embedding a color watermark into a chosen part of the quaternion hypercomplex spectrum have been developed in the last decade [23–27]. The Figure 4.11 shows a general image watermarking scheme using the QFT. The watermark can be embedded in the Quaternion Fourier spectrum coefficients of the original image according to the chosen *watermark embedding algorithm*.

Figure 4.12 shows the proposed general QFT-image watermarking detection scheme. The idea is based on the comparison of QFTs to the original and watermarked images, and as a result, the extraction of a watermark.

This idea is often combined with other methods, as with the Quaternion Singular Value Decomposition [27, 28]. The QFT is also used in color image enhancement [18, 29], and in pattern recognition and texture analysis [30–32]. The hypercomplex Fourier methods have also found some applications in physics [33, 34].

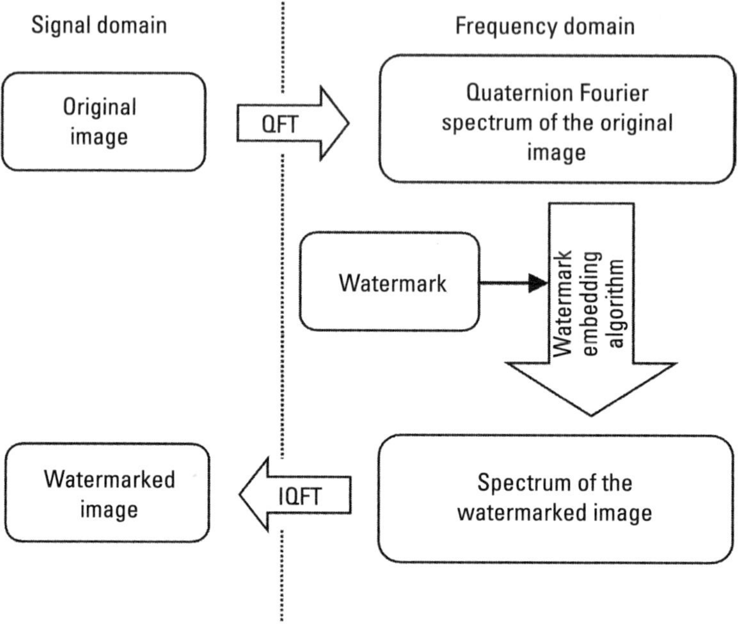

Figure 4.11   A general QFT-image watermarking scheme.

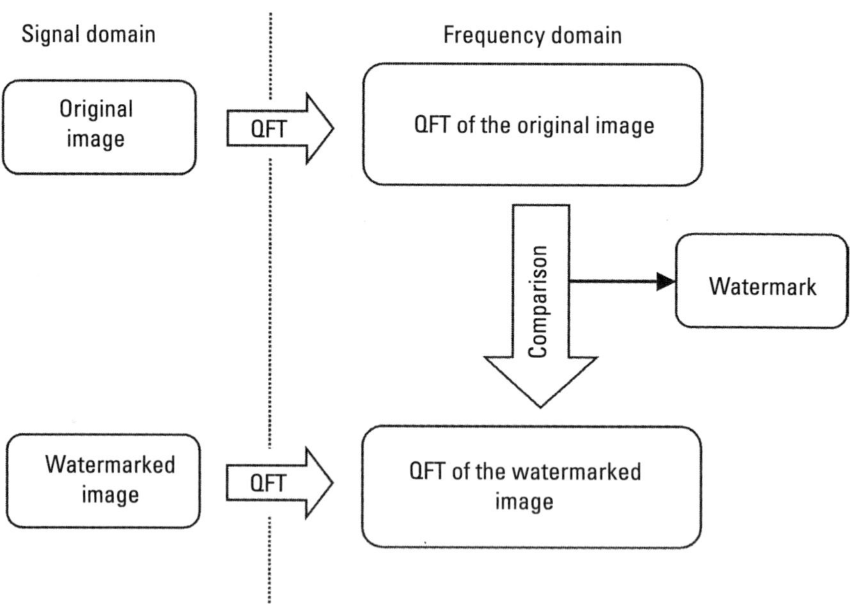

Figure 4.12   A general watermark detection scheme based on Quaternion Fourier transformation.

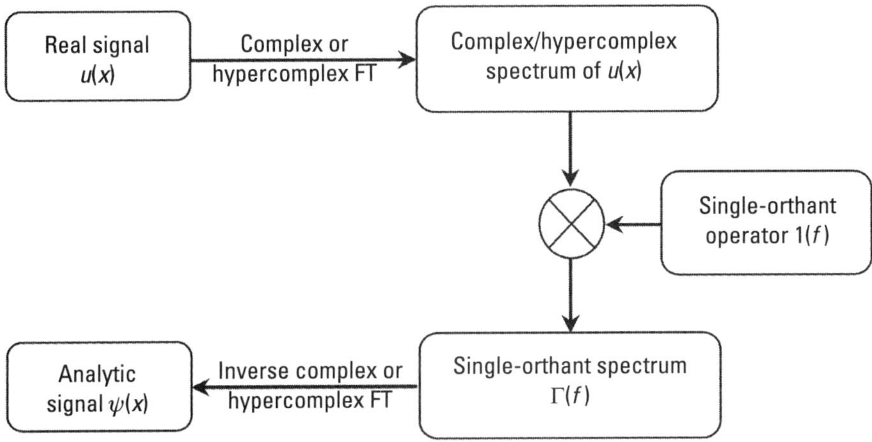

Figure 4.13 A general scheme of generation of the n-D analytic signal via the frequency domain.

4.4.1 Applications in the Domain of Analytic Signals

As will be shown in Chapter 5, the complex and hypercomplex Fourier transformation is a theoretical basis of the frequency-domain definitions of multidimensional analytic signals [35, 36]. The n-D complex/hypercomplex analytic signal is defined as the inverse complex/hypercomplex Fourier transform of the spectrum multiplied by the single-orthant operator (1.11). Figure 4.13 presents a general scheme of generation of the n-D complex/hypercomplex analytic signal $\psi(\boldsymbol{x})$.

4.5 Summary

In this chapter, we recalled basics of the theory of complex and hypercomplex Fourier transforms of n-D real signals with emphasis on 1-D, 2-D, and 3-D cases. We expressed different spectra as complex or hypercomplex sums of components differing in even/odd parity regarding corresponding variables. This allowed us to notice that, in the complex case, the full information about the frequency content of a n-D real signal is in a half space $f_1 > 0$. Differently, using the hypercomplex approach, we saw that a single orthant of the n-D frequency space contains full information about the frequency structure of n-D real signals. The formulas relating complex and hypercomplex FTs are very important from a practical point of view. They can be used to derive quaternion and octonion Fourier spectra starting with 2-D and 3-D complex FTs.

References

[1]   Mertins, A., *Signal Analysis*, Chichester, England: John Wiley & Sons Ltd., 1999.

[2]   Poularikas, A. D. (Ed.), *The Transforms and Applications Handbook, Second Edition*, Boca Raton, FL: CRC Press, IEEE Press, 2000.

[3]   Walker, J. S., *Fourier Analysis*, Oxford University Press, 1988.

[4]   Bülow, T., "Hypercomplex spectral signal representation for the processing and analysis of images," in *Bericht Nr. 99-3, Institut für Informatik und Praktische Mathematik*, Christian-Albrechts-Universität Kiel, August 1999.

[5]   Hahn, S. L., and K. M. Snopek, "The Unified Theory of n-Dimensional Complex and Hypercomplex Analytic Signals," *Bull. Polish Ac. Sci., Tech. Sci.*, Vol. 59, No. 2, 2011, pp. 167–181.

[6]   Bülow, T., and G. Sommer, "The Hypercomplex Signal—A Novel Extension of the Analytic Signal to the Multidimensional Case," *IEEE Trans. Signal Processing*, Vol. 49, No. 11, November 2001, pp. 2844–2852.

[7]   Ell, T. A., and S. J. Sangwine, "Quaternion involutions and anti-involutions," *Computers and Mathematics with Applications*, Vol. 53, 2007, pp. 137–143.

[8]   Hahn, S. L., and K. M. Snopek, "Comparison of Properties of Analytic, Quaternionic and Monogenic 2-D Signals," *WSEAS Transactions on Computers*, Issue 3, Vol. 3, July 2004, pp. 602–611.

[9]   Snopek, K.M., *Studies on Complex and Hypercomplex Multidimensional Analytic Signals*, prace naukowe, Elektronika z. 190, Oficyna Wydawnicza Politechniki Warszawskiej, Warsaw, 2013.

[10]   Folland, G. B., *Fourier Analysis and its Applications*, Pacific Grove, CA: Wadworth and Brooks/Cole, 1992.

[11]   Qian, S., and D. Chen, *Joint Time-Frequency Analysis*, Upper Saddle River, NJ: Prentice Hall PTR, 1966.

[12]   Flandrin, P., *Time-Frequency/Time-Scale Analysis*, San Diego, CA: Academic Press, 1999.

[13]   Allen, R. L., and D. W. Mills, *Signal Analysis—Time, Frequency, Scale, and Structure*, Piscataway, NJ: IEEE Press, Wiley-Interscience, 2004.

[14]   Gonzalez, R. C., and P. Wintz, *Digital Image Processing, Second Edition*, Boston, MA: Addison Wesley Publishing Company, 1987.

[15]   Sangwine, S. J., "Fourier transforms of colour images using quaternion or hypercomplex numbers," *Electron. Lett.*, Vol. 32, No. 21, October 1996, pp. 1979–1980.

[16]   Ell, T. A., and S. J. Sangwine, "Hypercomplex Fourier Transforms of Color Images," *IEEE Trans. Image Processing*, Vol. 16, No.1, January 2007, pp. 22–35.

[17]　Alexiadis, D. S., and G. D. Sergiadis, "Estimation of Motions in Color Image Sequences Using Hypercomplex Fourier Transforms," *IEEE Trans. Image Processing*, Vol. 18, No. 1, January 2009, pp. 168–187.

[18]　Khalil, M. I., "Applying Quaternion Fourier Transforms for Enhancing Color Images," *Int. J. of Image, Graphics and Signal Processing*, MECS, Vol. 4, No. 2, 2012, pp. 9–15.

[19]　Fernandez-Maloigne, C. (Ed.), *Advanced Color Image Processing and Analysis*, New York: Springer Science+Bussiness Media, 2013.

[20]　Venkatramana, R. B. D., and P. T. Yayachandra, *Color Image Processing Techniques using Quaternion Fourier Transforms*, Saarbrücken, Germany: LAP LAMBERT Academic Publishing, 2013.

[21]　Ell, T., and S. Sangwine, "Decomposition of 2D Hypercomplex Fourier Transforms into Pairs of Fourier Transforms," *Proc. EUSIPCO*, Vol. II, Tampere, Finland, Sept. 2000, pp. 1061–1064.

[22]　Felsberg, M., T. Bülow, G. Sommer, and V. M. Chernov, "Fast Algorithms of Hypercomplex Fourier Transforms," in *Geometric Computing with Clifford Algebras*, G. Sommer (Ed.), Berlin Heidelberg: Springer-Verlag, 2001, pp. 231–254.

[23]　Bas, P., N. Le Bihan, and J.-M.Chassery, "Color Image Watermarking Using Quaternion Fourier Transform," *Proc. ICASSP*, Hong Kong, 2003.

[24]　Tsui, T. K., X.-P. Zhang, and D. Androutsos, "Quaternion Image Watermarking using the Spatio-Chromatic Fourier Coefficients Analysis," *14th Annual ACM Int. Conf. on Multimedia, MM'2006*, Santa Barbara, CA, October 23–27, 2006, pp. 149–152.

[25]　Ma, X., Y. Xu, and X. Yang, "Color image watermarking using local quaternion Fourier spectra analysis," *IEEE Int. Conf. Multimedia and Expo*, Hannover, June 23–26, 2008, pp. 233–236.

[26]　Li, C., B. Li, L. Xiao, Y. Hu, and L. Tian, "A Watermarking Method Based on Hypercomplex Fourier Transform and Visual Attention," *J. Information & Computational Science*, Vol. 9, No. 15, 2012, pp. 4485–4492.

[27]　Wang, X., C. Wang, H. Yang, and P. Niu, "A robust blind color image watermarking in quaternion Fourier transform domain," *J. Systems and Software*, Vol. 86, No. 2, February 2013, pp. 255–277.

[28]　Sun, J., J. Yang, and D. Fu, "Color Images Watermarking Algorithm Based on Quaternion Frequency Singular Value Decomposition, " *Information and Control*, Vol. 40, No. 6, Shenyang, 2011, pp. 813–818.

[29]　Jin, L., H. Liu, X. Xu, and E. Song, "Quaternion-based color image filtering for impulsive noise suppression," *J. Electronic Imaging*, Vol. 19, No. 4, October–December, 2010.

[30]　Redfield, S. A., and Q. Q. Huynh, "Hypercomplex Fourier transforms applied to detection for side-scan sonar," *OCEANS'02 MTS/IEEE*, Vol. 4, October 29–31, 2002, pp. 2219–2224.

[31]   Assefa, D., L. Mansinha, K. F. Tiampo, H. Rasmussen, and K. Abdella, "Local quaternion Fourier transform and color texture analysis," *Signal Processing*, Vol. 60, No.
6, June 2010, pp. 1825–1835.

[32]   Martin, C. S., and S.-W. Kim (Eds.), "Progress in Pattern Recognition, Image Analysis, Computer Vision and Applications," *Proceedings of the 16th Iberoamerican Congress
CIARP'2011*, Pucón, Chile, November, 2011.

[33]   Mukundan, R., "Quaternions: From Classical Mechanics to Computer Graphics, and
Beyond," *Proc. 7th ATCM Conf*, Meleka, Malaysia, December 17–21, 2002, pp. 97–106.

[34]   Girard, P. R., *Quaternions, Clifford Algebras and Relativistic Physics*, Boston, MA:
Birkhäuser, 2008.

[35]   Hahn, S. L., "Multidimensional Complex Signals with Single-Orthant Spectra," *Proc.
IEEE*, Vol. 80, No. 8, 1992, pp. 1287–1300.

[36]   Hahn, S. L., *Hilbert Transforms in Signal Processing*, Norwood, MA: Artech House,
1996.

5

Complex and Hypercomplex Analytic Signals

This chapter presents the theory of multidimensional complex and hypercomplex analytic signals. The starting point is a real n-D signal $u(\boldsymbol{x})$, a function of the n-D Cartesian variable $\boldsymbol{x} = (x_1, x_1, \ldots, x_n)$. In one dimension, we usually deal with functions of time $u(t)$. The 1-D Fourier spectrum $U(f)$ may be a complex function $U(f) = \text{Re}(f) + j\text{Im}(f)$ of the 1-D frequency variable f. The support of U has a part at positive frequencies and a part at negative frequencies. The spectrum of the complex analytic signal is obtained by canceling the part at negative frequencies and doubling the part at positive frequencies. The inverse Fourier transform of this one-sided spectrum yields the analytic signal of the form $\psi(t) = u(t) + jv(t)$, where $v(t)$ is the Hilbert transform of u. It should be noted that the term *analytic signal* is not identical to the term *analytic function* (see next point) [1, 2].

In the case of 2-D real signals $u(x_1, x_2)$, the corresponding spectrum may be defined by a complex or hypercomplex Fourier transform (a quaternion Fourier transform) [3]. We have a 2-D complex or hypercomplex spectra with a support in all four quadrants. The corresponding analytic signals are defined by the inverse Fourier transform of a single-quadrant spectrum. The multiplication of U by a 2-D frequency domain unit step function yields a single quadrant spectrum. The generalization of this procedure is straightforward and presented in this chapter. The important polar forms of analytic signals are described in Chapter 8. A special kind of 2-D analytic signals introduced by Felsberg and Sommer [4] is called *monogenic signals*. The term *analytic* does not have the same interpretation with regard to analytic signals with single-orthant spectra for monogenic signals.

5.1　1-D Analytic Signals as Boundary Distributions of 1-D Analytic Functions

The term *analytic signal* has been applied by Gabor [5] to a complex function of a real variable x (usually a time variable $x = t$) of the form $\psi(x) = u(x) + jv(x)$. This signal is defined as a boundary distribution of a complex function $\psi(z)$ of a complex variable $z = x + jy$ of the form.

$$\psi(z) = \psi(x, y) = u(x, y) + jv(x, y) \tag{5.1}$$

Usually, $x = t$ and $y = \tau$ are time variables. Not every complex function is analytic. It should satisfy the conditions of analyticity. Let us recall the notion of the Cauchy integrals. Consider a Cartesian complex plane $(x, jy) \subset \mathbb{C}$ and a region $\mathbb{D} \subset \mathbb{C}$. For an analytic function, the circular integral for each closed path in $\mathbb{C}$ equals zero.

$$\oint_C f(z)\, dz = 0 \tag{5.2}$$

This integral is called the *first Cauchy integral*. Let us define a new function $f(z)/(z - z_0)$. This function has a pole at the point z_0. The contour integral of this function is

$$\oint_C \frac{f(z)}{z - z_0}\, dz = 2\pi j(z_0) \tag{5.3}$$

yielding the second Cauchy integral

$$f(z_0) = \frac{1}{2\pi j} \oint_C \frac{f(z)}{z - z_0}\, dz \tag{5.4}$$

An analytic function (5.1) should satisfy the Cauchy-Riemann partial derivatives

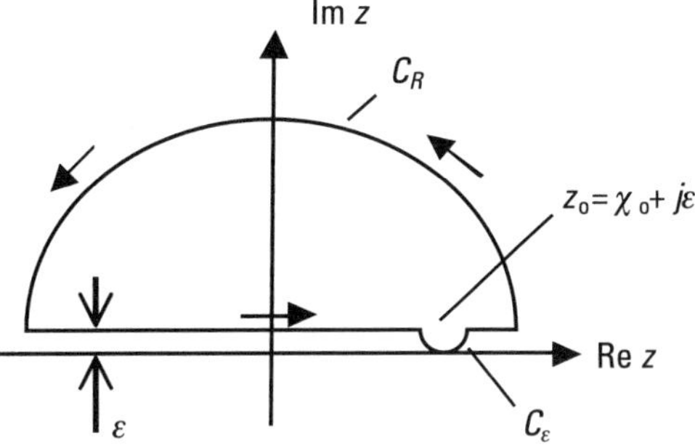

Figure 5.1   Contour of integration in the complex plane.

$$\frac{\partial u}{\partial x} = \frac{\partial v}{\partial y} \quad ; \quad \frac{\partial u}{\partial y} = -\frac{\partial v}{\partial x} \tag{5.5}$$

Let us take the contour presented in Figure 5.1. It contains three parts: a large half-circle of a radius C_R, a small half-circle around the pole z_0, and a line parallel to the x-axis, $y = y - \varepsilon$.

Using the counterclockwise contour integration, the Cauchy integral takes the form

$$f(z_0) = \frac{1}{2\pi j}\left\{ \int_{-R}^{x_0-\varepsilon} \frac{f(z)}{z-z_0}\,dz + \int_{x_0+\varepsilon}^{R} \frac{f(z)}{z-z_0}\,dz + \int_{C_R} \frac{f(z)}{z-z_0} + \int_{C_\varepsilon} \frac{f(z)}{z-z_0}\,dz \right\} \tag{5.6}$$

The first two integrals define in the limit $R \to \infty$, the Hilbert transform of a real function (signal) $u(x)$ of the form

$$v(x) = -\frac{1}{\pi} P \int_{-\infty}^{\infty} \frac{u(\eta)}{\eta - x}\,d\eta = \mathrm{H}[u(x)] \tag{5.7}$$

The symbol H denotes the Hilbert transform. For analytic functions, the third contour integral vanishes for $R \to \infty$, and the last integral in the limit $\varepsilon \to 0$ restores the function $u(x)$. The limit defines the boundary distribution of the function (5.1) of the form

$$\psi(x) = u\left(x, 0^+\right) + jv\left(x, 0^+\right) \tag{5.8}$$

and is called an *analytic signal*. The notation 0^+ indicates the limit at the positive side of the *x*-axis (upper half-plane) and is not used in applications. Equation (5.7) may be conveniently written in the form of the convolution

$$v(x) = u(x) * \frac{1}{\pi x} = H\big[u(x)\big] \tag{5.9}$$

The original signal $u(x)$ can also be written in the form of a convolution with the delta distribution $\delta(x)$

$$u(x) = u(x) * \delta(x) = I\big[u(x)\big] \tag{5.10}$$

The operator I is called identity operator. Using this operator the analytic signal (5.8) can be written in the form:

$$\psi(x) = I\big[u(x)\big] + jH\big[u(x)\big] = \left[\delta(x) + j\frac{1}{\pi x}\right] * u(x) = \psi_\delta(x) * u(x) \tag{5.11}$$

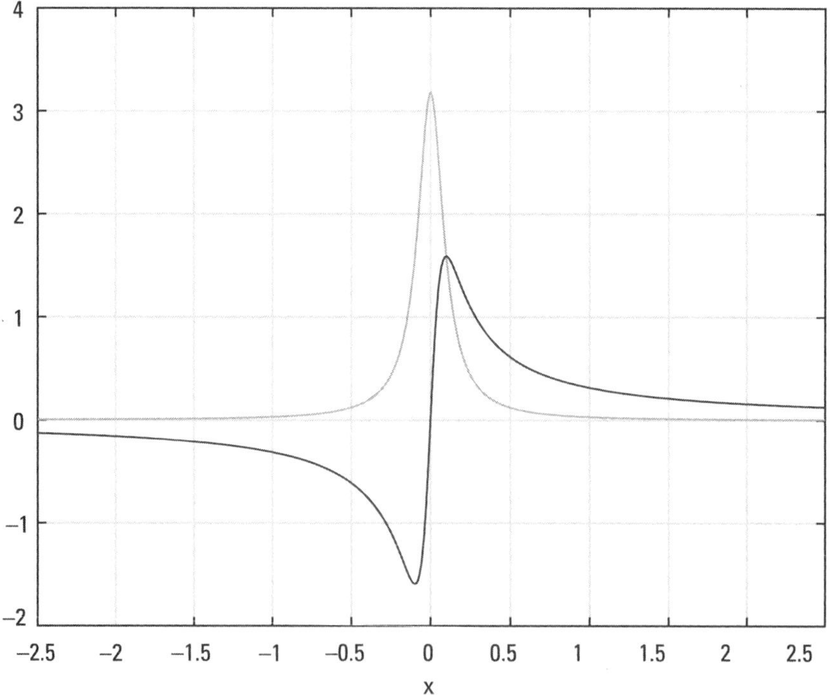

Figure 5.2   The approximating functions of the complex delta distribution, $a = 0.1$.

The expression in parenthesis is called *complex delta distribution* [10]. Of course, the imaginary part is also a distribution. It should be noted that $H[\delta(x)] = 1/(\pi x)$ (i.e., the complex delta distribution has the structure of an analytic signal). The approximating function of this distribution in (5.11) may conveniently have the form of a Cauchy analytic signal

$$\psi_\delta(x) = \lim_{a \to 0} \left\{ \frac{1}{\pi} \frac{a}{a^2 + x^2} + j\frac{1}{\pi} \frac{x}{a^2 + x^2} \right\} \qquad (5.12)$$

The complex delta distribution (5.12) can be used to shift the terms of the analytic signals using the shifted version

$$\psi(x - x_0) = \left[\delta(x - x_0) + j\frac{1}{\pi(x - x_0)} \right] * u(x) \qquad (5.13)$$

In computer simulations, the Cauchy analytic signal (5.12) can be conveniently applied using a small value of the constant a (see Figure 5.2) The Fourier transform of the complex delta distribution yields the frequency domain unit step (see Chapter 3)

$$F\left[\delta(x) + j\frac{1}{\pi x} \right] = 1 + \mathrm{sgn}(f) = 2 \cdot \mathbf{1}(f) \qquad (5.14)$$

where $\mathbf{1}(f)$ is the unit step distribution [1, 8]. Let us denote the Fourier transform (complex spectrum) of the signal $u(x)$ by $U(f)$. Because convolution in the signal domain corresponds to a product in the frequency domain, the spectrum of the analytic signal has the form

$$\Psi(f) = F[\psi(x)] = \left[1 + \mathrm{sgn}(f) \right] U(f) \qquad (5.15)$$

This is the one-sided spectrum of the analytic signal at positive frequencies. It may be shown that the conjugate analytic signal has a one-sided spectrum at negative frequencies. The real signal and the Hilbert transform can be written in the form:

$$u(x) = \frac{\psi(x) + \psi^*(x)}{2} \qquad (5.16)$$

$$v(x) = \frac{\psi(x) - \psi^*(x)}{2j} \qquad (5.17)$$

Example 5.1

In this simple example, the real cosine signal and its Fourier transform are:

$$u(x) = \cos(2\pi f_0 x)$$

$$= 0.5\left[e^{j2\pi f_0 x} + e^{-j2\pi f_0 x} \right] \overset{F}{\Leftrightarrow} 0.5\left[\delta(f - f_0) + \delta(f + f_0) \right]$$

The Hilbert transform is $v(x) = \sin(\omega x)$ and $\psi(x) = e^{j\omega x}$, $\omega = 2\pi f_o$. The instantaneous amplitude $A(x) = \sqrt{u^2 + v^2} = 1$ and the instantaneous phase $\varphi(x) = \tan^{-1}(v/u) = \omega x$. This example is presented to get comparisons with similar examples for 2-D and 3-D signals.

Example 5.2

This example with a simple Gaussian pulse is an introduction to similar examples for 2-D and 3-D signals. Consider the Gaussian signal and its Fourier spectrum

$$u(t) = e^{-\pi t^2} \overset{F}{\Leftrightarrow} U(f) = e^{-\pi f^2} \tag{5.18}$$

The corresponding Hilbert pair is

$$u(t) = e^{-\pi t^2} \overset{H}{\Leftrightarrow} v(t) = \int_0^\infty e^{-\pi f^2} \sin(2\pi ft)\, df \tag{5.19}$$

The signal u is an even function and the Hilbert transform v is an odd function. However, evenness and oddness are relative. Let us shift the signal by t_0. The shifted signal and its Fourier spectrum are

$$u(t - t_0) = e^{-\pi(t - t_0)^2} \overset{F}{\Leftrightarrow} U(f) = e^{-\pi f^2} e^{-j2\pi ft_o}$$

$$= e^{-\pi f^2} \cos(2\pi ft_0) - je^{-\pi f^2} \sin(2\pi ft_0) = U_e - jU_o \tag{5.20}$$

The shifted signal is now a union of an even and odd term $u(t - t_o) = u_e + u_o$ (see Chapter 3)

$$u_e = \frac{u(t - t_0) + u(-t - t_0)}{2} = e^{-\pi(t^2 + t_0^2)} \cosh(2\pi tt_0) \tag{5.21}$$

$$u_o = \frac{u(t - t_0) - u(-t - t_0)}{2} = e^{-\pi(t^2 + t_0^2)} \sinh(2\pi tt_0) \tag{5.22}$$

The corresponding Hilbert transform is

$$v\left(t - t_0\right) = \int_0^\infty e^{-\pi f^2} \sin\left[2\pi f\left(t - t_0\right)\right] df = v_e + v_o \qquad (5.23)$$

The waveforms illustrating this example are displayed in Figure 5.3.

We now recall Vakman's conditions (criteria), which the operator of the Hilbert transform and analytic signals must satisfy [30, 31].

- *First condition:* The operator H of the Hilbert transform must be continuous:

$$H[u + \partial u] \Rightarrow H[u] \text{ for } \|\partial u\| \to 0$$

- *Second condition:* Phase independence of scaling and homogeneity. Let the signal $u(t)$ be replaced by $\varepsilon u(t)$ for a small constant $\varepsilon > 0$. Then its instantaneous phase and frequency must remain the same (see polar representation in Chapter 7).
- *Third condition:* Harmonic independence: The Hilbert transform of the cosine signal $A\cos(\omega t + \varphi)$ must be $A\sin(\omega t + \varphi)$.

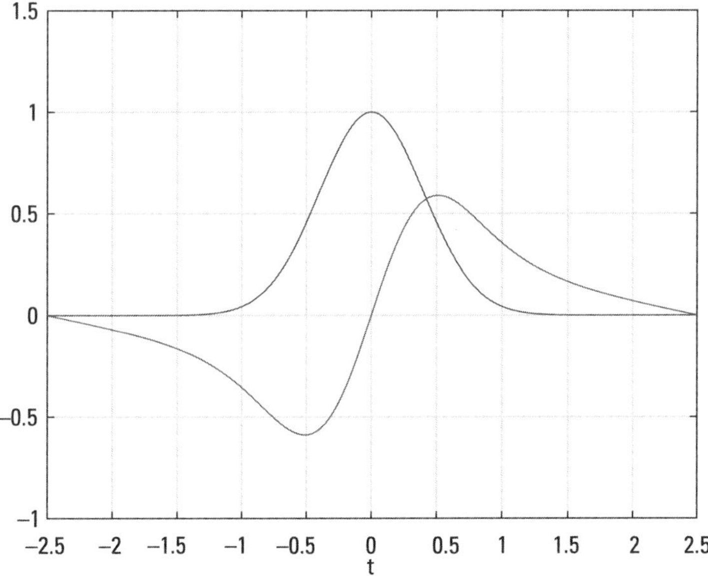

Figure 5.3 Waveforms illustrating Example 5.2.

5.2 The n-D Analytic Signal

The 1-D analytic signals are boundary distributions of 1-D analytic functions. The extension for n-D functions (signals) is straightforward. Consider the n-D complex space $\mathbb{C}^n$ of complex variables $z = (z_1, z_2, \ldots, z_n)$; $z_k = x_k + jy_k$. The space $\mathbb{C}$ is a Cartesian product of n complex planes. We define a complex-valued n-D function analytic in the region $D^n = D_1 \times D_2 \times \ldots \times D_n$, $D^n \subset \mathbb{C}^n$, $D_k \subset \mathbb{C}_k$. The function $f(z)$ is represented by the n-D Cauchy integral:

$$f(z) = \frac{1}{(2\pi j)^n} \oint_{\partial D_1} \cdots \oint_{\partial D_n} \frac{f(\xi_1, \ldots, \xi_n) d\xi_n}{(\xi_1 - z_1) \ldots (\xi_n - z_n)} \tag{5.24}$$

where δD_k are closed contours each (as shown in Figure 5.1). The procedure of integration for each contour is the same as in the 1-D case. In consequence, the 1-D signal in (5.11) is replaced by the product

$$\psi(\mathbf{x}) = \frac{1}{2^n} \prod_{k=1}^{n} \left\{ j\mathrm{H}\left[u(x_k)\right] + \mathrm{I}\left[u(x_k)\right] \right\} \tag{5.25}$$

or using the convolution notation

$$\psi(\mathbf{x}) = \frac{1}{2^n} \prod_{k=1}^{n} \left\{ \delta(x_k) + j\frac{1}{\pi x_k} \right\} * \ldots * u(\mathbf{x}) \tag{5.26}$$

Again, the use of the factor $1/2^n$ is a matter on convention and usually is omitted. Note that the n-D complex delta distribution [10] is a product of 1-D complex delta distributions.

5.2.1 The 2-D Complex Analytic Signals

The 2-D complex analytic signals with the single quadrant spectrum in the first quadrant is

$$\psi_1(x_1, x_2) = \left\{ \left[\delta(x_1) + j\frac{1}{\pi x_1} \right] \times \left[\delta(x_2) + j\frac{1}{\pi x_2} \right] \right\} * * u(x_1, x_2) \tag{5.27}$$

The developed form is

$$\psi_1(x_1, x_2) = u(x_1, x_2) - v(x_1, x_2) + j\left[v_1(x_1, x_2) + v_2(x_1, x_2) \right] \tag{5.28}$$

Note that the convolution with the delta distribution restores the signal u

$$u(x_1,x_2) = \delta(x_1) * \delta(x_2) * u(x_1,x_2) = \delta(x_1,x_2) * *u(x_1,x_2) \quad (5.29)$$

The second term of the real part

$$v(x_1,x_2) = \frac{1}{\pi^2 x_1 x_2} * *u(x_1,x_2) \quad (5.30)$$

is the total Hilbert transform of u with regard to both variables. The terms of the imaginary part are the partial Hilbert transforms with regard to the variables x_1 and x_2 [6, 7]:

$$v_1(x_1,x_2) = \frac{\delta(x_2)}{\pi x_1} * *u(x_1,x_2), \quad v_2(x_1,x_2) = \frac{\delta(x_1)}{\pi x_2} * *u(x_1,x_2) \quad (5.31)$$

The 2-D extension of (5.15) yields the single quadrant spectrum of the 2-D analytic signal

$$\Psi_1(f_1,f_2) = F\big[\psi_1(x_1,x_2)\big] = \big[1 + \mathrm{sgn}(f_1)\big] \times \big[1 + \mathrm{sgn}(f_2)\big] U(f_1,f_2) \quad (5.32)$$

The subscript 1 denotes the single quadrant spectrum in the first quadrant. The single quadrant frequency operator in (5.32) (see Chapter 3) can be written more compactly in the form of a 2-D unit step

$$4 \times \mathbf{1}(f) = 1 + s_1 + s_2 + s_1 s_2; \ s_1 = \mathrm{sgn}(f_1); \ s_2 = \mathrm{sgn}(f_2) \quad (5.33)$$

The signal (5.28) has the single quadrant spectrum in the first quadrant (see Chapter 3). The corresponding signals and its frequency operators for the quadrants 1, 2, 3, and 4 are displayed in Table 5.1.

Note that we have two pairs of conjugate signals: $\psi_1 = \psi_4^*$ and $\psi_2 = \psi_3^*$. This is the consequence of Hermitian symmetry of the complex spectrum of the real signal u. The insertion in the Fourier transform of the four terms of the signal $u(x_1,x_2) = u_{ee} + u_{oe} + u_{eo} + u_{oo}$ yields

$$U(f_1,f_2) = \int\limits_{-\infty}^{\infty}\int\limits_{-\infty}^{\infty} u(x_1,x_2) e^{-j2\pi(f_1 x_1 + f_2 x_2)} dx_1\, dx_2 = U_{ee} - U_{oo} - j(U_{oe} + U_{eo})$$

$$(5.34)$$

Table 5.1
2-D Complex Signals with Single Quadrant Spectra $s_1 = \mathrm{sgn}(f_1)$, $s_2 = \mathrm{sgn}(f_2)$

Quadrant n	Signal $\psi_n(x_1, x_2)$	Operator $4 \times 1(f_1, f_2)$
1	$\psi_1 = u - v + j(v_1 + v_2)$	$1 + s_1 + s_2 + s_1 s_2$
2	$\psi_2 = u + v - j(v_1 - v_2)$	$1 - s_1 + s_2 - s_1 s_2$
3	$\psi_3 = u + v + j(v_1 - v_2)$	$1 + s_1 - s_2 - s_1 s_2$
4	$\psi_4 = u - v - j(v_1 + v_2)$	$1 - s_1 - s_2 + s_1 s_2$

Table 5.2
Four Options of the Spectrum (5.32)

Quadrant	Spectrum
1	$U(f_1, f_2) = U_{ee} - U_{oo} - j(U_{oe} + U_{eo})$
2	$U(-f_1, f_2) = U_{ee} + U_{oo} + j(U_{oe} - U_{eo})$
3	$U(f_1, -f_2) = U_{ee} + U_{oo} - j(U_{oe} - U_{eo})$
4	$U(-f_1, -f_2) = U_{ee} - U_{oo} + j(U_{eoe} + U_{eo})$

Table 5.2 contains the four options of this complex spectrum. There are two conjugate pairs

$$U(f_1, f_2) = U^*(-f_1, -f_2) \text{ and } U(-f_1, f_2) = U^*(f_1, -f_2) \quad (5.35)$$

Consequently, the spectral information in the half-plane $f_1 < 0$ is redundant. The signal u can be recovered by knowledge of the spectrum in the half-plane $f_1 > 0$. The signal u and its Hilbert transforms can be recovered using the formulae:

$$u(x_1, x_2) = \frac{\psi_1 + \psi_2 + \psi_3 + \psi_4}{4} = \frac{\psi_1 + \psi_1^* + \psi_3 + \psi_3^*}{4} \quad (5.36)$$

$$v(x_1, x_2) = \frac{-\psi_1 + \psi_2 + \psi_3 - \psi_4}{4} = \frac{-\psi_1 - \psi_1^* + \psi_3 + \psi_3^*}{4} \quad (5.37)$$

$$v_1(x_1, x_2) = \frac{\psi_1 - \psi_2 + \psi_3 - \psi_4}{4j} = \frac{\psi_1 - \psi_1^* + \psi_3 - \psi_3^*}{4j} \quad (5.38)$$

$$v_2(x_1,x_2) = \frac{\psi_1 + \psi_2 - \psi_3 - \psi_4}{4j} = \frac{\psi_1 - \psi_1^* - \psi_3 + \psi_3^*}{4j} \qquad (5.39)$$

Note that the energies of the signals ψ_1 and ψ_4 differ from the energies of ψ_2 and ψ_3 (see Chapter 7). The squared norm of ψ_1 is

$$\begin{aligned}
\left\|\psi_1\right\|^2 &= \psi_1 \times \psi_1^* = (u-v)^2 + \left(v_1 + v_2\right)^2 \\
&= u^1 + v^2 + v_1^2 + v_2^2 + 2\left(uv - v_1v_2\right)
\end{aligned} \qquad (5.40)$$

and of ψ_4

$$\begin{aligned}
\left\|\psi_4\right\|^2 &= \psi_4 \times \psi_4^* = (u+v)^2 + \left(v_1 - v_2\right)^2 \\
&= u^1 + v^1 + v_1^2 + v_2^2 - 2\left(uv - v_1v_2\right)
\end{aligned} \qquad (5.41)$$

The energies are different because the integrals of the squared norms differ. However, for 2-D separable functions, $u(x_1,x_2) = u_1(x_1)u_2(x_2)$, the energies are equal since $uv = v_1v_2$, and consequently,

$$\left\|\psi_i\right\|^2 = u^2 + v^2 + v_1^2 + v_2^2 \; ; i = 1,2,3,4 \qquad (5.42)$$

Example 5.3

This example presents the extension of the Example 5.1 for 2-D. *Consider* the real 2-D cosine and its Fourier transform

$$u(x_1,x_2) = \cos\left[2\pi\left(f_{10}x_1 + f_{20}x_2\right)\right]$$

$$\overset{2F}{\Longleftrightarrow} 0.5\left[\delta\left(f_1 - f_{10}\right)\delta\left(f_2 - f_{20}\right) + \delta\left(f_1 + f_{10}\right)\delta\left(f_2 + f_{20}\right)\right]$$

Note the two quadrant support of the spectrum in the first and fourth quadrant. The total Hilbert transform is $v(x_1,x_2) = -\cos[2\pi(f_{10}x_1 + f_{20}x_2)]$ and the two partial Hilbert transforms are $v_1(x_1,x_2) = \sin[2\pi(f_{10}x_1 + f_{20}x_2)]$ and $v_2(x_1,x_2) = \sin[2\pi(f_{10}x_1 + f_{20}x_2)]$. The analytic signal with spectrum in the first quadrant is $\psi_1(x_1,x_2) = u - v + j(v_1 + v_2) = \exp[j2\pi(f_{10}x_1 + f_{20}x_2)]$. Its amplitude equals 2 and the phase $\phi(x_1,x_2) = 2\pi(f_{10}x_1 + f_{20}x_2)$.

Example 5.4

This example is a continuation of the Example 5.2 extended for 2-D signals. Consider the 2-D Gaussian signal and its Fourier spectrum

$$
u(x_1,x_2) = \frac{1}{\sqrt{1-\rho^2}}\exp\left[\frac{-\pi}{1-\rho^2}\left(x_1^2 + x_2^2 - 2\rho x_1 x_2\right)\right]
$$

$$
\overset{F}{\Leftrightarrow} U(f_1,f_2) = \exp\left[-\pi\left(f_1^2 + f_2^2 + 2\rho f_1 f_2\right)\right]; \quad 0 \le \rho < 1
\tag{5.43}
$$

Using (3.11)–(3.15) (Chapter 3), it is shown that the signal and its spectrum have only even-even and odd-odd terms. The two signal terms are

$$
u(x_1,x_2) = u_{ee} + u_{oo}
$$

$$
= \frac{1}{\sqrt{1-\rho^2}}\exp\left[\frac{\pi}{1-\rho^2}\right]\left[\cosh\left(\frac{2\pi\rho}{1-\rho^2}x_1 x_2\right) + \sinh\left(\frac{2\pi\rho}{1-\rho^2}x_1 x_2\right)\right]
\tag{5.44}
$$

The spectrum is a real function and its two terms are

$$
U(f_1,f_2) = U_{ee} - U_{oo}
$$

$$
= e^{-\pi\left(f_1^2 + f_2^2\right)}\cosh\left(2\pi\rho f_1 f_2\right) - e^{-\pi\left(f_1^2 + f_2^2\right)}\sinh\left(2\pi\rho f_1 f_2\right)
\tag{5.45}
$$

For $\rho = 0$ it is an even-even function. The corresponding Fourier transforms are

$$
U_{ee}(f_1,f_2) = \int\limits_{-\infty}^{\infty}\int\limits_{-\infty}^{\infty} u_{ee}(x_1 x_2)\cos(2\pi f_1 x_1)\cos(2\pi f_2 x_2)\,dx_1\,dx_2
\tag{5.46}
$$

$$
U_{oo}(f_1,f_2) = \int\limits_{-\infty}^{\infty}\int\limits_{-\infty}^{\infty} u_{oo}(x_1 x_2)\sin(2\pi f_1 x_1)\sin(2\pi f_2 x_2)\,dx_1\,dx_2
\tag{5.47}
$$

Due to symmetry, the integrals can be calculated from zero to infinity and by multiplication by four. The total Hilbert and partial Hilbert transforms can be calculated in the signal domain using the convolutions (5.30), (5.31), or in the frequency domain using the integrals

$$
v(x_1,x_2) = v_{ee} + v_{oo}
$$

$$
= \int\limits_{-\infty}^{\infty}\int\limits_{-\infty}^{\infty} -\mathrm{sgn}(f_1)\mathrm{sgn}(f_2)U(f_1,f_2)\cos\left[2\pi\left(f_1 x_1 + f_2 x_2\right)\right]df_1\,df_2
\tag{5.48}
$$

$$v_1\left(x_1,x_2\right) = v_{1oe} + v_{1eo}$$

$$= \int\limits_{-\infty}^{\infty} \int\limits_{-\infty}^{\infty} \mathrm{sgn}\left(f_1\right) U\left(f_1,f_2\right) \sin\left[2\pi\left(f_1 x_1 + f_2 x_2\right)\right] df_1\, df_2 \tag{5.49}$$

$$v_2\left(x_1,x_2\right) = v_{2oe} + v_{2eo}$$

$$= \int\limits_{-\infty}^{\infty} \int\limits_{-\infty}^{\infty} \mathrm{sgn}\left(f_2\right) U\left(f_1,f_2\right) \sin\left[2\pi\left(f_1 x_1 + f_2 x_2\right)\right] df_1\, df_2 \tag{5.50}$$

Due to the symmetries the integrals can be calculated one-sided from 0 to ∞ multiplying by 4.

As in Example 5.1, let us investigate the signal (5.43) shifted in the signal plane (x_1,x_2) to the point (a,b)

$$u\left[x_1 - a,\ x_2 - b\right] = \frac{1}{\sqrt{1-\rho^2}}\, e^{\frac{-\pi}{1-\rho^2}\left[\left(x_1-a\right)^2 + \left(x_2-b\right)^2 - 2\rho\left(x_1-a\right)\left(x_2-b\right)\right]} \ ;\ \ 0 \le \rho < 1$$

$$\tag{5.51}$$

The spectrum of this signal is

$$U\left(f_1,f_2\right) = e^{-\pi\left(f_1^2 + f_2^2 + 2\rho f_1 f_2\right)} e^{-j2\pi\left[f_1 a + f_2 b\right]} \tag{5.52}$$

This spectrum is complex, with two real terms and two imaginary terms. Assuming that the positive shift variables (a,b) are sufficiently large, we get a single quadrant spectrum of an analytic 2-D signal. Such a signal is called *quasi-analytic* (see Chapter 8). The shifted signal is a union of all four terms with regard to evenness or oddness. The formulae for these terms can be derived. However, they are lengthy and in implementations it is easier to use computer calculations. In Figure 5.4 2-D images illustrating this example are shown.

5.2.2 3-D Complex Analytic Signals

The 3-D real signal $u(x_1,x_2,x_3)$ can have eight terms with regard to evenness and oddness. The insertion of these terms in the 3-D Fourier transform

$$U\left(f_1,f_2,f_3\right) = \int\limits_{-\infty}^{\infty} \int\limits_{-\infty}^{\infty} \int\limits_{-\infty}^{\infty} u\left(x_1,x_2,x_3\right) e^{-j2\pi\left(f_1 x_1 + f_2 x_2 + f_3 x_3\right)}\, dx_1\, dx_2\, dx_3 \tag{5.53}$$

yields eight terms of the spectrum displayed in the first line of Table 5.3.

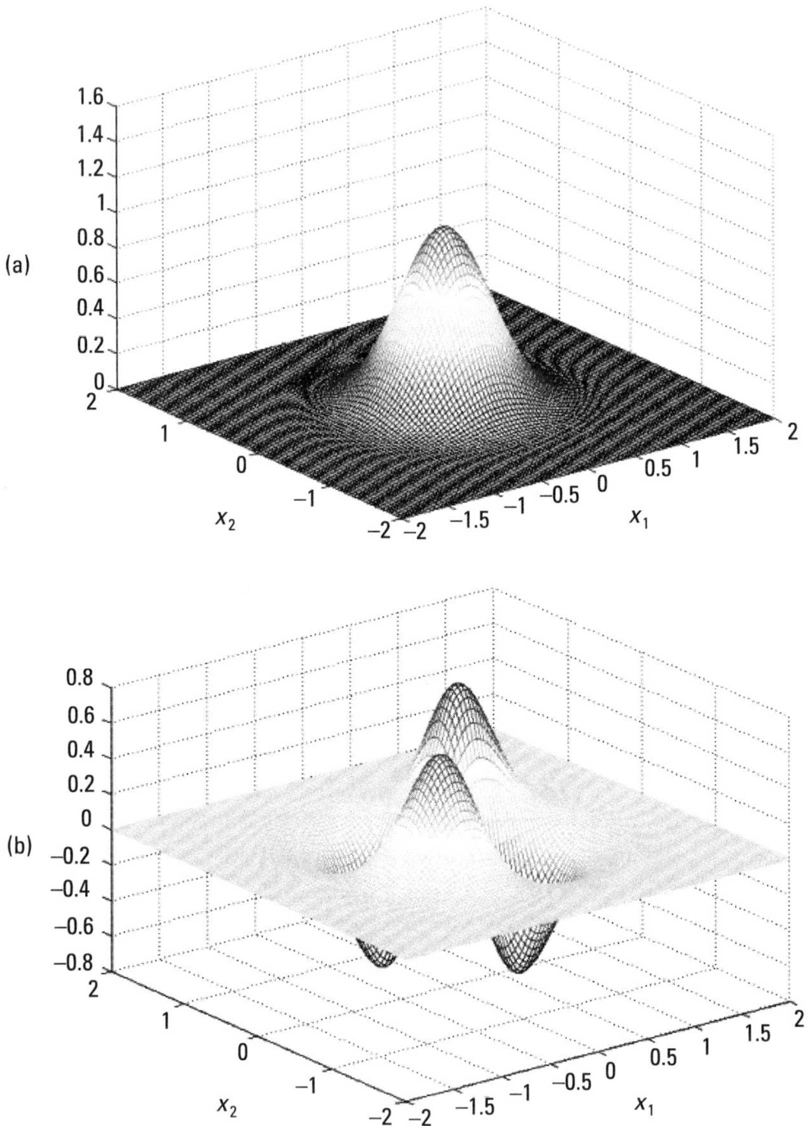

Figure 5.4 (a) The Gaussian signal (5.43), $\rho = 0$. (b) The total Hilbert transform $v(x_1, x_2)$. (c) The partial Hilbert transform $v_1(x_1, x_2)$. (d) $v_2(x_1, x_2)$.

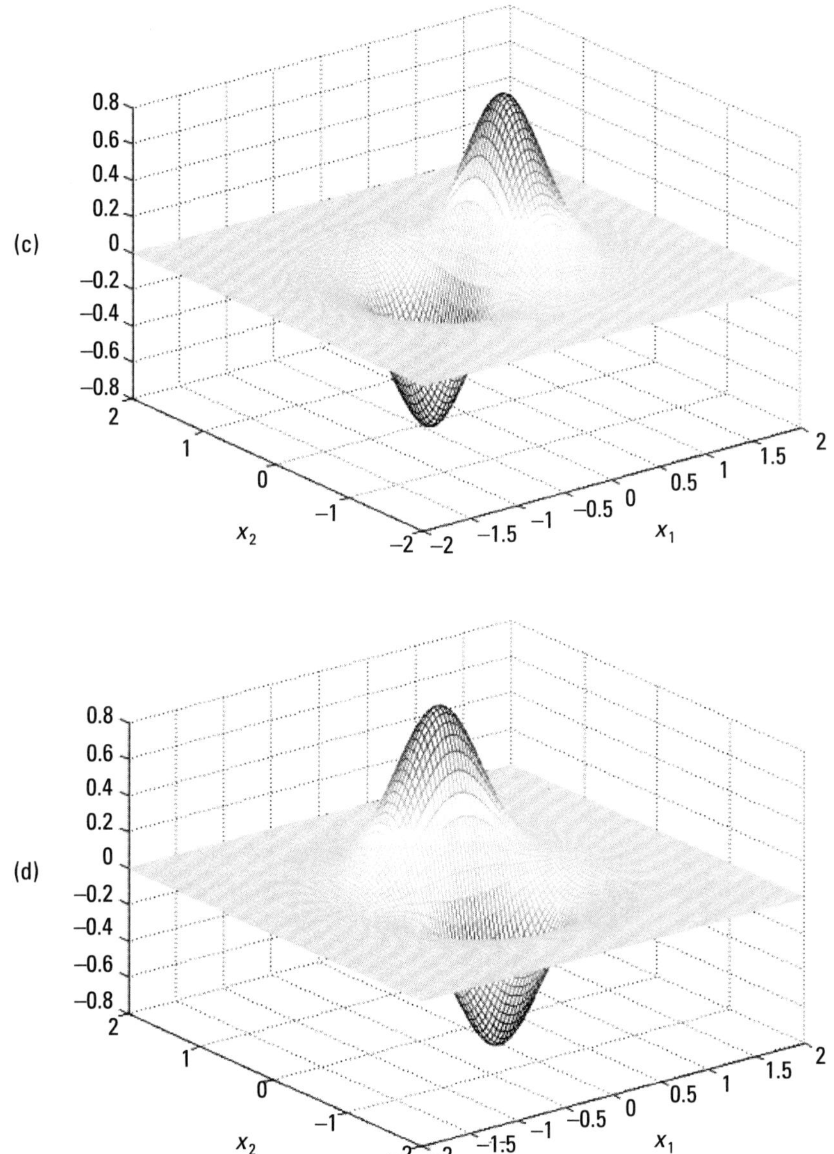

Figure 5.4 Continued

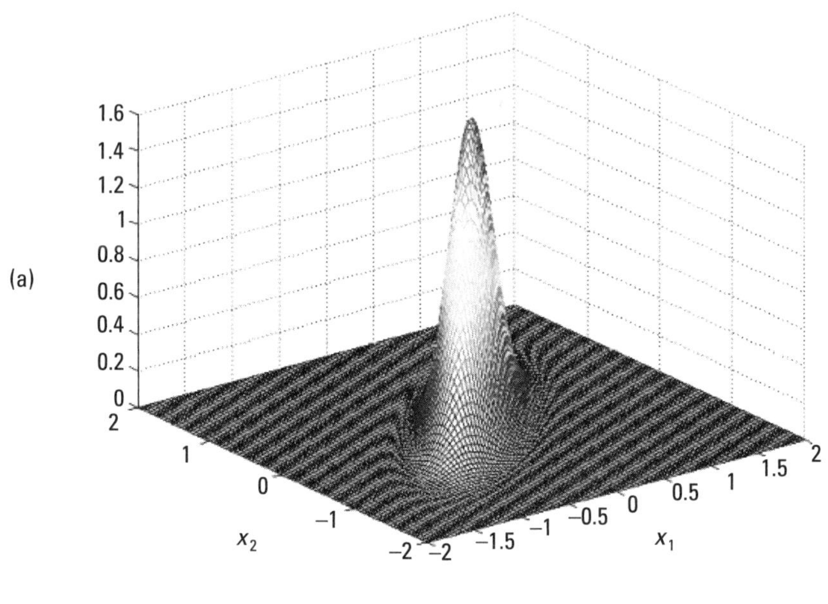

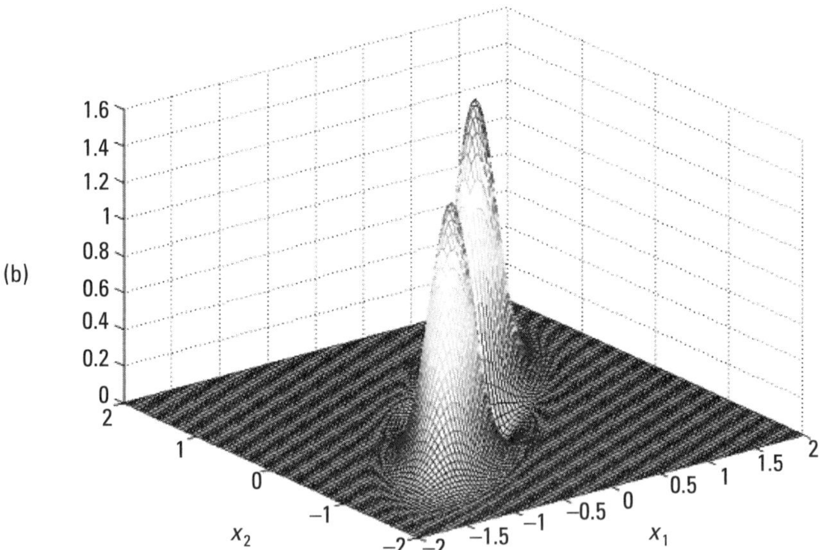

Figure 5.5   (a) The Gaussian signal (5.43), $\rho = 0.8$. (b) The total Hilbert transform $v(x_1, x_2)$. (c) The partial Hilbert transform $v_1(x_1, x_2)$. (d) $v_2(x_1, x_2)$.

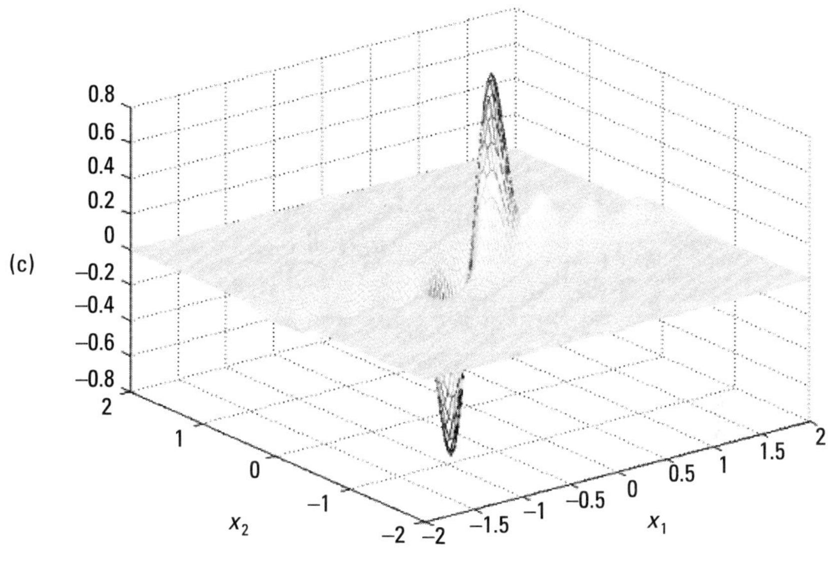

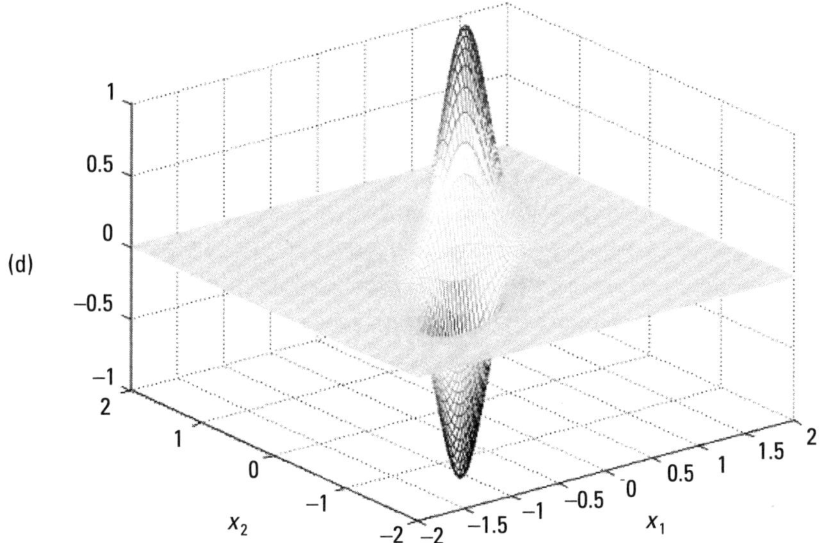

Figure 5.5 Continued

Table 5.3
Signs of the Terms of 3-D Single Octant Spectra in Successive Octants

Octant	Eight Terms of a 3-D Spectrum with Single Octant Support
1	$U\left(f_1,f_2,f_3\right) = U_{eee} - U_{ooe} - U_{oeo} - U_{eoo} - j\left(U_{oee} + U_{eoe} + U_{eeo} - U_{ooo}\right)$
2	$U\left(-f_1,f_2,f_3\right) = U_{eee} + U_{ooe} + U_{oeo} - U_{eoo} - j\left(U_{oee} + U_{eoe} - U_{eeo} + U_{ooo}\right)$
3	$U\left(f_1,-f_2,f_3\right) = U_{eee} + U_{ooe} - U_{oeo} + U_{eoo} - j\left(+U_{oee} - U_{eoe} + U_{eeo} + U_{ooo}\right)$
4	$U\left(-f_1,-f_2,f_3\right) = U_{eee} - U_{ooe} + U_{oeo} + U_{eoo} - j\left(+U_{oee} - U_{eoe} - U_{eeo} - U_{ooo}\right)$
5	$U\left(f_1,f_2,-f_3\right) = U_{eee} - U_{ooe} + U_{oeo} + U_{eoo} - j\left(-U_{oee} + U_{eoe} + U_{eeo} + U_{ooo}\right)$
6	$U\left(-f_1,f_2,-f_3\right) = U_{eee} + U_{ooe} - U_{oeo} + U_{eoo} - j\left(-U_{oee} + U_{eoe} - U_{eeo} - U_{ooo}\right)$
7	$U\left(f_1,-f_2,-f_3\right) = U_{eee} + U_{ooe} + U_{oeo} - U_{eoo} - j\left(-U_{oee} - U_{eoe} + U_{eeo} - U_{ooo}\right)$
8	$U\left(-f_1,-f_2,-f_3\right) = U_{eee} - U_{ooe} - U_{oeo} - U_{eoo} - j\left(-U_{oee} - U_{eoe} - U_{eeo} + U_{ooo}\right)$

The spectra in the next lines illustrate the 3-D Hermitian symmetry.

$$U\left(f_1,f_2,f_3\right) = U^*\left(-f_1,-f_2,-f_3\right), \ U\left(-f_1,f_2,f_3\right) = U^*\left(f_1,-f_2,-f_3\right)$$
$$U\left(f_1,-f_2,f_3\right) = U^*\left(-f_1,+f_2,-f_3\right), \ U\left(-f_1,-f_2,f_3\right) = U^*\left(f_1,f_2,-f_3\right)$$

$$(5.54)$$

In the signal domain the extension of the 2-D (5.27) to 3-D is

$$\psi_1\left(x_1,x_2,x_3\right)$$

$$= \left\{\left[\delta\left(x_1\right) + j\frac{1}{\pi x_1}\right] \times \left[\delta\left(x_2\right) + j\frac{1}{\pi x_2}\right]\left[\delta\left(x_3\right) + j\frac{1}{\pi x_3}\right]\right\} *** u\left(x_1,x_2,x_3\right)$$

$$(5.55)$$

and the single octant spectrum in the first octant is

$$\Psi_1\left(f_1,f_2,f_3\right) = F\left[\psi\left(x_1,x_2,x_3\right)\right]$$
$$= \left[1+\text{sgn}\left(f_1\right)\right]\times\left[1+\text{sgn}\left(f_2\right)\right]\times\left[1+\text{sgn}\left(f_3\right)\right]U\left(f_1,f_2,f_3\right)$$

(5.56)

The inverse Fourier transform of this spectrum

$$U\left(f_1,f_2,f_3\right) = \int\limits_{-\infty}^{\infty}\int\limits_{-\infty}^{\infty}\int\limits_{-\infty}^{\infty}\Psi_1\left(f_1,f_2,f_3\right)e^{j2\pi\left(f_1x_1+f_2x_2+f_3x_3\right)}df_1\,df_2\,df_3$$

(5.57)

yields the 3-D analytic signal with single octant spectrum in the first octant; this is displayed in the first line of Table 5.4. The Hilbert transforms of the seven terms of this signal can be calculated in the signal domain using the following convolutions. The partial Hilbert transforms with regard to a single variable are

$$v_1\left(x_1,x_2,x_3\right) = \frac{\delta\left(x_2\right)\delta\left(x_3\right)}{\pi x_1}\,{*}{*}{*}\,u\left(x_1,x_2,x_3\right)$$

(5.58)

$$v_2\left(x_1,x_2,x_3\right) = \frac{\delta\left(x_1\right)\delta\left(x_3\right)}{\pi x_2}\,{*}{*}{*}\,u\left(x_1,x_2,x_3\right)$$

(5.59)

$$v_3\left(x_1,x_2,x_3\right) = \frac{\delta\left(x_1\right)\delta\left(x_2\right)}{\pi x_3}\,{*}{*}{*}\,u\left(x_1,x_2,x_3\right)$$

(5.60)

The three partial Hilbert transforms with regard to two variables

$$v_{12}\left(x_1,x_2,x_3\right) = \frac{\delta\left(x_3\right)}{\pi^2 x_1 x_2}\,{*}{*}{*}\,u\left(x_1,x_2,x_3\right)$$

(5.61)

$$v_{13}\left(x_1,x_2,x_3\right) = \frac{\delta\left(x_2\right)}{\pi^2 x_1 x_3}\,{*}{*}{*}\,u\left(x_1,x_2,x_3\right)$$

(5.62)

$$v_{23}\left(x_1,x_2,x_3\right) = \frac{\delta\left(x_1\right)}{\pi^2 x_2 x_3}\,{*}{*}{*}\,u\left(x_1,x_2,x_3\right)$$

(5.63)

Table 5.4
Eight Analytic Signals with Single Octant Spectra in Succesive Octants and the Corresponding Unit Step Frequency Operators

Octant n	Signal $\psi_n(x_1, x_2, x_3)$	Frequency Domain Operator $8 \times 1(f_1, f_2, f_3)$
1	$\psi_1 = u - v_{12} - v_{13} - v_{23} + j\left[v_1 + v_2 + v_3 - v\right]$	$1 + s_1 + s_2 + s_3 + s_1 s_2 + s_1 s_3 + s_2 s_3 + s_1 s_2 s_3$
2	$\psi_2 = u + v_{12} + v_{13} - v_{23} + j\left[-v_1 + v_2 + v_3 + v\right]$	$1 - s_1 + s_2 + s_3 - s_1 s_2 - s_1 s_3 + s_2 s_3 - s_1 s_2 s_3$
3	$\psi_3 = u + v_{12} - v_{13} + v_{23} + j\left[v_1 - v_2 + v_3 + v\right]$	$1 + s_1 - s_2 + s_3 - s_1 s_2 + s_1 s_3 - s_2 s_3 - s_1 s_2 s_3$
4	$\psi_4 = u - v_{12} + v_{13} + v_{23} + j\left[-v_1 - v_2 + v_3 - v\right]$	$1 - s_1 - s_2 + s_3 + s_1 s_2 - s_1 s_3 - s_2 s_3 + s_1 s_2 s_3$
5	$\psi_5 = u - v_{12} + v_{13} + v_{23} - j\left[-v_1 - v_2 + v_3 - v\right]$	$1 + s_1 + s_2 - s_3 + s_1 s_2 - s_1 s_3 - s_2 s_3 - s_1 s_2 s_3$
6	$\psi_6 = u + v_{12} - v_{13} + v_{23} - j\left[v_1 - v_2 + v_3 + v\right]$	$1 - s_1 + s_2 - s_3 - s_1 s_2 + s_1 s_3 - s_2 s_3 + s_1 s_2 s_3$
7	$\psi_7 = u + v_{12} + v_{13} - v_{23} - j\left[-v_1 + v_2 + v_3 + v\right]$	$1 + s_1 - s_2 - s_3 - s_1 s_2 - s_1 s_3 + s_2 s_3 + s_1 s_2 s_3$
8	$\psi_8 = u - v_{12} - v_{13} - v_{23} - j\left[v_1 + v_2 + v_3 - v\right]$	$1 - s_1 - s_2 - s_3 + s_1 s_2 + s_1 s_3 + s_2 s_3 - s_1 s_2 s_3$

and the total Hilbert transform

$$v\left(x_1, x_2, x_3\right) = \frac{1}{\pi^3 x_1 x_2 x_3} * * * u\left(x_1, x_2, x_3\right) \qquad (5.64)$$

In computer implementations, it is advisable to calculate the original signal using the convolution

$$u\left(x_1, x_2, x_3\right) = \delta\left(x_1\right)\delta\left(x_2\right)\delta\left(x_3\right) * * * u\left(x_1, x_2, x_3\right) \qquad (5.65)$$

using a suitable approximation of the delta distribution, for example given by (5.12). The Hilbert transforms can be calculated using the inverse Fourier transforms and the relations (5.66)

$$v_{ik} = F^{-1}\left[-\text{sgn}\left(f_i\right)\text{sgn}\left(f_k\right)U\right]; \quad i,k = (1,2),\ (1,3),\ (2,3) \qquad (5.66)$$

$$v = F^{-1}\left[j\,\text{sgn}\left(f_1\right)\text{sgn}\left(f_2\right)\text{sgn}\left(f_3\right)U\right] \qquad (5.67)$$

where F^{-1} denotes the 3-D inverse Fourier transform defined by (4.8). The original signal should be calculated using the inverse Fourier transform for the same reason as in the signal domain.

The original signal can be reconstructed using the signals from Table 5.4 using

$$u(x_1,x_2,x_3) = \frac{\psi_1 + \psi_1^* + \psi_3 + \psi_3^* + \psi_5 + \psi_5^* + \psi_7 + \psi_7^*}{8}. \qquad (5.68)$$

As in the 2-D case, the energies of the signals displayed in Table 5.4 differ. There are four pairs of conjugated signals of different energy. Again, for separable 3-D signals, all four energies are equal and the square of the norm is

$$\left\|\psi_i\right\|^2 = u^2 + v^2 + v_1^2 + v_2^2 + v_3^2 + v_{12}^2 + v_{13}^2 + v_{23}^2 \; ; \; i = 1,2,3,\ldots,8 \qquad (5.69)$$

Example 5.5

This example is a 3-D continuation of Examples 5.1 and 5.3. Consider the real 3-D harmonic signal

$$u(x_1,x_2,x_3) = \cos\left[2\pi\left(f_{10}x_1 + f_{20}x_2 + f_{30}x_3\right)\right]$$

$$U(f_1,f_2,f_3) = 0.5\left[\delta(f_1 - f_{10})\delta(f_2 - f_{20})\delta(f_3 - f_{30}) + \delta(f_1 + f_{10})\delta(f_2 + f_{20})\delta(f_3 + f_{30})\right]$$

Let us omit presentation of the simple but lengthy trigonometric formulas for the partial and total Hilbert transforms of the analytic signal $\psi_1(x_1,x_2,x_3)$. The insertion of these functions in the first formula in Table 5.4 yields $\psi_1(x_1,x_2,x_3) = 4e^{j2\pi(f_{10}x_1 + f_{20}x_2 + f_{30}x_3)}$. Its amplitude is 4 and the phase is $\phi = 2\pi\left(f_{10}x_1 + f_{20}x_2 + f_{30}x_3\right)$. Note the two octant support of the real signal in the octants 1 and 8.

Example 5.6

This example represents the extension of Examples 5.2 and 5.4 for 3-D Gaussian signals. The 3-D simplified *Gaussian* signal with all variances equal $(2\pi)^{-0.5}$ and equal all values of the correlation coefficients is

$$u(x_1,x_2,x_3) = \frac{1}{(2\pi)^{3/2}\sqrt{1 + 2\rho^3 - 3\rho^2}}$$

$$\exp\left[\frac{-1}{2(1 + 2\rho^3 - 3\rho^2)}\left[(1 - \rho^2)(x_1^2 + x_2^2 + x_3^2) + 2\rho(1 - \rho)(x_1x_2 + x_1x_3 + x_2x_3)\right]\right]$$

$$(5.70)$$

In order to achieve a more compact notations, let us write this signal in the form

$$u\left(x_1,x_2 x_3\right) = F_1\left(x_1^2 + x_2^2 + x_3^2\right)\exp\left[C \times \left(x_1 x_2 + x_1 x_3 + x_2 x_3\right)\right]$$

where

$$F_1 = \frac{1}{(2\pi)^{3/2}\sqrt{1+2\rho^3-3\rho^2}}\exp\left[\frac{-\left(1-\rho^2\right)}{2\left(1+2\rho^3-3\rho^2\right)}\left(x_1^2 + x_2^2 + x_3^2\right)\right]$$

and

$$C = \frac{-\rho(1-\rho)}{1+3\rho^3-3\rho^2}$$

The signal is a union of four terms with regard to evenness and oddness, $\alpha = x_1 x_2$, $\beta = x_1 x_3$, $\gamma = x_2 x_3$.

$$u_{eee}\left(x_1,x_2,x_3\right)$$
$$= F_1\left\{\begin{matrix}\cosh\left[C(\alpha+\beta+\gamma)\right]+\cosh\left[C(-\alpha+\beta+\gamma)\right]\\+\cosh\left[C(\alpha-\beta+\gamma)\right]+\cosh\left[C(-\alpha-\beta+\gamma)\right]\end{matrix}\right\}$$

$$u_{eeo}\left(x_1,x_2,x_3\right)$$
$$= F_1\left\{\begin{matrix}\sinh\left[C(\alpha+\beta+\gamma)\right]+\sinh\left[C(-\alpha+\beta+\gamma)\right]\\+\sinh\left[C(\alpha-\beta+\gamma)\right]+\sinh\left[C(-\alpha-\beta+\gamma)\right]\end{matrix}\right\}$$

$$u_{oee}\left(x_1,x_2,x_3\right)$$
$$= F_1\left\{\begin{matrix}\cosh\left[C(\alpha+\beta+\gamma)\right]-\cosh\left[C(-\alpha+\beta+\gamma)\right]\\+\cosh\left[C(\alpha-\beta+\gamma)\right]-\cosh\left[C(-\alpha-\beta+\gamma)\right]\end{matrix}\right\}$$

$$u_{ooe}\left(x_1,x_2,x_3\right)$$
$$= F_1\left\{\begin{matrix}\cosh\left[C(\alpha+\beta+\gamma)\right]-\cosh\left[C(-\alpha+\beta+\gamma)\right]\\-\cosh\left[C(\alpha-\beta+\gamma)\right]+\cosh\left[C(-\alpha-\beta+\gamma)\right]\end{matrix}\right\}$$

The Fourier spectrum of this signal is a real function

$$U\left(f_1,f_2,f_3\right)=\exp\left[-2\pi^2\left(f_1^2+f_2^2+f_3^2\right)-\rho\left(f_1f_2+f_1f_3+f_2f_3\right)\right] \quad (5.71)$$

and is a union of four terms (the four terms of the imaginary part all equal zero)

$$U\left(f_1,f_2,f_3\right)=U_{eee}-U_{eoe}-U_{oeo}-U_{eoo} \quad (5.72)$$

Following the 1-D and 2-D examples, the shift of the signal (5.70) to a point (a,b,c) in the signal space produces a complex spectrum with all eight terms with regard to evenness and oddness.

$$u\left(x_1-a,x_2-b,x_3-c\right)\overset{3F}{\Leftrightarrow}U\left(f_1,f_2,f_3\right)e^{-j2\pi\left(f_1a+f_2b+f_3c\right)} \quad (5.73)$$

5.3 Hypercomplex n-D Analytic Signals

Hypercomplex n-D analytic signals are defined in exactly the same way as in the complex case; that is the boundary distributions of hypercomplex analytic functions. The difference is in the application of the algebra of hypercomplex numbers. Their single-orthant spectra are calculated in the form of a product of a single orthant operator (see (5.56)) and the spectrum is calculated using the hypercomplex version of the complex Fourier transform. The notion of the hypercomplex Fourier transform of a real n-D signal is not unique and depends on the choice of the algebra, for example, Cayley-Dickson or Clifford (see Chapter 2). As in the complex case, the starting point is the hypercomplex version of the Cauchy integral

$$f(\mathbf{z})=\frac{1}{\left(2\pi e_1\right)}\oint_{\partial D_1}\frac{1}{\left(2\pi e_2\right)}\oint_{\partial D_2}\frac{1}{\left(2\pi e_4\right)}\oint_{\partial D_3}\cdots\frac{1}{\left(2\pi e_m\right)}\oint_{\partial D_n}\frac{f\left(\xi,\ldots,\xi_n\right)d\xi_n}{\left(\xi_1-z_1\right)\ldots\left(\xi_n-z_n\right)}$$

$$(5.74)$$

This representation is not unique and depends of the algebra of multiplication of the imaginary units e_1. In the case of the Cayley-Dickson algebra $m=2^{N-1}$, we have a sequence with subscripts 1, 2, 4, 8..., and in the case of the Clifford algebra we have a sequence with subscripts 1, 2, 3, 4,... . It should be noted that the signal domain definition of the n-D analytic signal in the form of (5.26) representing the n-fold convolution of a real signal $u(\pmb{x})$ with the

n-D complex delta distribution. Analogously, the n-D hypercomplex analytic signal is defined by the n-fold convolution of $u(\boldsymbol{x})$ with the n-D hypercomplex delta distribution. The Cayley-Dickson version is

$$\Psi_{hyp}(\mathbf{x}) = \prod_{k=1}^{m}\left[\delta(x_k) + \frac{e_k}{\pi x_k}\right] * \ldots * u(\mathbf{x}) \; ; \; k = 1,2,4,8\ldots \tag{5.75}$$

and the version with Clifford algebra is

$$\Psi_{hyp}(\mathbf{x}) = \prod_{k=1}^{m}\left[\delta(x_k) + \frac{e_k}{\pi x_k}\right] * \ldots * u(\mathbf{x}) \; ; \; k = 1,2,3,\ldots \tag{5.76}$$

In 1-D, the complex and hypercomplex functions are the same.

5.3.1 2-D Quaternion Signals

In 2-D, the difference between the Cayley-Dickson approach and the Clifford approach is purely formal. The imaginary units e_1, e_2, e_3 are replaced by e_1, e_2, $(e_{12} = e_1 e_2)$. The definition (5.75) defines a quaternion-valued analytic signal of the form

$$\Psi_{q1}(x_1,x_2) = \left[\delta(x_1,x_2) + e_1\frac{\delta(x_2)}{\pi x_1} + e_2\frac{\delta(x_1)}{\pi x_2} + e_3\frac{1}{\pi^2 x_1 x_2}\right] * * u(x_1,x_2) \tag{5.77}$$

We get the following form of the quaternion analytic signal

$$\Psi_{q1}(x_1,x_2) = u + e_1 v_1 + e_2 v_2 + e_3 v \tag{5.78}$$

where u, v_1, v_2, and v are the same functions as defined for the 2-D complex analytic signal, i.e., the original signal and its partial and total Hilbert transforms (see (5.30) and (5.31)). The subscript $q1$ denotes that this signal has a quaternion single quadrant spectrum in the first quadrant. In Table 5.5, we see the signals with single quadrant spectra in all four quadrants and the corresponding frequency domain unit step operators (see (5.33)).

Note that the quaternion Fourier transform of u called QFT (see Chapter 4):

$$U_q(f_1,f_2) = \int_{-\infty}^{\infty}\int_{-\infty}^{\infty} e^{-e1\cdot 2\pi f_1 x_1} u(x_1,x_2) e^{-e_2 2\pi f_2 x_2}\, dx_1 dx_2 \tag{5.79}$$

Table 5.5
Quaternion Signals and Unit Step Operators in Successive Quadrants

Quadrant n	$\psi_{qi}(x_1, x_2)$	Unit Step Operator
1	$\psi_{q1} = u + e_1 v_1 + e_2 v_2 + e_3 v$	$4 \times 1(f_1, f_2) = 1 + \operatorname{sgn}(f_1) + \operatorname{sgn}(f_2) + \operatorname{sgn}(f_1)\operatorname{sgn}(f_2)$
2	$\psi_{q2} = u - e_1 v_1 + e_2 v_2 - e_3 v$	$4 \times 1(-f_1, f_2) = 1 - \operatorname{sgn}(f_1) + \operatorname{sgn}(f_2) - \operatorname{sgn}(f_1)\operatorname{sgn}(f_2)$
3	$\psi_{q3} = u + e_1 v_1 - e_2 v_2 - e_3 v$	$4 \times 1(f_1, -f_2) = 1 + \operatorname{sgn}(f_1) - \operatorname{sgn}(f_2) - \operatorname{sgn}(f_1)\operatorname{sgn}(f_2)$
4	$\psi_{q4} = u - e_1 v_1 - e_2 v_2 + e_3 v$	$4 \times 1(-f_1, -f_2) = 1 - \operatorname{sgn}(f_1) - \operatorname{sgn}(f_2) + \operatorname{sgn}(f_1)\operatorname{sgn}(f_2)$

Because the 2-D signal is a real function, it can be inserted in front or at the end with no change of signs of the result. The insertion of u as a union of even and odd terms (see Chapter 3, (3.11)–(3.15)) and using the Cayley-Dickson algebra (see the multiplication Table 2.2) presents the quaternion spectra U_q of four signals displayed in Table 5.6. In the first line, the operator with the single quadrant support in the first quadrant is shown. In the following lines, the spectra for next quadrants are shown. They are calculated directly by appropriate change of signs of the first quadrant operators. The same result yields the calculation of the involution.

The inverse QFT is

$$u(x_1, x_2) = \int\limits_{-\infty}^{\infty} \int\limits_{-\infty}^{\infty} e^{e_2 2\pi f_1 x_2} U_q(f_1, f_2) e^{e_1 2\pi f_2 x_1}\, df_1 df_2 \qquad (5.80)$$

Since the spectrum U_q is a complex function, the order of the integrand cannot be changed. All four analytic signals in Table 5.5 have the same norm given by the norm of separable complex signals (see (5.42)).

Table 5.6
Illustration of the Quaternionic Hermitian Symmetry

Quadrant	Quaternion Spectrum	Involution
1	$U_q(f_1, f_2) = U_{ee} - e_1 U_{oe} - e_2 U_{eo} + e_3 U_{oo}$	$U_q(f_1, f_2)$
2	$U_q(-f_1, f_2) = U_{ee} + e_1 U_{oe} - e_2 U_{eo} - e_3 U_{oo}$	$-e_2 U_q(f_1, f_2) e_2$
3	$U_q(f_1, -f_2) = U_{ee} - e_1 U_{oe} + e_2 U_{eo} - e_3 U_{oo}$	$-e_1 U_q(f_1, f_2) e_1$
4	$U_q(-f_1, -f_2) = U_{ee} + e_1 U_{oe} + e_2 U_{eo} + e_3 U_{oo}$	$-e_3 U_q(f_1, f_2) e_3$

Example 5.7

The quaternion valued representation for the real signal of Example 5.3 is

$$\psi_q(x_1, x_2) = \cos\left[2\pi\left(f_{10}x_1 + f_{20}x_2\right)\right] + e_1 \sin\left[2\pi\left(f_{10}x_{1+} + f_{20}x_2\right)\right]$$
$$+ e_2 \sin\left[2\pi\left(f_{10}x_1 + f_{20}x_2\right)\right] - e_3 \cos\left[2\pi\left(f_{10}x_1 + f_{20}x_2\right)\right]$$

Its amplitude equals 2 and the phase functions are described in Chapter 8.

5.3.2   3-D Hypercomplex Analytic Signals

The 3-D analytic signals are called octonions. Differently to the 2-D case we describe two kinds of octonions, one defined using the Cayley-Dickson algebra (see Table 2.3) and the second using the Clifford algebra Cl_{03} (see Table 2.4). Using the Cayley-Dickson approach, the signal domain definition of the octonion analytic signal is

$$\psi_1^{CD}\left(x_1, x_2, x_3\right)$$

$$= \left\{\left[\delta\left(x_1\right) + e_1 \frac{1}{\pi x_1}\right] \times \left[\delta\left(x_2\right) + e_2 \frac{1}{\pi x_2}\right] \times \left[\delta\left(x_3\right) + e_4 \frac{1}{\pi x_3}\right]\right\} *** u\left(x_1, x_2, x_3\right)$$

$$(5.81)$$

All signals in the Table 5.7 have the same norm

$$\psi_i^{CD} \times \left(\psi_i^{CD}\right)^* = u^2 + v_1^2 + v_2^2 + v_{12}^2 + v_3^2 + v_{13}^2 + v_{23}^2 + v^2 \qquad (5.82)$$

Table 5.7
The Eight Octonion Signals Defined by the Cayley-Dickson Algebra

Octant	Cayley-Dickson Octonions
1	$\psi_1^{CD}\left(x_1, x_2, x_3\right) = u + e_1v_1 + e_2v_2 + e_3v_{12} + e_4v_3 + e_5v_{13} + e_6v_{23} + e_7v$
3	$\psi_3^{CD}\left(x_1, x_2, x_3\right) = u + e_1v_1 - e_2v_2 - e_3v_{12} + e_4v_3 + e_5v_{13} - e_6v_{23} - e_7v$
5	$\psi_5^{CD}\left(x_1, x_2, x_3\right) = u + e_1v_1 + e_2v_2 + e_3v_{12} - e_4v_{13} - e_6v_{23} - e_7v$
7	$\psi_7^{CD}\left(x_1, x_2, x_3\right) = u + e_1v_1 - e_2v_2 - e_3v_{12} - e_4v_3 - e_5v_{13} + e_6v_{23} + e_7v$
2	$\psi_2^{CD}\left(x_1, x_2, x_3\right) = u - e_1v_1 + e_2v_2 - e_3v_{12} + e_4v_3 - e_5v_{13} + e_6v_{23} - e_7v$
4	$\psi_4^{CD}\left(x_1, x_2, x_3\right) = u - e_1v_1 - e_2v_2 + e_3v_{12} + e_4v_3 - e_5v_{13} - e_6v_{23} + e_7v$
6	$\psi_6^{CD}\left(x_1, x_2, x_3\right) = u - e_1v_1 + e_2v_2 - e_3v_{12} - e_4v_3 + e_5v_{13} - e_6v_{23} + e_7v$
8	$\psi_8^{CD}\left(x_1, x_2, x_3\right) = u - e_1v_1 - e_2v_2 + e_3v_{12} - e_4v_3 + e_5v_{13} + e_6v_{23} - e_7v$

The real signal u can be reconstructed using all eight signals

$$u\left(x_1,x_2,x_3\right) = \frac{\psi_1^{CD} + \psi_2^{CD} + \psi_3^{CD} + \psi_4^{CD} + \psi_5^{CD} + \psi_6^{CD} + \psi_7^{CD} + \psi_8^{CD}}{8} \qquad (5.83)$$

Let us recall, that for 3-D complex signals the reconstruction due to the Hermitian symmetry of the corresponding spectra needs only four signals (see(5.68)).

5.4 Monogenic 2-D Signals

The quaternion-valued hypercomplex signals have three imaginary terms defined by two partial Hilbert transforms and the total Hilbert transform. The monogenic 2-D hypercomplex signal has been proposed by Sommer and Felsberg [4] in the form

$$\psi_M\left(x_1,x_2\right) = u\left(x_1,x_2\right) + e_1 v_{r1}\left(x_1,x_2\right) + e_2 v_{r2}\left(x_1,x_2\right) \qquad (5.84)$$

where v_{r1} and v_{r2} are Riesz transforms of u:

$$v_{r1}\left(x_1,x_2\right) = u\left(x_1,x_2\right) ** r_1\left(x_1,x_2\right) \qquad (5.85)$$

$$v_{r2}\left(x_1,x_2\right) = u\left(x_1,x_2\right) ** r_2\left(x_1,x_2\right) \qquad (5.86)$$

The Riesz kernels have the form

$$r_1\left(x_1,x_2\right) = \frac{x_1}{2\pi\left(\sqrt{x_1^2 + x_2^2}\right)^3} \qquad (5.87)$$

$$r_2\left(x_1,x_2\right) = \frac{x_2}{2\pi\left(\sqrt{x_1^2 + x_2^2}\right)^3} \qquad (5.88)$$

and their 2-D Fourier transforms

$$U_{r1}\left(f_1,f_2\right) = \frac{-e_1 f_1}{\sqrt{f_1^2 + f_2^2}} \qquad (5.89)$$

$$U_{r2}\left(f_1, f_2\right) = \frac{-e_1 f_2}{\sqrt{f_1^2 + f_2^2}} \qquad (5.90)$$

The corresponding QFT is given by (5.79) and can be calculated using (see Chapter 4, (4.76))

$$QFT\left(f_1, f_2\right) = U\left(f_1, f_2\right)\frac{1-e_3}{2} + U\left(f_1, -f_2\right)\frac{1+e_3}{2} \qquad (5.91)$$

The insertion of (5.79) yields

$$QFT\left[r_1\right] = \frac{-e_1 f_1}{\sqrt{f_1^2 + f_2^2}} \qquad (5.92)$$

and the insertion of (5.91) yields

$$QFT\left[r_2\right] = \frac{-e_1 f_2}{\sqrt{f_1^2 + f_2^2}} \qquad (5.93)$$

In conclusion, the monogenic signal (5.84) can be written in the form of the convolution of the real signal with the Riesz kernel

$$\psi_M\left[x_1, x_2\right] = u\left(x_1, x_2\right) ** \left[\delta\left(x_1, x_2\right) + e_1 r_1\left(x_1, x_2\right) + e_2 r_2\left(x_1, x_2\right)\right] \qquad (5.94)$$

The frequency domain of this relation has the form of the product

$$QFT\left(\psi_M\right) = QFT(u) \times \left(1 + \frac{f_1 + f_2}{\sqrt{f_1^2 + f_2^2}}\right) \qquad (5.95)$$

It should be noted that, in the case of analytic and quaternion 2-D signals, the frequency domain operator has the form $[1 + \mathrm{sgn}(f_1)][1 + \mathrm{sgn}(f_2)]$. Figure 5.6 shows both operators.

Evidently, the monogenic operator is not a single quadrant operator. We observe that in the fourth quadrant it changes the sign. Asymptotically, in the first quadrant for large values of f_1 and f_2, it equals 3 while the operator (a) equals 4. The polar form of the monogenic signal is presented in Chapter 7. The amplitude equals the norm of (5.84) and Felsberg and Sommer [4] defined two angles called orientation angle and phase angle. The energies

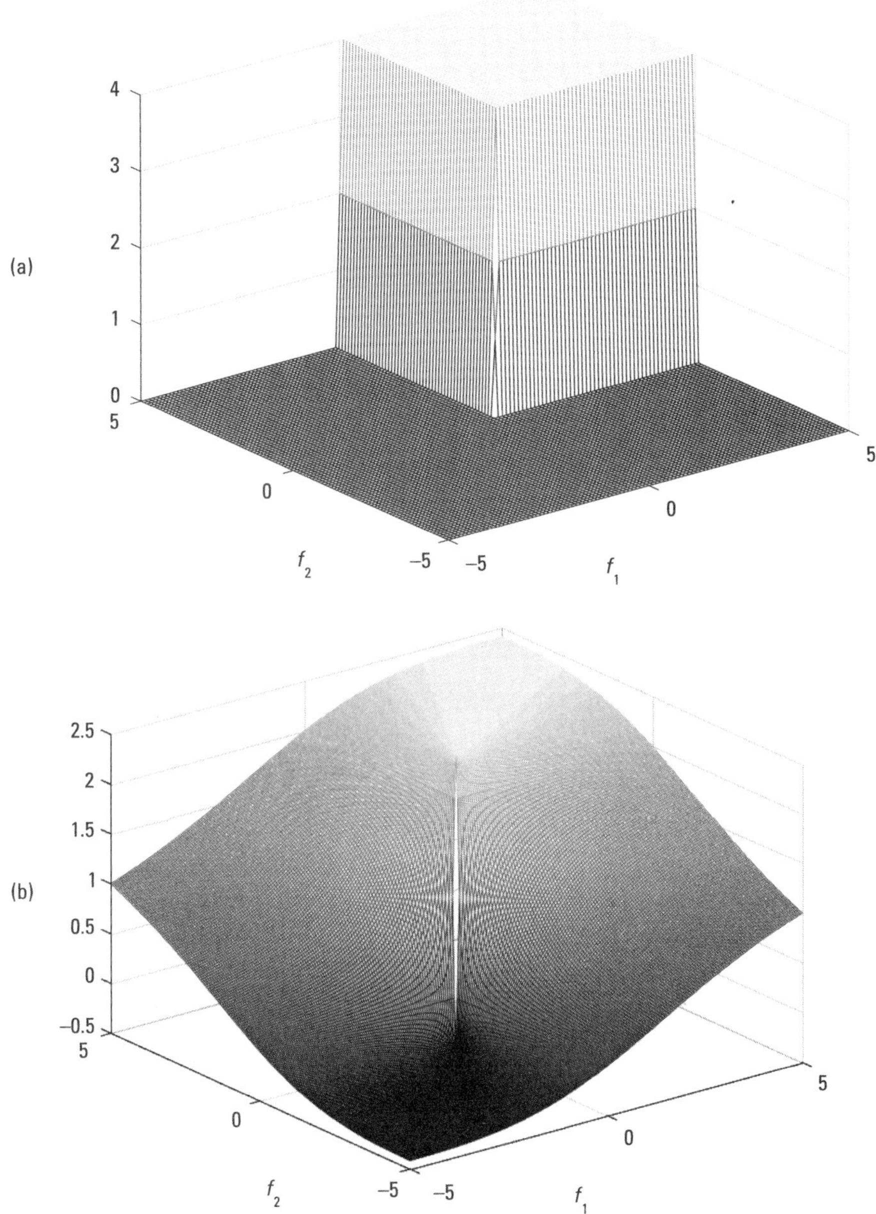

Figure 5.6 (a) The operator $[1 + \mathrm{sgn}(f_1)][1 + \mathrm{sgn}(f_2)]$, and (b) the operator $1 + \dfrac{f_1 + f_2}{\sqrt{f_1^2 + f_2^2}}$.

of two Riesz terms in (5.84) may differ. Their sum equals the energy of the real term u. Therefore, the energy of the monogenic signal equals two energies of the real term. Let us briefly discuss the term *analytic signal*, which was used by Felsberg and Sommer [4]. This term has been introduced to the signal theory by Gabor [5], who defined the notion of 1-D analytic signal. It should be noted that analytic signals are boundary distributions of analytic functions. These boundary distributions are defined by the Cauchy integrals (see (5.24)). However, Felsberg and Sommer [4] define the monogenic signal as *a boundary 2-D distribution of a 3-D Laplace equation*. Therefore, they use the term "analytic" defined differently from the analytic signals with single-orthant spectra. The Riesz transform has been described by Larkin et al. [11], who use the phrase *spiral phase quadrature transform*.

Example 5.8

This example is a continuation of the Example 5.3 for the monogenic signal. Let

$$u(x_1, x_2) = \cos\left[2\pi(f_{10}x_1 + f_{20}x_2)\right] \overset{2F}{\Longleftrightarrow} U(f_1, f_2)$$

$$= 0.5\left[\delta(f_1 - f_{10})\delta(f_2 - f_{20}) + \delta(f_1 + f_{10})\delta(f_2 + f_{20})\right].$$

The spectrum of the term v_{r1} in (5.84) is

$$V_{r1}(f_1, f_2) = \frac{-e_1 f_1}{\sqrt{f_1^2 + f_2^2}} 0.5\left[\delta(f_1 - f_{10})\delta(f_2 - f_{20}) + \delta(f_1 + f_{10})\delta(f_2 + f_{20})\right]$$

Using the relations

$$f(x)\delta(f - a) = f(a)\delta(f - a) \text{ and } f(x)\delta(f + a) = f(-a)\delta(f + a)$$

we get

$$V_{r1}(f_1, f_2) = \frac{-e_1 f_{10}}{\sqrt{f_{10}^2 + f_{20}^2}} 0.5\left[\delta(f_1 - f_{10})\delta(f_2 - f_{20}) - \delta(f_1 + f_{10})\delta(f_2 + f_{20})\right]$$

Note the change of sign of the second delta terms. The inverse QFT of this spectrum yields

$$v_{r1}(x_1, x_2) = \frac{f_{10}}{\sqrt{f_{10}^2 + f_{20}^2}} \sin\left[2\pi(f_{10}x_1 + f_{20}x_2)\right]$$

In exactly the same way, we get

$$v_{r2}(x_1,x_2) = \frac{f_{20}}{\sqrt{f_{10}^2 + f_{20}^2}}\sin\left[2\pi\left(f_{10}x_1 + f_{20}x_2\right)\right]$$

The insertion $f_{20} = kf_{10}$ yields the following form of the monogenic signal (5.84)

$$\psi_M(x_1,x_2) = \cos\left[2\pi\left(f_{10}x_1 + f_{20}x_2\right)\right] + e_1\frac{1}{\sqrt{1+k^2}}\sin\left[2\pi\left(f_{10}x_1 + f_{20}x_2\right)\right]$$
$$+ e_2\frac{k}{\sqrt{1+k^2}}\sin\left[2\pi\left(f_{10}x_1 + f_{20}x_2\right)\right]$$

Evidently, the energies of the two imaginary terms are equal only for $k = 1$. However, the sum of the energies is independent of k and equal to the energy of the real term. This statement applies for any monogenic signal.

Example 5.9

This example presents the monogenic version of Example 5.4. The 2-D Gaussian signal is given by (5.43), $\rho = 0.8$. The three terms of the monogenic signal are displayed in Figure 5.7.

In this example, the energies of the Riesz terms are equal (computer integration) and their sum equals the energy of the real signal u.

5.5 A Short Survey of the Notions of Analytic Signals with Single Orthant Spectra

In this chapter, we presented the definitions and properties of n-D complex and hypercomplex analytic signals with single orthant Fourier spectra. This summary recalls the basic properties of complex and hypercomplex analytic signals, including comments about the polar representation of these signals, which will be presented in detail in Chapter 7. The 2-D monogenic hypercomplex signal is not a member of the family of analytic signals with single orthant spectra. We include this signal, bearing in mind its importance in signal processing.

1. All analytic signals with single orthant spectra have the form of an extension of the Gabor's 1-D analytic signal to higher dimensions.

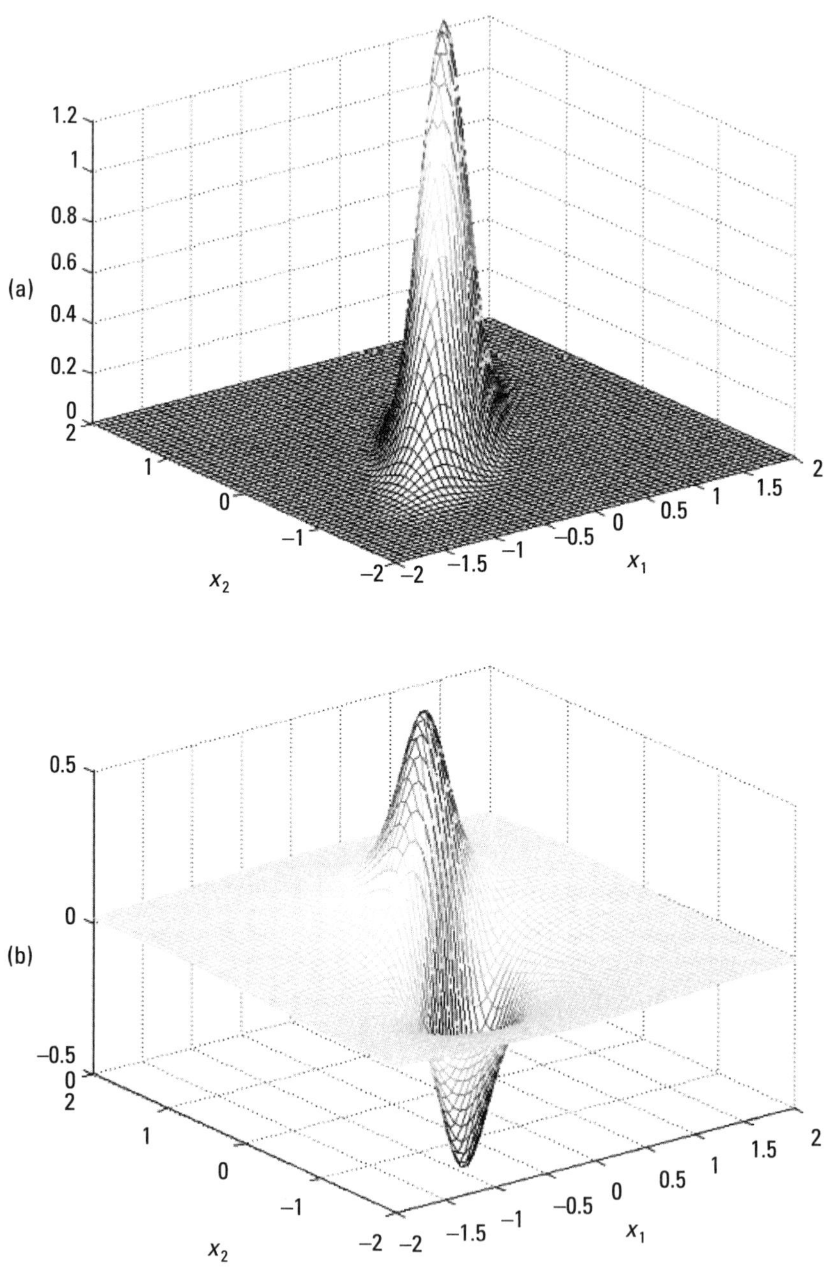

Figure 5.7   (a)The Gaussian 2-D signal (5.43), $\rho = 0.8$, (b) the Riesz term v_{r1}, (c) v_{r2}.

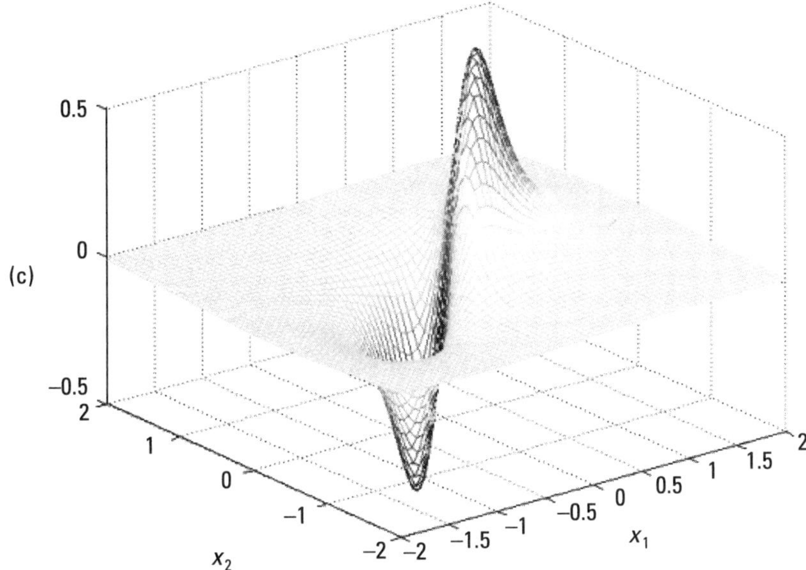

Figure 5.7 Continued

The Gabor's analytic signal is the boundary distribution of the 1-D analytic function. In the signal domain, it is defined by a convolution of a real signal u with the 1-D complex delta distribution. In the frequency domain, its spectrum is given by the product of the spectrum of u by a 1-D unit step distribution. The polar representation defines a single amplitude and single phase function (i.e., two functions (2^1)).

2. All higher dimensional *complex analytic signals* are defined in the signal domain by the convolution of the real signal u with the n-D complex delta distribution. It should be noted that this distribution has the form of a product of 1-D distributions (see (5.26)). Analogously, in the frequency domain, the spectrum of the complex analytic signal has the form of a product of the n-D spectrum of u by the n-D unit step distribution. Again, it has the form of a product of 1-D unit step distributions. The polar representation of n-D complex analytic signals defines 2^n functions, 2^{n-1} amplitudes, and 2^{n-1} phase functions (see Chapter 7). The reconstruction of the real signal using the amplitude-phase representation requires the knowledge of a half-space (and not a single orthant) spectrum.

3. All higher dimensional *hypercomplex analytic signals* are defined in the signal domain by the convolution of the real signal u with the n-D *hypercomplex delta* distribution. It should be noted that this distribution has the form of a product of 1-D distributions (see (5.75) and (5.76)). Analogously, in the frequency domain, the spectrum of the complex analytic signal has the form of a product of the n-D hypercomplex spectrum of u by the n-D unit step distribution. Again, it has the form of a product of 1-D unit step distributions. It should be noted that the hypercomplex spectrum depends on the choice of the algebra of the hypercomplex numbers. The polar representation of the hypercomplex signals differs with regard to the complex case. The total number of functions is the same. However, in 2-D, the quaternion analytic signal is represented by a single amplitude and three phase functions. The real signal can be reconstructed using the knowledge of single quadrant spectrum. However, there are formulae enabling the calculation the amplitude and three quaternion phase functions starting with the two amplitudes and two phase functions of the complex representation. The exact polar representation of an octonion hypercomplex analytic signal is still unknown. An approximate solution defines a single amplitude and seven phase functions.

4. The monogenic 2-D quaternion hypercomplex signal has the form of a convolution of the real signal u with the monogenic delta distribution $\psi_{\delta M} = \delta(x_1,x_2) + e_1 r_1(x_1,x_2) + e_2 r_2(x_1,x_2)$. Evidently, it is not a product of 1-D distributions. Its polar form defines a single amplitude and two phase functions. An extension of the monogenic signal to higher dimensions was not described in [4].

5.6 Survey of Application of n-D Analytic Signals

For technical reasons (volume), in this book we are not able to present a detailed description of applications. Instead, this survey is a guide to selected references describing applications.

5.6.1 Applications Presented in Other Chapters of this Book

In Chapter 9, we will describe Wigner distributions and ambiguity functions of analytic and monogenic signals.

5.6.2 Applications Described in Hahn's Book on Hilbert Transforms

In [7], there are descriptions of the role of analytic signals in modulation theory, amplitude modulation, frequency modulation, single-sideband modulation, and compatible single-sideband modulation. An extension for 2-D signals in the form of single-quadrant modulation is presented. Various kinds of Hilbert filters are described in detail. Analytic signals are used to define the power of signals.

5.6.3 Selected Applications

Hypercomplex wavelets are described in [16] and [23]. The octonic representation of electromagnetic fields is described in [19, 26] and field theory is described in [20, 26]. Applications in physics are numerous. For example, applications in relativistic quantum mechanics can be found in [21] and [22]. Olhede and Metikas described the notion of a homomorphic 1-D analytic signal [23]. The notion of entropy of analytic signals is introduced in [28]. Analytic waves are described by Vakman in [30, 31]. Applications in medicine (i.e., the processing of medical images) are presented in [18, 32].

References

[1] Antosik, P., Mikusiński, J., and Sikorski, P., *Theory of Distributions—The Sequential Approach,* Warsaw, Poland: PWN, 1973.

[2] Bedrosian, E., "The Analytic Signal Representation of Modulated Waveforms," *Proc. IRE,* October 1962, pp. 2071–2076.

[3] Bülow, T., and G. Sommer, "Hypercomplex Signals—A Novel Extension of the Analytic Signal to the Multidimensional Case," *IEEE Trans. Sign. Proc.,* Vol. 49, No. 11, Nov. 2001, pp. 2844–2852.

[4] Felsberg, M., and G. Sommer, "The Monogenic Signal," *IEEE Trans Sign. Proc.,* Vol. 49, No. 12, Dec. 2001, pp. 3136–3144.

[5] Gabor, D., "Theory of Communications" *Trans. Inst. Electr. Eng.,* Vol. 3, 1946, pp. 429–456.

[6] Hahn, S. L., "Multidimensional Complex Signals with Single-Orthant Spectra," *Proc. IEEE,* Vol. 80, No. 8, August 1992, pp. 1287–1300.

[7] Hahn, S. L., *Hilbert Transforms in Signal Processing,* Norwood, MA: Artech House, 1996.

[8]   Hahn, S. L., and K. M. Snopek, "The Unified Theory of n-Dimensional Complex and Hypercomplex Analytic Signals, " *Bull. Polish Ac. Sci., Tech. Sci.*, Vol. 59, No. 2, 2011, pp. 167–181.

[9]   Hahn, S. L., and K. M. Snopek, "Comparison of Properties of Analytic, Quaternionic and Monogenic 2D Signals," *WSEAS Transactions on Computers*, Issue 3, Vol. 3, July 2004, pp. 602–611.

[10]   Hahn, S. L., "The n-dimensional complex delta distribution," *IEEE Trans. Sign. Proc.*, Vol. 44, No. 7, July 1996, pp.1833–1837.

[11]   Larkin, K. G., D. Bone, and M. A. Oldfield, "Natural demodulation of 2D fringe patterns: General Background of the spiral phase quadrature transform," *J. Opt.Soc. Am.*, Vol. 18, No. 8, August 2001, pp.1862–1870.

[12]   Schwartz, L., *Méthodes Mathématiques pour les Science Physique*, Paris, France: Hermann, 1965.

[13]   Snopek, K. M., *Studies on Complex and Hypercomplex Multidimensional Analytic Signals*, Prace naukowe, Elektronika z. 190, Oficyna Wydawnicza Politechniki Warszawskiej, Warsaw, 2013.

[14]   Vakman, D., "On the analytic signal, the Teager-Kaiser energy algorithm, and other methods for defining amplitude and frequency," *IEEE Trans. Signal Process.*, Vol. 44, No. 4, 1996, pp. 791–797.

[15]   Bülow, T., "Hypercomplex spectral signal representation for the processing and analysis of images," in Bericht Nr. 9903, Institut für Informatik und Praktische Mathematik, Christian-Albrechts-Universität Kiel, 1999, http://www.cis.upenn.edu/~thomasbl/thesis.html.

[16]   Chan, W. L., H. Choi, and B. Baraniuk B, "Coherent multidcale image processing using dual-tree quaternion wavelets," *IEEE Trans. Image Process*, Vol. 17, No. 7, July 2008, pp. 1069–1082.

[17]   Cerejeiras, P., M. Ferierra, and U. Kähler, "Monogenic wavelets over the unit ball," *Zeitschrift fürAnalysis und ihre Anwendunden*, Vol. 24, 2005, pp. 841–852.

[18]   Wang, L., "Myocardial Motion Estimation from 2-D Analytic Phases and Preliminary Study on the Hypercomplex Signal," Medical Imaging INSA de Lyon, 2014.

[19]   Mironov, V. L., and S. V. Mironov, "Octonic representation of electromagnetic field equations," *Journal of Mathematical Physics*, Vol. 50, No. 012901, 2009.

[20]   Mironov, V. L., and S. V. Mironov, "Noncommutative sedeons and their application in field theory," *Institute for physics and microstructures*, RAS, November 2011.

[21]   Mironov, V. L., and S. V. Mironov, "Octonic second order equations of relativistic quantum mechanics," *JMP*, Vol. 50, No. 012302, 2009.

[22]   Mironov, V. L., and S. V. Mironov, "Octonic relativistic quantum mechanics," 2008, http:// arXiv.org./abs/0803.0375.

[23] Olhede, S. C., and G. Metikas, "The hyperanalytic wavelet transform," *Imperial College Statistics*, Technical Report TR-06-02, February 2008, pp. 1–49.

[24] Poletti, M., "The homomorphic analytic signal," *IEEE Trans. Signal Processing*, Vol. 45, 1997, pp.1943–1953.

[25] Poularikas, A. D. (ed.), *The Transforms and Applications Handbook, Second Edition*, Boca Raton, FL: CRC Press, IEEE Press, 2000.

[26] Tolan, T., K. Ozdas, and M. Tanisli, "Reformulation of electromagnetism with octonions," *Il Nuovo Cimento*, Vol. 121 B, No. 1, 2005, pp. 43–55.

[27] Ulrich, S., "Conformal relativity with hyper complex variables," *Proceedings of the Royal Society A*, Vol. 470, 2014, http://dx.doi.org/10.1098/rspa.2014.0027.

[28] Man'ko, M. A., "Entropy of an Analytic Signal," *Journal of Russian Laser Research*, Vol. 45, 1997, pp. 1943–1953.

[29] Selesnick, W. I., R. G. Baraniuk, and N. G. Kingsbury, "The dual-tree complex wavelet transform," *IEEE Signal Processing Magazine*, November 2005, pp. 123–150.

[30] Vakman, D., "Analytic waves," *International Journal of Theoretical Physics*, Nol.36, No.1, 1997.

[31] Vakman, D., *Signals, Oscillations and Waves, A Modern Approach*, Norwood, MA: Artech House, 1998.

[32] Washinger, C., T. Klein, and N. Navaly, "The 2D analytic signal on RF and B-Mode Ultra Sound Images," in *Information Processing in Medical Imaging* [IPMI], Vol. 121 B, No.1, 2011, pp.43–56.

6

Ranking of Analytic Signals

As shown in Chapters 4 and 5, the Fourier frequency support of an n-D complex or hypercomplex analytic signal is limited to a single orthant. For example, for 3-D signals, this support is called an octant and for 2-D signals—a quadrant. For a given n-D real signal, the corresponding analytic signal is a union of 2^n terms. For example, the octonion analytic signal ($n = 3$) is a union of 8 terms: a real signal u and 7 various Hilbert transforms (6 partial and a total one). For $n = 2$ (quaternion analytic signals), we have a sum of four terms (see Appendix H, Hilbert quadruples).

Of course, the 1-D analytic signal is a union of two terms: the real signal and its Hilbert transform. Analyzing forms of complex and hypercompex analytic signals in 1-D, 2-D, and 3-D, we notice partially conjugate pairs of signals, which inspires us to introduce a kind of systematic (ranking/hierarchy) in the n-D frequency space.

The definition of *ranking* is simple. We assign to the analytic signal with a single orthant spectrum support the highest rank equal n; that is, for 3-D signals, the highest rank is 3, and for 2-D signals, it is 2. The lower rank signal is obtained by enlarging the frequency support joining two neighbor orthants (and forming a so-called *suborthant*), for example, two octants separated by a common plane or two quadrants separated by a common line.

In this chapter, it will be shown that the doubling of the frequency domain support reduces the number of terms. For $n = 3$, the signal of the highest rank has eight terms: the signals of rank 2 have four terms and the lowest rank (equal 1) only two terms, the same as the number of terms of the 1-D Gabor's analytic signal. The difference is that the lowering of rank reduces the number of terms with no change of signal dimensions. For example, for 3-D signals, the signal of the lowest rank has the form $\psi(x_1,x_2,x_3) = u(x_1,x_2,x_3) + jH_1(x_1,x_2,x_3)$, where H_1 denotes the partial Hilbert transform with regard to x_1 as a consequence of the choice of the frequency domain support of the lowest rank signal in the half-plane f_1. Regarding terminology, the use of the term *ranking* may be questioned. An alternative choice could be *hierarchy*.

6.1   Definition of a Suborthant

The *orthant*, as shown in Chapter 3, is a strictly determined part of the n-D frequency space [1, 2]. Evidently, this notion can be used in any n-D space.

Let us recall that in 1-D, the orthant is a *half-axis*. Therefore, in 1-D we have two (2^1) orthants: $f > 0$ and $f < 0$ (see Figure 6.1(a)). Further, the 2-D plane is divided into four (2^2) *quadrants*: $(f_1 > 0, f_2 > 0)$, $(f_1 < 0, f_2 > 0)$, $(f_1 > 0, f_2 < 0)$, and $(f_1 < 0, f_2 < 0)$, numbered as shown in Figure 6.1(b). The 3-D space is modeled as a union of eight (2^3) *octants*:

$$\left(f_1 > 0, f_2 > 0, f_3 > 0\right), \left(f_1 < 0, f_2 > 0, f_3 > 0\right),$$
$$\left(f_1 > 0, f_2 < 0, f_3 > 0\right), \left(f_1 < 0, f_2 < 0, f_3 > 0\right),$$
$$\left(f_1 > 0, f_2 > 0, f_3 < 0\right), \left(f_1 < 0, f_2 > 0, f_3 < 0\right),$$
$$\left(f_1 > 0, f_2 < 0, f_3 < 0\right), \left(f_1 < 0, f_2 < 0, f_3 < 0\right)$$

labeled with 1, 2, …, 8 as shown in Figure 6.2. It should be noted that the orthant's numbering differs from that which is commonly used by mathematicians. For example in 2-D, the standard mathematical convention applies the numeration of quadrants as follows: No. 1. for $(f_1 > 0, f_2 > 0)$, No. 2. for $(f_1 < 0, f_2 > 0)$, No. 3. for $(f_1 < 0, f_2 < 0)$, and No. 4 for $(f_1 > 0, f_2 < 0)$.

In this book, we apply the convention introduced by Hahn in [3] and used in all his further works concerning multidimensional analytic signals with single-orthant spectra [4–6]. The advantage of such a label is that all higher dimensional orthants in the half-space $f_1 > 0$ are always odd-numbered. It should be noted that the same indexing applies for all analytic signals used in this book (see Chapter 5).

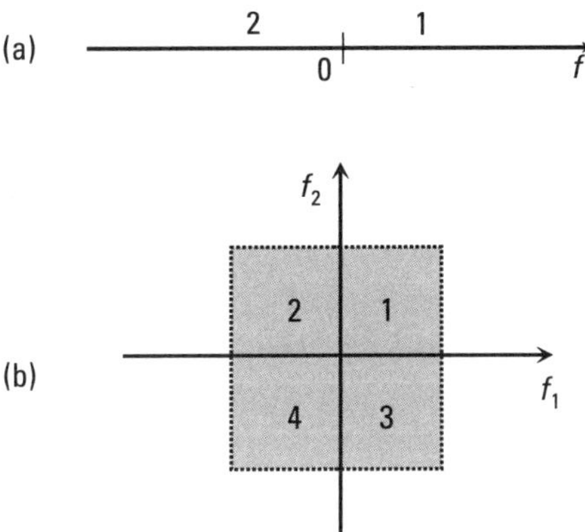

Figure 6.1 Labeling of orthants in 1-D and 2-D.

Consider the n-D space, composed of 2^n orthants. We assign the order 1 to a single orthant of the n-D space. A family of *suborthants* of higher order is defined as follows:

A *suborthant* of order 2 is formed of two adjacent orthants of order 1 having a *common hyperplane*. The number of suborthants of the second order is $2^n/2 = 2^{n-1}$. Continuing, we define a suborthant of order 3 as a union of two adjacent second order suborthants with a common hyperplane. The number of the third order suborthants is 2^{n-2}. The process is continued and finally we

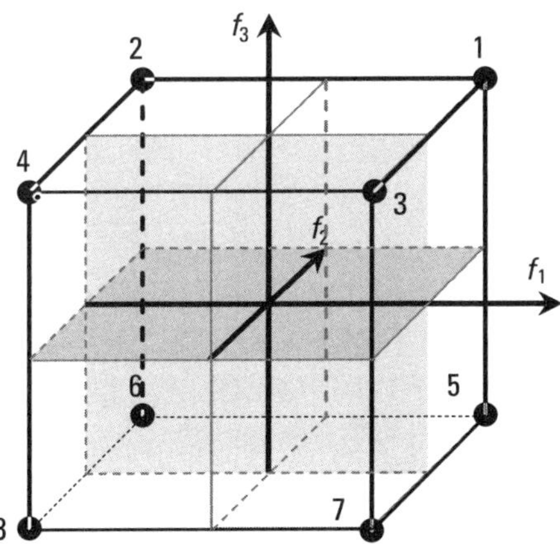

Figure 6.2 Labeling of octants in 3-D.

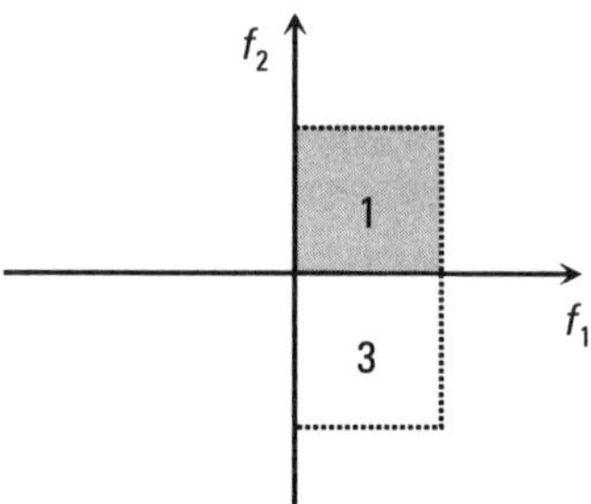

Figure 6.3 Subquadrant (1,3) in the half plane $f_1 > 0$.

get two suborthants of order n (i.e., two half-spaces $\left(f_1 > 0, f_2, \ldots, f_n\right)$ and $\left(f_1 < 0, f_2, \ldots, f_n\right)$, each composed of two suborthants of order $n - 1$).

Let us illustrate the above notions in 2-D and 3-D for $f_1 > 0$.

6.1.1   Subquadrants in 2-D

We call a *subquadrant* (1,3) (of order 2) in $f_1 > 0$, a half-plane composed of two neighboring quadrants 1 and 3 (of order 1) (see Figure 6.3). They have a *common half-line*.

6.1.2   Suboctants in 3-D

In the half-space $f_1 > 0$ of the 3-D space, a *suboctant of order 2* is defined as a union of two adjacent octants of order 1 having a *common half-plane* (i.e.,

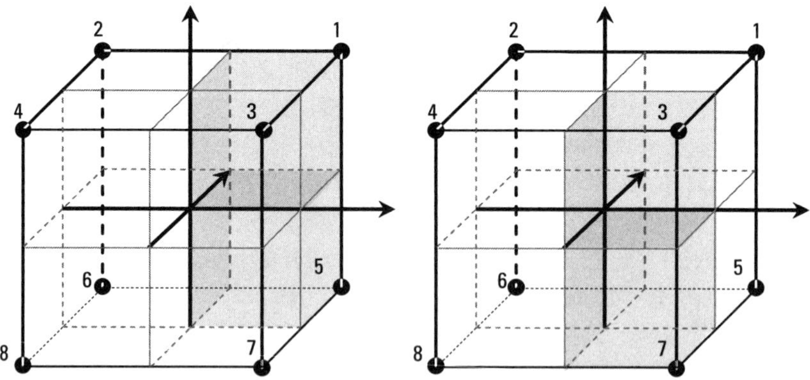

Figure 6.4 Suboctants (1,5) and (3,7) in the half-space $f_1 > 0$.

(1,5)–(3,7) as in Figure 6.4, or (1,3)–(5,7)). The *suboctant of order 3* is the half-space $f_1 > 0$.

6.2 Ranking of Complex Analytic Signals

Let us consider the n-dimensional signal space, $x = (x_1, x_1, \ldots, x_n)$ is the n-D signal domain variable. Let $R = n$ be the highest *rank* assigned to the n-D complex analytic signal with a single-orthant spectrum. For example, for octonion signals, $R = 3$ and such a signal is a union of eight terms. We define *lower rank* n-D complex signals by applying the step-by-step procedure ending with a signal of rank $R = 1$. The lowest rank R signal always has two terms, just as the Gabor's analytic signal. In the case of octonion signals, a signal of rank $R = 2$ is a union of four terms, and a signal of rank $R = 1$ has only two components. The ranking procedure uses the notion of a suborthant introduced in Section 6.1. It should be noted that only the number of terms is reduced and not the dimension of a signal. Let us study in detail cases $n = 2$ and $n = 3$.

6.2.1 Ranking of 2-D Complex Analytic Signals

Consider two complex analytic signals with single-quadrant spectra in the $f_1 > 0$ plane (see Table 5.1 and Figure 6.1(b)):

$$\psi_1(x_1, x_2) = u(x_1, x_2) - v(x_1, x_2) + \left[v_1(x_1, x_2) + v_2(x_1, x_2) \right] \cdot e_1 \quad (6.1)$$

$$\psi_3(x_1, x_2) = u(x_1, x_2) + v(x_1, x_2) + \left[v_1(x_1, x_2) - v_2(x_1, x_2) \right] \cdot e_1 \quad (6.2)$$

The rank of the above signals is $R = 2$. We define a 2-D *complex signal of lower rank* $R = 1$ as a mean value of signals ψ_1 and ψ_3 given by (6.1) and (6.2) as follows

$$\psi_{(f_1 > 0)}(x_1, x_2) = \frac{\psi_1 + \psi_3}{2} = u(x_1, x_2) + v_1(x_1, x_2) \cdot e_1 \quad (6.3)$$

The signal (6.3) is sometimes called a *partial analytic signal* [7]. The process of ranking in the signal domain is illustrated in Figure 6.5. The signal (6.3) can equivalently be defined in the frequency domain as the inverse 2-D FT of a spectrum limited to the half-plane $(f_1 > 0, f_2)$ equals the subquadrant (1,3) (see Figure 6.6).

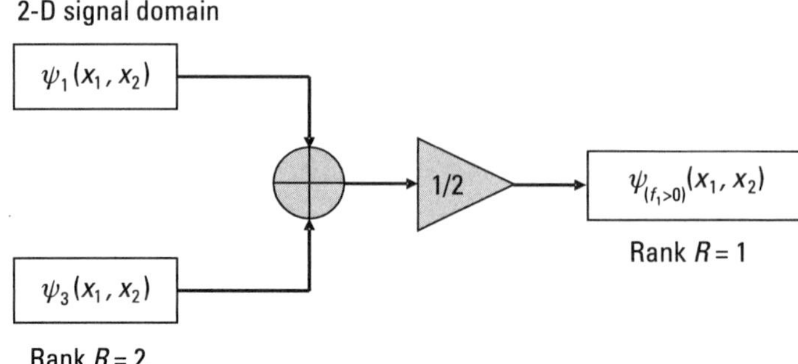

Figure 6.5   Ranking of 2-D complex signals in the signal domain.

Example 6.1

Let us consider a 2-D real signal $u(x_1,x_2) = \cos(2\pi f_{10}x_1)\cos(2\pi f_{20}x_2)$, $f_{10}, f_{20} > 0$ and its complex analytic signal of rank 2 given by

$$\psi_1(x_1,x_2) = \exp\left[e_1 2\pi\left(f_{10}x_1 + f_{20}x_2\right)\right]$$
$$= \cos\left[2\pi\left(f_{10}x_1 + f_{20}x_2\right)\right] + \sin\left[2\pi\left(f_{10}x_1 + f_{20}x_2\right)\right]\cdot e_1$$

It can equivalently be written in the form (6.1) with $v(x_1,x_2) = \sin(2\pi f_{10}x_1)\sin(2\pi f_{20}x_2)$, $v_1(x_1,x_2) = \sin(2\pi f_{10}x_1)\cos(2\pi f_{20}x_2)$, $v_2(x_1,x_2) = \cos(2\pi f_{10}x_1)\sin(2\pi f_{20}x_2)$. The 2-D complex analytic signal with a spectrum in quadrant 3 is $\psi_3(x_1,x_2) = \exp[e_1 2\pi(f_{10}x_1 - f_{20}x_2)]$.

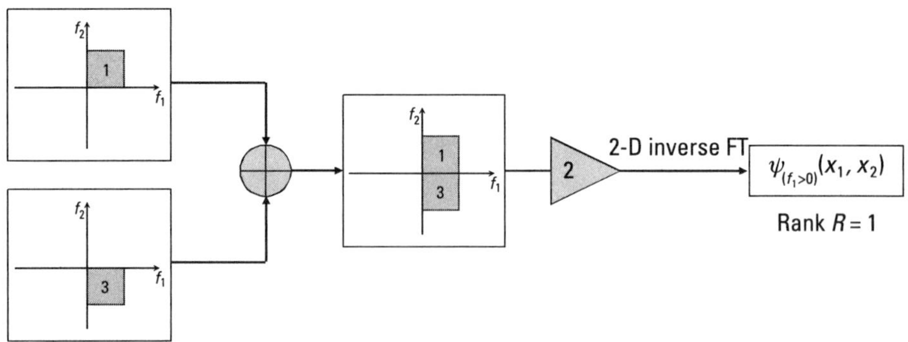

Figure 6.6   Ranking of 2-D complex signals in the frequency domain.

Then, a complex signal of rank $R = 1$ is $\psi_{(f_i>0)}(x_1,x_2) = \exp(e_1 2\pi f_{10}x_1)$ $\cos(2\pi f_{20}x_2)$. So, we see that $\psi_{(f_i>0)}(x_1,x_2)$ has exactly the form (6.3), that is

$$\psi_{(f_i>0)}(x_1,x_2) = \cos(2\pi f_{10}x_1)\cos(2\pi f_{20}x_2) + \sin(2\pi f_{10}x_1)\cos(2\pi f_{20}x_2)\cdot e_1$$

$$= u(x_1,x_2) + v_1(x_1,x_2)\cdot e_1$$

Moreover, its spectrum composed of two delta pulses in the subquadrant (1.3) is given by the formula

$$F\left\{\psi_{(f_i>0)}(x_1,x_2)\right\} = \frac{1}{2}\delta(f_1 - f_{10})\delta(f_2 - f_{20}) + \frac{1}{2}\delta(f_1 - f_{10})\delta(f_2 + f_{20})$$

6.2.2 Ranking of 3-D Complex Analytic Signals

Let us explain the process of ranking in 3-D. We recall the signal-domain definitions of 3-D complex analytic signals with spectra in $f_1 > 0$ (see Chapter 5, Table 5.4):

$$\psi_1(x_1,x_2,x_3) = u - v_{12} - v_{13} - v_{23} + (v_1 + v_2 + v_3 - v)\cdot e_1 \qquad (6.4)$$

$$\psi_3(x_1,x_2,x_3) = u + v_{12} - v_{13} + v_{23} + (v_1 - v_2 + v_3 + v)\cdot e_1 \qquad (6.5)$$

$$\psi_5(x_1,x_2,x_3) = u - v_{12} + v_{13} + v_{23} + (v_1 + v_2 - v_3 + v)\cdot e_1 \qquad (6.6)$$

$$\psi_7(x_1,x_2,x_3) = u + v_{12} + v_{13} - v_{23} + (v_1 - v_2 - v_3 - v)\cdot e_1 \qquad (6.7)$$

The process of ranking of 3-D complex signals in the (x_1,x_2,x_3)–domain is illustrated in Figure 6.7. We assign the highest rank $R = 3$ to all complex analytic signals $\psi_i(x_1,x_2,x_3)$, $i = 1,3,5,7$ given by (6.4)–(6.7). Then, we calculate a mean value (Step 1) of two 3-D analytic signals (having a frequency support in suboctants (1,5) and (3,7)) getting two signals of rank $R = 2$:

$$\psi_{(1,5)}(x_1,x_2,x_3) = \frac{\psi_1 + \psi_5}{2} = u - v_{12} + (v_1 + v_2)\cdot e_1 \qquad (6.8)$$

$$\psi_{(3,7)}(x_1,x_2,x_3) = \frac{\psi_3 + \psi_7}{2} = u + v_{12} + (v_1 - v_2)\cdot e_1 \qquad (6.9)$$

It should be noted that the structure of (6.8)–(6.9) is similar to (6.1) and (6.2), respectively. Continuing the ranking process (Step 2), we come to a 3-D complex signal of the lowest rank $R = 1$:

$$\Psi_{(f_1>0)}\left(x_1,x_2,x_3\right) = \frac{\Psi_{(1,5)} + \Psi_{(3,7)}}{2} = u\left(x_1,x_2,x_3\right) + v_1\left(x_1,x_2,x_3\right)\cdot e_1 \qquad (6.10)$$

having a structure similar to (6.4). As in the 2-D case, the equivalent process can be performed in the frequency domain as shown in Figure 6.8.

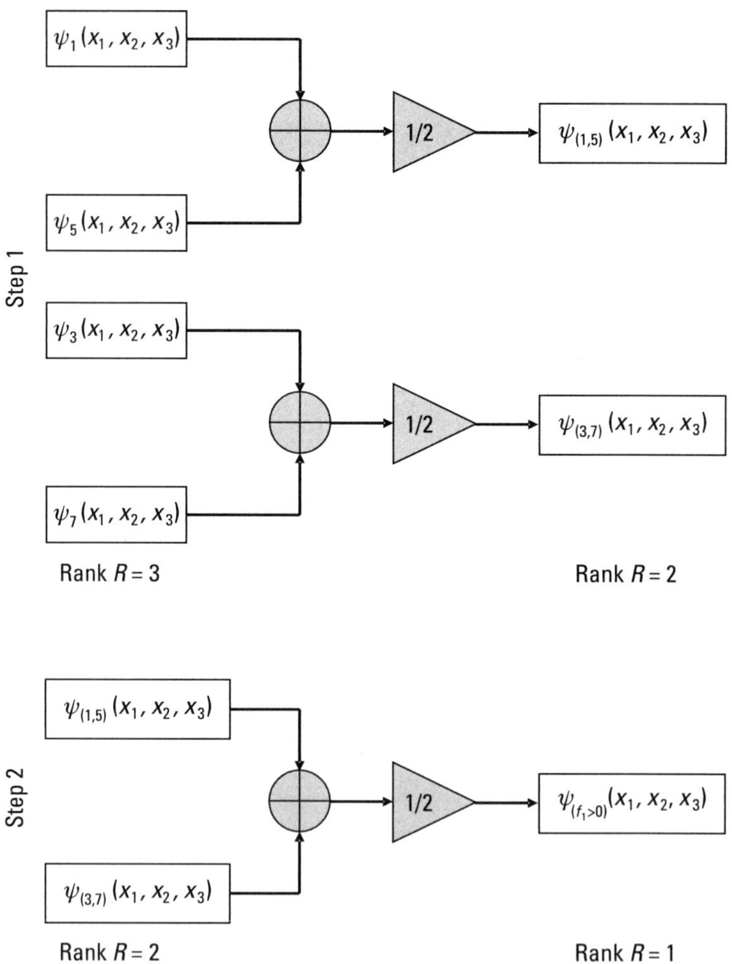

Figure 6.7 Ranking of 3-D complex signals in the signal domain.

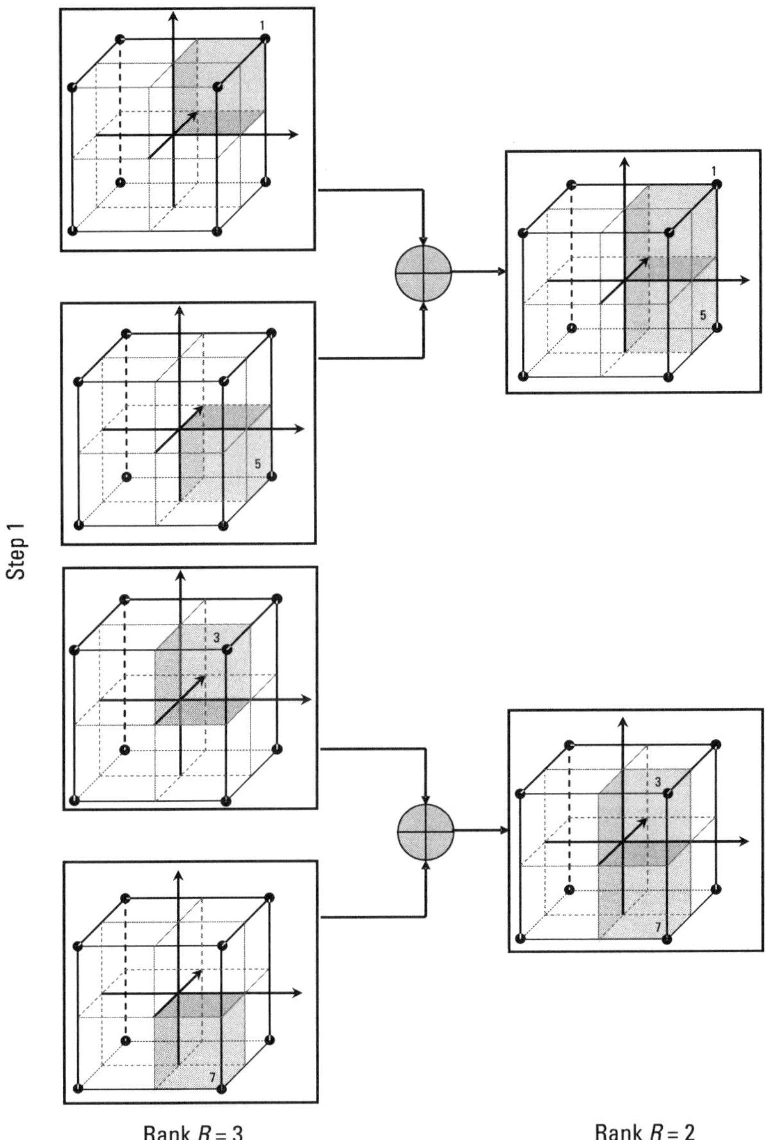

Figure 6.8 Ranking of 3-D complex signals in the frequency domain.

We obtain the signal (6.10) as the 3-D inverse FT of a spectrum limited to a half-space $(f_1 > 0, f_2, f_3)$ composed of two second-order suboctants (1,5) and (3,7).

Example 6.2

Let us consider the 3-D real signal $u(x_1, x_2, x_3) = \cos(2\pi f_{10}x_1)\cos(2\pi f_{20}x_2)\cos(2\pi f_{30}x_3)$, $f_{10}, f_{20}, f_{30} > 0$ and its 3-D analytic signal of rank 3 given by

$$\psi_1\left(x_1, x_2, x_3\right) = \exp\left[e_1 2\pi\left(f_{10}x_1 + f_{20}x_2 + f_{30}x_3\right)\right]$$
$$= \cos\left[2\pi\left(f_{10}x_1 + f_{20}x_2 + f_{30}x_3\right)\right]$$
$$+ \sin\left[2\pi\left(f_{10}x_1 + f_{20}x_2 + f_{30}x_3\right)\right]\cdot e_1$$

To simplify the notation, let us substitute $\omega_i = 2\pi f_{i0}$, $i = 1,2,3$. The above signal can equivalently be written in the form (6.4) with:

$$v_1\left(x_1, x_2, x_3\right) = \sin\left(\omega_1 x_1\right)\cos\left(\omega_2 x_2\right)\cos\left(\omega_3 x_3\right)$$
$$v_2\left(x_1, x_2, x_3\right) = \cos\left(\omega_1 x_1\right)\sin\left(\omega_2 x_2\right)\cos\left(\omega_3 x_3\right)$$
$$v_3\left(x_1, x_2, x_3\right) = \cos\left(\omega_1 x_1\right)\cos\left(\omega_2 x_2\right)\sin\left(\omega_3 x_3\right)$$
$$v_{12}\left(x_1, x_2, x_3\right) = \sin\left(\omega_1 x_1\right)\sin\left(\omega_2 x_2\right)\cos\left(\omega_3 x_3\right)$$
$$v_{13}\left(x_1, x_2, x_3\right) = \sin\left(\omega_1 x_1\right)\cos\left(\omega_2 x_2\right)\sin\left(\omega_3 x_3\right)$$
$$v_{23}\left(x_1, x_2, x_3\right) = \cos\left(\omega_1 x_1\right)\sin\left(\omega_2 x_2\right)\sin\left(\omega_3 x_3\right)$$
$$v\left(x_1, x_2, x_3\right) = \sin\left(\omega_1 x_1\right)\sin\left(\omega_2 x_2\right)\sin\left(\omega_3 x_3\right)$$

Other 3-D complex analytic signals with spectra in octants 3, 5, 7 (respectively) are

$$\psi_3\left(x_1, x_2, x_3\right) = \exp\left[e_1 2\pi\left(f_{10}x_1 - f_{20}x_2 + f_{30}x_3\right)\right]$$
$$\psi_5\left(x_1, x_2, x_3\right) = \exp\left[e_1 2\pi\left(f_{10}x_1 + f_{20}x_2 - f_{30}x_3\right)\right]$$
$$\psi_7\left(x_1, x_2, x_3\right) = \exp\left[e_1 2\pi\left(f_{10}x_1 - f_{20}x_2 - f_{30}x_3\right)\right]$$

The 3-D complex signals of rank 2 given by (6.8)–(6.9) respectively are

$$\psi_{(1,5)}\left(x_1,x_2,x_3\right) = \exp\left[e_1 2\pi\left(f_{10}x_1 + f_{20}x_2\right)\right]\cos\left(2\pi f_{30}x_3\right)$$

$$\psi_{(3,7)}\left(x_1,x_2,x_3\right) = \exp\left[e_1 2\pi\left(f_{10}x_1 - f_{20}x_2\right)\right]\cos\left(2\pi f_{30}x_3\right)$$

Finally, the 3-D complex signal of rank 1 is

$$\psi_{(f_i>0)}\left(x_1,x_2,x_3\right) = \exp\left(e_1 2\pi f_{10}x_1\right)\cos\left(2\pi f_{20}x_2\right)\cos\left(2\pi f_{30}x_3\right)$$

As expected, the signal $\psi_{(f_i>0)}(x_1,x_2,x_3)$ has exactly the form (6.10), that is

$$\psi_{(f_i>0)}\left(x_1,x_2,x_3\right) = \cos\left(2\pi f_{10}x_1\right)\cos\left(2\pi f_{20}x_2\right)\cos\left(2\pi f_{30}x_3\right)$$
$$+ \sin\left(2\pi f_{10}x_1\right)\cos\left(2\pi f_{20}x_2\right)\cos\left(2\pi f_{30}x_3\right)\cdot e_1$$

and its spectrum is limited to the suboctant (1,5)–(3,7).

Examples 6.1 and 6.2 clearly visualize the procedure of ranking in 2-D and 3-D. We assemble all of the results in Table 6.1, showing the hierarchy of n-D complex analytic signals $n = 1, 2, 3$ of rank $R \le n$.

6.3 Ranking of Hypercomplex Analytic Signals

In this section, we apply the ranking procedure described in Section 6.2 to construct n-D hypercomplex signals of rank $R < n$. The ranking of quaternion and octonion analytic signal will end with exactly the same result as in case of complex analytic signals.

Similarly as above, let $R = n$ be the highest rank assigned to the n-D Cayley-Dickson analytic signals with a single-orthant spectrum support. Let us start from the simplest case of quaternion analytic signals ($n = 2$) and then pass to octonion analytic signals ($n = 3$).

6.3.1 Ranking of 2-D Cayley-Dickson Analytic Signals

Let us recall the signal-domain definitions of 2-D Cayley-Dickson (quaternion) analytic signals (see Chapter 5, Table 5.7)

$$\psi_1^{CD}\left(x_1,x_2\right) = u + v_1\cdot e_1 + v_2\cdot e_2 + v\cdot e_3 \tag{6.11}$$

$$\psi_3^{CD}\left(x_1,x_2\right) = u + v_1\cdot e_1 - v_2\cdot e_2 - v\cdot e_3 \tag{6.12}$$

Table 6.1

Hierarchy of n-D Complex Analytic Signals Related to the Real Signal $u(\boldsymbol{x}) = \displaystyle\prod_{i=1}^{n} \cos(\omega_i x_i)$, $n = 1,2,3$

	$n=1$		$n=2$		$n=3$	
$R=3$	——		——		1	$\exp\!\left[e_1\!\left(\omega_1 x_1 + \omega_2 x_2 + \omega_3 x_3\right)\right]$
					3	$\exp\!\left[e_1\!\left(\omega_1 x_1 - \omega_2 x_2 + \omega_3 x_3\right)\right]$
					5	$\exp\!\left[e_1\!\left(\omega_1 x_1 + \omega_2 x_2 - \omega_3 x_3\right)\right]$
					7	$\exp\!\left[e_1\!\left(\omega_1 x_1 - \omega_2 x_2 - \omega_3 x_3\right)\right]$
$R=2$	——		1	$\exp\!\left[e_1\!\left(\omega_1 x_1 + \omega_2 x_2\right)\right]$	$(1,5)$	$\exp\!\left[e_1\!\left(\omega_1 x_1 + \omega_2 x_2\right)\right]\cos(\omega_3 x_3)$
$R=1$	1	$\exp(e_1\omega_1 x_1)$	3	$\exp\!\left[e_1\!\left(\omega_1 x_1 - \omega_2 x_2\right)\right]$	(3.7)	$\exp\!\left[e_1\!\left(\omega_1 x_1 - \omega_2 x_2\right)\right]\cos(\omega_3 x_3)$
			$(1,3)$	$\exp(e_1\omega_1 x_1)\cos(\omega_2 x_2)$	$(1,5)\text{–}(3,7)$	$\exp(e_1\omega_1 x_1)\cos(\omega_2 x_2)\cos(\omega_3 x_3)$
$u(\boldsymbol{x})$	$\cos(\omega_1 x_1)$		$\cos(\omega_1 x_1)\cos(\omega_2 x_2)$		$\cos(\omega_1 x_1)\cos(\omega_2 x_2)\cos(\omega_3 x_3)$	

All of the above analytic signals have rank $R = 2$ and form partially conjugate pairs. We define two 2-D signals of rank $R = 1$ by calculating the mean value of signals with spectra in the subquadrant (1,3)

$$\psi_{(f_i>0)}^{CD}\left(x_1,x_2\right) = \frac{\psi_1^{CD} + \psi_3^{CD}}{2} = u\left(x_1,x_2\right) + v_1\left(x_1,x_2\right) \cdot e_1 \qquad (6.13)$$

We notice that (6.13) is exactly the same as (6.4).

Example 6.3

Let us consider a 2-D real signal $u(x_1,x_2) = \cos(2\pi f_{10}x_1)\cos(2\pi f_{20}x_2)$ and its (total and partial) Hilbert transforms: $v(x_1,x_2) = \sin(2\pi f_{10}x_1)\sin(2\pi f_{20}x_2)$, $v_1(x_1,x_2) = \sin(2\pi f_{10}x_1)\cos(2\pi f_{20}x_2)$, $v_2(x_1,x_2) = \cos(2\pi f_{10}x_1)\sin(2\pi f_{20}x_2)$.

The quaternion analytic signals of rank 2 are given by (6.11)–(6.12). To simplify the notation, let us write: $c_i = \cos(\omega_i x_i)$, $s_i = \sin(\omega_i x_i)$, $i = 1,2$.

$$\begin{aligned}
\psi_1^{CD}\left(x_1,x_2\right) &= c_1c_2 + s_1c_2 \cdot e_1 + c_1s_2 \cdot e_2 + s_1s_2 \cdot e_3 \\
&= c_1c_2 + s_1c_2 \cdot e_1 + \left(c_1s_2 + s_1s_2 \cdot e_1\right) \cdot e_2 \\
&= \left(c_1 + s_1 \cdot e_1\right)c_2 + \left[\left(c_1 + s_1 \cdot e_1\right)s_2\right] \cdot e_2 \\
&= \left(c_1 + s_1 \cdot e_1\right)\left(c_2 + s_2 \cdot e_2\right) = \exp\left(e_1\omega_1 x_1\right)\exp\left(e_2\omega_2 x_2\right)
\end{aligned}$$

$$\begin{aligned}
\psi_3^{CD}\left(x_1,x_2\right) &= c_1c_2 + s_1c_2 \cdot e_1 - c_1s_2 \cdot e_2 - s_1s_2 \cdot e_3 \\
&= c_1c_2 + s_1c_2 \cdot e_1 - \left(c_1s_2 + s_1s_2 \cdot e_1\right) \cdot e_2 \\
&= \left(c_1 + s_1 \cdot e_1\right)c_2 - \left[\left(c_1 + s_1 \cdot e_1\right)s_2\right] \cdot e_2 \\
&= \left(c_1 + s_1 \cdot e_1\right)\left(c_2 - s_2 \cdot e_2\right) = \exp\left(e_1\omega_1 x_1\right)\exp\left(-e_2\omega_2 x_2\right)
\end{aligned}$$

The signal of rank 1 is calculated as a mean value given by (6.13) of the above signals:

$$\begin{aligned}
\psi_{(f_i>0)}^{CD}\left(x_1,x_2\right) &= \frac{1}{2}\left[\exp\left(e_1\omega_1 x_1\right)\exp\left(e_2\omega_2 x_2\right) + \exp\left(e_1\omega_1 x_1\right)\exp\left(-e_2\omega_2 x_2\right)\right] \\
&= \exp\left(e_1\omega_1 x_1\right)\cos\left(\omega_2 x_2\right).
\end{aligned}$$

It should be noted that the signal $\psi_{(f_i>0)}^{CD}(x_1,x_2)$ is exactly the same as the complex analytic signal of rank 1 from Example 6.1. This result confirms once again equivalence of complex and hypercomplex approaches. We will show that this conclusion can be generalized to higher dimensional Cayley-Dickson analytic signals.

6.3.2 Ranking of 3-D Cayley-Dickson Analytic Signals

In this section, we will show that the ranking process of 3-D Cayley-Dickson (octonion) analytic signals will give the same exact result as for the 3-D complex analytic signals shown in Section 6.2.2.

The octonion analytic signals of the highest rank $R = 3$ (see Table 5.7, Chapter 5) are given by

$$\psi_1^{CD}\left(x_1,x_2,x_3\right) = u + v_1 \cdot e_1 + v_2 \cdot e_2 + v_{12} \cdot e_3 + v_3 \cdot e_4 + v_{13} \cdot e_5 + v_{23} \cdot e_6 + v \cdot e_7$$

$$(6.14)$$

$$\psi_3^{CD}\left(x_1,x_2,x_3\right) = u + v_1 \cdot e_1 - v_2 \cdot e_2 - v_{12} \cdot e_3 + v_3 \cdot e_4 + v_{13} \cdot e_5 - v_{23} \cdot e_6 - v \cdot e_7$$

$$(6.15)$$

$$\psi_5^{CD}\left(x_1,x_2,x_3\right) = u + v_1 \cdot e_1 + v_2 \cdot e_2 + v_{12} \cdot e_3 - v_3 \cdot e_4 - v_{13} \cdot e_5 - v_{23} \cdot e_6 - v \cdot e_7$$

$$(6.16)$$

$$\psi_7^{CD}\left(x_1,x_2,x_3\right) = u + v_1 \cdot e_1 - v_2 \cdot e_2 - v_{12} \cdot e_3 - v_3 \cdot e_4 - v_{13} \cdot e_5 + v_{23} \cdot e_6 + v \cdot e_7$$

$$(6.17)$$

The 3-D Cayley-Dickson signals of rank $R = 2$ are derived as the mean value of corresponding signals of rank 3 given by (6.14), (6.16), and (6.15), (6.17), respectively

$$\psi_{(1,5)}^{CD}\left(x_1,x_2,x_3\right) = \frac{\psi_1^{CD} + \psi_5^{CD}}{2} = u + v_1 \cdot e_1 + v_2 \cdot e_2 + v_{12} \cdot e_3 \quad (6.18)$$

$$\psi_{(3,7)}^{CD}\left(x_1,x_2,x_3\right) = \frac{\psi_3^{CD} + \psi_7^{CD}}{2} = u + v_1 \cdot e_1 - v_2 \cdot e_2 - v_{12} \cdot e_3 \quad (6.19)$$

We see that the above signals have a quaternion structure similar to (6.11) and (6.12). Finally, the 3-D signal of rank 1 (of a complex form) is a mean of (6.18) and (6.19):

$$\psi^{CD}_{(f_1>0)}\left(x_1,x_2,x_3\right) = \frac{\psi^{CD}_{(1,5)} + \psi^{CD}_{(3,7)}}{2} = u\left(x_1,x_2,x_3\right) + v_1\left(x_1,x_2,x_3\right)\cdot e_1 \qquad (6.20)$$

The above signal coincides with (6.10).

Example 6.4

Let us consider the 3-D real signal $u(x_1,x_2,x_3) = \cos(\omega_1 x_1)\cos(\omega_2 x_2)\cos(\omega_3 x_3)$, $\omega_i = 2\pi f_{i0}$, $i = 1,2,3$, and its total and partial Hilbert transforms:

$$v_1\left(x_1,x_2,x_3\right) = \sin\left(\omega_1 x_1\right)\cos\left(\omega_2 x_2\right)\cos\left(\omega_3 x_3\right)$$

$$v_2\left(x_1,x_2,x_3\right) = \cos\left(\omega_1 x_1\right)\sin\left(\omega_2 x_2\right)\cos\left(\omega_3 x_3\right)$$

$$v_3\left(x_1,x_2,x_3\right) = \cos\left(\omega_1 x_1\right)\cos\left(\omega_2 x_2\right)\sin\left(\omega_3 x_3\right)$$

$$v_{12}\left(x_1,x_2,x_3\right) = \sin\left(\omega_1 x_1\right)\sin\left(\omega_2 x_2\right)\cos\left(\omega_3 x_3\right)$$

$$v_{13}\left(x_1,x_2,x_3\right) = \sin\left(\omega_1 x_1\right)\cos\left(\omega_2 x_2\right)\sin\left(\omega_3 x_3\right)$$

$$v_{23}\left(x_1,x_2,x_3\right) = \cos\left(\omega_1 x_1\right)\sin\left(\omega_2 x_2\right)\sin\left(\omega_3 x_3\right)$$

$$v\left(x_1,x_2,x_3\right) = \sin\left(\omega_1 x_1\right)\sin\left(\omega_2 x_2\right)\sin\left(\omega_3 x_3\right)$$

Let us use the simplified notation: $c_i = \cos(\omega_i x_i)$, $s_i = \sin(\omega_i x_i)$, $i = 1,2,3$. The octonion analytic signals of rank 3 are given by (6.14)–(6.17). Using (6.18) and (6.19), we obtain signals of rank 2:

$$\psi^{CD}_{(1,5)}\left(x_1,x_2,x_3\right) = c_1 c_2 c_3 + s_1 c_2 c_3 \cdot e_1 + c_1 s_2 c_3 \cdot e_2 + s_1 s_2 c_3 \cdot e_3$$

$$= c_1 c_2 c_3 + s_1 c_2 c_3 \cdot e_1 + \left(c_1 s_2 c_3 + s_1 s_2 c_3 \cdot e_1\right)\cdot e_2$$

$$= \left(c_1 + s_1 \cdot e_1\right)c_2 c_3 + \left[\left(c_1 + s_1 \cdot e_1\right)s_2 c_3\right]\cdot e_2$$

$$= \left(c_1 + s_1 \cdot e_1\right)\left(c_2 + s_2 \cdot e_2\right)c_3 = e^{e_1 \omega_1 x_1} e^{e_2 \omega_2 x_2} \cos\left(\omega_3 x_3\right)$$

$$\psi^{CD}_{(3,7)}\left(x_1,x_2,x_3\right) = c_1 c_2 c_3 + s_1 c_2 c_3 \cdot e_1 - c_1 s_2 c_3 \cdot e_2 - s_1 s_2 c_3 \cdot e_3$$

$$= c_1 c_2 c_3 + s_1 c_2 c_3 \cdot e_1 - \left(c_1 s_2 c_3 + s_1 s_2 c_3 \cdot e_1\right)\cdot e_2$$

$$= \left(c_1 + s_1 \cdot e_1\right)c_2 c_3 - \left[\left(c_1 + s_1 \cdot e_1\right)s_2 c_3\right]\cdot e_2$$

$$= \left(c_1 + s_1 \cdot e_1\right)\left(c_2 - s_2 \cdot e_2\right)c_3 = e^{e_1 \omega_1 x_1} e^{-e_2 \omega_2 x_2} \cos\left(\omega_3 x_3\right)$$

Table 6.2
Ranking of n-D Cayley-Dickson Analytic Signals Related to the Real Signal $u(x) = \prod_{i=1}^{n} \cos(\omega_i x_i)$

	$n=1$		$n=2$		$n=3$	
$R=3$	——		——		1	$\exp(e_1\omega_1 x_1)\exp(e_2\omega_2 x_2)\exp(e_3\omega_3 x_3)$
					3	$\exp(e_1\omega_1 x_1)\exp(-e_2\omega_2 x_2)\exp(e_3\omega_3 x_3)$
					5	$\exp(e_1\omega_1 x_1)\exp(e_2\omega_2 x_2)\exp(-e_3\omega_3 x_3)$
					7	$\exp(e_1\omega_1 x_1)\exp(-e_2\omega_2 x_2)\exp(-e_3\omega_3 x_3)$
$R=2$	——		1	$\exp(e_1\omega_1 x_1)\exp(e_2\omega_2 x_2)$	(1,5)	$\exp(e_1\omega_1 x_1)\exp(e_2\omega_2 x_2)\cos(\omega_3 x_3)$
$R=1$	1	$\exp(e_1\omega_1 x_1)$	3	$\exp(e_1\omega_1 x_1)\exp(-e_2\omega_2 x_2)$	(3,7)	$\exp(e_1\omega_1 x_1)\exp(-e_2\omega_2 x_2)\cos(\omega_3 x_3)$
			(1,3)	$\exp(e_1\omega_1 x_1)\cos(\omega_2 x_2)$	(1,5)–(3,7)	$\exp(e_1\omega_1 x_1)\cos(\omega_2 x_2)\cos(\omega_3 x_3)$
$u(x)$	$\cos(\omega_1 x_1)$		$\cos(\omega_1 x_1)\cos(\omega_2 x_2)$		$\cos(\omega_1 x_1)\cos(\omega_2 x_2)\cos(\omega_3 x_3)$	

Finally, the signal of rank 1 defined by (6.20) is

$$\psi^{CD}_{(f_1>0)}\left(x_1,x_2,x_3\right) = \frac{1}{2}\left[e^{e_1\omega_1 x_1} e^{e_2\omega_2 x_2}\cos\left(\omega_3 x_3\right) + e^{e_1\omega_1 x_1} e^{-e_2\omega_2 x_2}\cos\left(\omega_3 x_3\right)\right]$$

$$= e^{e_1\omega_1 x_1}\cos\left(\omega_2 x_2\right)\cos\left(\omega_3 x_3\right)$$

We obtained the same result as shown in Example 6.2. This proves once again the equivalence of both the complex and hypercomplex approach. To synthesize obtained results, we collected them in Table 6.2, showing forms of n-D hypercomplex analytic signals $n = 1, 2, 3$ of rank $R \le n$ for signals from Examples 6.3 and 6.4.

6.4 Summary

In this chapter, we discussed the idea of *lower-rank n-D analytic signals*. The process of *ranking* is based on enlarging the support of the spectrum by formation suborthants of n-D space, in form of a union of two neighboring (adjacent) orthants (see Section 6.1). The inverse Fourier transform of the *enlarged* spectrum yields an n-D analytic signal with half the components. The lowest rank R signal has always two terms, just as the Gabor's analytic signal. For $n > 2$, the step-by-step ranking procedure produces more than one lower rank signal because there are different options in creating suborthants. The results obtained in Sections 6.2 and 6.3 can be generalized to the n-D case:

$$\psi_{(f_1>0)}(\boldsymbol{x}) = u(\boldsymbol{x}) + jv_1(\boldsymbol{x}) \tag{6.21}$$

$$\psi_{(f_1<0)}(\boldsymbol{x}) = u(\boldsymbol{x}) - jv_1(\boldsymbol{x}) \tag{6.22}$$

So, we see that the n-dimensional complex and hypercomplex analytic signals of rank $R = 1$ are always n-D complex signals with the real part $u(\boldsymbol{x})$ and the imaginary part in form of the partial Hilbert transform $v_1(\boldsymbol{x})$ of u, [1, 2].

References

[1]   Hahn, S. L., and K. M. Snopek, "The Unified Theory of n-Dimensional Complex and Hypercomplex Analytic Signals," *Bull. Polish Ac. Sci., Tech. Sci.*, Vol. 59, No. 2, 2011, pp. 167–181.

[2]   Snopek, K. M., *Studies on Complex and Hypercomplex Multidimensional Analytic Signals*, Prace naukowe, Elektronika z. 190, Oficyna Wydawnicza Politechniki Warszawskiej, Warsaw, 2013.

[3]   Hahn, S. L., "Multidimensional Complex Signals with Single-Orthant Spectra," *Proc. IEEE*, Vol. 80, No. 8, August 1992, pp. 1287–1300.

[4]   Hahn, S., "Amplitudes, Phases and Complex Frequencies of 2D Gaussian Signals" *Bull. Polish Academy Sci., Tech. Sci.*, Vol. 40, No. 3, 1992, pp. 289–311.

[5]   Hahn, S. L., "The n-dimensional complex delta distribution," *IEEE Trans. Sign.Proc.*, Vol. 44, No. 7, July 1996, pp.1833–1837.

[6]   Hahn, S. L., *Hilbert Transforms in Signal Processing*, Norwood, MA: Artech House, 1996.

[7]   Bülow, T., M. Felsberg, and G. Sommer, "Non-Commutative Hypercomplex Fourier Transforms of Multidimensional Signals" in *Geometric Computing with Clifford Algebra*, G. Sommer (Ed.), Berlin: Springer-Verlag, 2001, pp. 187–207.

7

Polar Representation of Analytic Signals

7.1 Introduction

This chapter presents polar representation of complex and hypercomplex analytic signals. For a given real n-D signal $u(\mathbf{x})$, the corresponding complex analytic signal has the form $\psi(\mathbf{x}) = \mathrm{Re}(\mathbf{x}) + e_1\mathrm{Im}(\mathbf{x})$. Its polar form is $\psi(\mathbf{x}) = A(\mathbf{x})\exp[e_1\varphi(\mathbf{x})]$, where $A(\mathbf{x}) = \sqrt{[\mathrm{Re}\psi(\mathbf{x})]^2 + [\mathrm{Im}\psi(\mathbf{x})]^2}$ and $\mathrm{Tan}[\varphi(\mathbf{x})] = \mathrm{Im}[\psi(\mathbf{x})]/\mathrm{Re}[\psi(\mathbf{x})]$. The capital T denotes the multibranch tangent function. In the case of 1-D time signals: $A(t) = \sqrt{[\mathrm{Re}\psi(t)]^2 + [\mathrm{Im}\psi(t)]^2}$ and $\mathrm{Tan}[\varphi(t)] = \mathrm{Im}[\psi(t)]/\mathrm{Re}[\psi(t)]$. The real signal can be reconstructed form its polar form using the relation $u(t) = A(t)\cos[\varphi(t)]$.

Why, therefore, do we need the polar form?

Firstly, this form enables the definitions of instantaneous amplitude, instantaneous phase, and the instantaneous frequency of a signal. Secondly, in specific n-D applications (for example, in medical imaging), it is possible to extract some features not directly observable in the real signal. The polar form of hypercomplex signals differs from the polar form of complex signals. Sections 7.4 and 7.5 will show that the complex and hypercomplex polar representations are equivalent. There are closed formulae enabling the calculation of amplitudes and the phase functions of complex and hypercomplex signals.

The amplitudes and phase functions in the complex case are uniquely defined for any dimensions. However, the reconstruction needs a calculation of $n/2$ amplitudes and $n/2$ phase functions. The hypercomplex polar representation for $n = 3$ (octonions) is known only approximately and is unknown for $n = 4$ (sedenions). Details are presented in Sections 7.5.1 and 7.5.2. It should be noted that the Fourier spectrum of a real signal may be a complex function. The polar form of this spectrum also defines the amplitude and phase functions [14]. This problem is not described in this book.

7.2 Polar Representation of Complex Numbers

Consider a complex number $z_0 = x_0 + y_0 \cdot e_1$ (we use the notation e_1 and not the commonly used j). The polar form of this number is

$$z_0 = x_0 + y_0 \cdot e_1 = \left| z_0 \right| e^{e_1 \arg z_0} = \left| z_0 \right| e^{e_1 \varphi} \tag{7.1}$$

where the absolute value

$$\left| z_0 \right| = A_0 = \sqrt{x_0^2 + y_0^2} \tag{7.2}$$

is called an *amplitude* and the argument,

$$\arg\left(z_0\right) = \varphi = \operatorname{a\,tan2}\left(y_0, x_0\right) \tag{7.3}$$

is an *angle*. The notation (7.1) defines a static (nonrotating) phasor in the form of a scaled line attached to the origin of the (x,y)-plane and ending with an arrow at the point (x_0, y_0). The natural logarithm of (7.1),

$$\operatorname{Ln}\left(z_0\right) = \ln\left| z_0 \right| + \varphi \cdot e_1 = A_0 + \varphi \cdot e_1 \tag{7.4}$$

defines the *complex amplitude*. In electrical engineering, the phasor is a complex number representing a sinusoidal (harmonic) function whose amplitude A, angular frequency $\omega = 2\pi f$ and initial phase ϕ are time-invariant.

7.3 Polar Representation of 1-D Analytic Signals

Consider the Gabor's 1-D analytic signal $\psi(t) = u(t) + v(t) \cdot e_1$, where $v(t) = H[u(t)]$ is the Hilbert transform of u. Let us recall, that the Hilbert transform

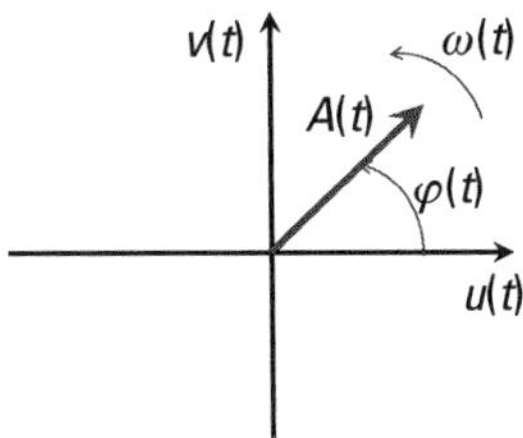

Figure 7.1   The rotating phasor of length $A(t)$ and angular frequency (velocity) $\omega(t)$.

is the only one satisfying the Vakman's three conditions: the amplitude continuity, phase independence of scale and harmonic correspondence [15, 16].

The polar form of the analytic signal is represented by a phasor rotating counterclockwise as shown in Figure 7.1.

The formula

$$\psi(t) = u(t) + v(t) \cdot e_1 = A(t)\exp\left[e_1\varphi(t)\right] \tag{7.5}$$

defines the *instantaneous amplitude*

$$A(t) = \sqrt{u^2(t) + v^2(t)} \tag{7.6}$$

and

$$\varphi(t) = \operatorname{atan2}\bigl(v(t), u(t)\bigr) \tag{7.7}$$

defines the *instantaneous phase*. The logarithm of the analytic signal (7.5) defines the *instantaneous complex phase* [7, 8]

$$\Phi_c(t) = \operatorname{Ln}\left[\psi(t)\right] = \operatorname{Ln}\left[A(t)\right] + \varphi(t) \tag{7.8}$$

The capital L denotes the multibranch character of the logarithm of a complex function. The time derivative of the complex phase defines the *instantaneous complex frequency*:

$$s(t) = \frac{d}{dt}\left\{\operatorname{Ln}\left[\psi(t)\right]\right\} = \frac{1}{A(t)}\frac{dA(t)}{dt} + \frac{d\varphi(t)}{dt}\cdot e_1 = \alpha(t) + \omega(t)\cdot e_1 \tag{7.9}$$

The function $\alpha(t)$ is called the *instantaneous radial velocity* (a measure of the time change of the amplitude) and $\omega(t)$ is the *instantaneous angular velocity* (the phasor rotates with a variable in time angular velocity). It can be written in the form (dots over u and v define corresponding time derivatives d/dt):

$$\omega(t) = \frac{d}{dt}\operatorname{atan2}\big[v(t),u(t)\big] = \frac{u(t)\dot{v}(t) - v(t)\dot{u}(t)}{u^2(t) + v^2(t)} \tag{7.10}$$

The *instantaneous frequency* is

$$f(t) = \frac{\omega(t)}{2\pi} \tag{7.11}$$

In conclusion, we have defined the notions of the instantaneous amplitude, instantaneous phase, and instantaneous frequency of the Gabor's analytic signal. Note that no definition is true or false. We simply agree to apply a specific name to the defined quantity. The aforementioned quantities are uniquely defined as properties of the Gabor's analytic signal. Cohen, Loughlin, and Vakman [3] believed that the notions of instantaneous amplitude and frequency are not unique. Hahn indicated that the Gabor's analytic signal defines uniquely the above notions [13].

In terms of the polar representation, the real signal and its Hilbert transform have the form

$$u(t) = A(t)\cos\big[\varphi(t)\big],\ \ v(t) = A(t)\sin\big[\varphi(t)\big] \tag{7.12}$$

Example 7.1

The Gabor's 1-D analytic signal has a representation in the form of a convolution of the real signal with the complex delta distribution (see Chapter 5, (5.11)). The approximating function of the complex delta distribution is (Chapter 5, (5.12))

$$\psi_\delta(t) = u(t) + v(t)\cdot e_1 = \frac{a}{\pi\big(a^2 + t^2\big)} + \frac{t}{\pi\big(a^2 + t^2\big)}\cdot e_1 \tag{7.13}$$

The polar form of this analytic signal is

$$\psi_\delta(t) = A(t)\exp\big[e_1\varphi(t)\big] = \frac{1}{\pi\sqrt{a^2 + t^2}}\exp\big[e_1\operatorname{atan2}(t,a)\big] \tag{7.14}$$

Therefore, the instantaneous complex phase is

$$\Phi_c(t) = \mathrm{Ln}\left(\frac{1}{\pi\sqrt{a^2 + t^2}}\right) + a\tan 2(t,a)\cdot e_1 \tag{7.15}$$

and the instantaneous complex angular frequency

$$s(t) = \dot{\Phi}_c(t) = \alpha(t) + \omega(t)\cdot e_1 = \frac{-t}{a^2 + t^2} + \frac{a}{a^2 + t^2}\cdot e_1 \tag{7.16}$$

In the limit $a \to 0$, we get the instantaneous complex phase of the complex delta distribution

$$\Phi_{c\delta}(t) = \mathrm{Ln}\left[\frac{1}{\pi|t|}\right] + 0.5\pi\,\mathrm{sgn}(t)\cdot e_1 \tag{7.17}$$

and the instantaneous complex angular frequency

$$s(t) = \dot{\Phi}_{c\delta}(t) = \alpha(t) + \omega(t)\cdot e_1 = -\frac{1}{t} + \pi\delta(t)\cdot e_1 \tag{7.18}$$

Example 7.2

Consider the analytic signal with the real part in the form of the interpolating function of the sampling theory (see 7.7 in [10]).

$$\psi(t) = \frac{\sin(at)}{at} + \frac{\sin^2(0.5at)}{0.5at}\cdot e_1 = A(t)\exp\left[e_1\varphi(t)\right] \tag{7.19}$$

We get

$$\psi(t) = \left|\frac{\sin(0.5at)}{0.5at}\right|\exp\left[\frac{e_1 at}{2}\right] \tag{7.20}$$

Therefore, the instantaneous complex phase is

$$\Phi_c(t) = \mathrm{Ln}\left|\frac{\sin(0.5at)}{0.5at}\right| + \frac{at}{2}\cdot e_1 \tag{7.21}$$

and the instantaneous complex angular frequency

$$s(t) = \frac{a}{2}\cot(0.5at) + \frac{a}{2}\cdot e_1 \tag{7.22}$$

In conclusion, the interpolating function is a signal with variable amplitude and a constant angular frequency.

Example 7.3

The classic representation of a frequency modulated harmonic carrier is [10]

$$\psi_{FM}(t) = A_0 \exp\left\{e_1\left[\omega_0 t + \Phi_0 + \varphi(t)\right]\right\}, \quad \Omega_0 = 2\pi f_0 \qquad (7.23)$$

where $\varphi(t)$ represents the angle modulation of the carrier f_0. This signal is not an analytic signal since real and imaginary parts do not form a Hilbert pair. However, for large values of the carrier frequency it is nearly analytic (*quasi-analytic*) (see Chapter 8).

7.3.1 Representation of the Instantaneous Complex Frequency using the Wigner Distribution

The time-frequency Wigner distribution $W(t,f)$ of analytic signals is described in Chapter 9. Let us show formulae defining terms of the instantaneous complex frequency using moments of the Wigner distribution. The spectral moments have the form

$$m_n(t) = \int\limits_{-\infty}^{\infty} \omega^n W(t,f)\,df \qquad (7.24)$$

The instantaneous amplitude is uniquely defined by the zero order moment $m_0(t) = A^2(t)$. The terms of the instantaneous complex frequency are

$$\alpha(t) = \frac{\dot{m}_0(t)}{2m_0(t)}, \quad \omega(t) = \frac{m_1(t)}{m_0(t)}; \quad \dot{m}_0 = dm_0/dt \qquad (7.25)$$

7.4 Polar Representation of 2-D Analytic Signals

7.4.1 2-D Complex Analytic Signals with Single-Quadrant Spectra

The 2-D complex signals with single quadrant spectra are represented by two amplitudes and two phase functions [9]. Due to the Hermitian symmetry (see Chapter 5, Table 5.2), we have a pair of analytic signals with spectra in the first and third quadrant. These quadrants form the half-plane $f_1 > 0$. The two analytic signals are

$$\psi_1(x_1,x_2) = u(x_1,x_2) - v(x_1,x_2) + \left[v_1(x_1,x_2) + v_2(x_1,x_2)\right] \cdot e_1$$
$$= \mathrm{Re}_1 + \mathrm{Im}_1 \cdot e_1 \tag{7.26}$$

$$\psi_3(x_1,x_2) = u(x_1,x_2) + v(x_1,x_2) + \left[v_1(x_1,x_2) - v_2(x_1,x_2)\right] \cdot e_1$$
$$= \mathrm{Re}_2 + \mathrm{Im}_2 \cdot e_1. \tag{7.27}$$

Their polar form is

$$\psi_1(x_1,x_2) = \mathrm{Re}_1 + \mathrm{Im}_1 \cdot e_1 = A_1(x_1,x_2) e^{e_1\varphi_1(x_1,x_2)} \tag{7.28}$$

$$\psi_3(x_1,x_2) = \mathrm{Re}_2 + \mathrm{Im}_2 \cdot e_1 = A_2(x_1,x_2) e^{e_1\varphi_2(x_1,x_2)} \tag{7.29}$$

The two amplitude functions are

$$A_1(x_1,x_2) = \sqrt{\mathrm{Re}_1^2 + \mathrm{Im}_1^2} = \sqrt{u^2 + v^2 + v_1^2 + v_2^2 + 2(uv + v_1 v_2)} \tag{7.30}$$

$$A_2(x_1,x_2) = \sqrt{\mathrm{Re}_2^2 + \mathrm{Im}_2^2} = \sqrt{u^2 + v^2 + v_1^2 + v_2^2 - 2(uv + v_1 v_2)} \tag{7.31}$$

and the two phase functions

$$\varphi_1(x_1,x_2) = \mathrm{atan}2(\mathrm{Im}_1,\mathrm{Re}_1) = \mathrm{atan}2(v_1 + v_2, u - v) \tag{7.32}$$

$$\varphi_2(x_1,x_2) = \mathrm{atan}2(\mathrm{Im}_2,\mathrm{Re}_2) = \mathrm{atan}2(v_1 - v_2, u + v) \tag{7.33}$$

The reconstruction formula of a 2-D real signal from its polar representation is

$$u(x_1,x_2) = \frac{A_1 \cos \varphi_1 + A_2 \cos \varphi_2}{2} \tag{7.34}$$

Note, that the two amplitudes (7.30) and (7.31) can be written as

$$A_1(x_1,x_2) = \sqrt{A_q^2 + d^2}, \; A_2(x_1,x_2) = \sqrt{A_q^2 - d^2} \tag{7.35}$$

where $A_q(x_1,x_2) == \sqrt{u^2 + v^2 + v_1^2 + v_2^2}$ is the amplitude of the quaternion analytic signal (see next Section) and $d^2 = 2(uv + v_1 v_2)$.

7.4.1.1 The Case of 2-D Separable Real Signals

The 2-D real signal is a separable function if it is a product of 1-D functions of the form $u(x_1,x_2) = g_1(x_1)g_2(x_2)$. The total Hilbert transform is $v(x_1,x_2) = h_1(x_1) h_2(x_2)$ and the partial Hilbert transforms of such a signal is $v_1(x_1,x_2) = h_1(x_1) g_2(x_2)$, $v_2(x_1,x_2) = g_1(x_1)h_2(x_2)$, where h denotes a 1-D Hilbert transform of g. These formulae show that for separable signals $d^2 = 0$ and both amplitudes (7.30) and (7.31) have the same value. The polar 1-D representations of 1-D analytic signals $\psi_{g1}(x_1) = g_1(x_1) + h_1(x_1) \cdot e_1 = a_1(x_1)\exp[e_1\alpha_1(x_1)]$ and $\psi_{g2}(x_2) = g_2(x_2) + h_2(x_2) \cdot e_1 = a_2(x_2)\exp[e_1\alpha_2(x_2)]$ show that the two phase angles (7.32) and (7.33) for separable signals have the form

$$\varphi_1\left(x_1,x_2\right) = \alpha_1\left(x_1\right) + \alpha_2\left(x_2\right), \quad \varphi_2\left(x_1,x_2\right) = \alpha_1\left(x_1\right) - \alpha_2\left(x_2\right) \quad (7.36)$$

Evidently, for separable signals, the quaternion phase angles are given by the sum and difference of the complex phase angles.

7.4.2 2-D Hypercomplex Quaternion Analytic Signals with Single-Quadrant Spectra

The 2-D quaternion analytic signal (see Chapter 5, (5.78)) is [1, 5, 6]

$$\psi_q\left(x_1,x_2\right) = u\left(x_1,x_2\right) + v_1\left(x_1,x_2\right) \cdot e_1 + v_2\left(x_1,x_2\right) \cdot e_2 + v\left(x_1,x_2\right) \cdot e_3 \quad (7.37)$$

The polar form of this signal is

$$\psi_q\left(x_1,x_2\right)$$
$$= A_q\left(x_1,x_2\right)\exp\left[e_1\phi_1\left(x_1,x_2\right)\right]\exp\left[e_3\phi_3\left(x_1,x_2\right)\right]\exp\left[e_2\phi_2\left(x_1,x_2\right)\right] \quad (7.38)$$

where the amplitude A_q is given by the norm

$$A_q\left(x_1,x_2\right) = \sqrt{u^2 + v_1^2 + v_2^2 + v^2} \quad (7.39)$$

and three phase functions are defined by selected terms of the Rodriquez matrix [1]

$$\mathfrak{M}(q) = \begin{pmatrix} R_{11} = u^2 + v_1^3 - v_2^2 - v^2 & R_{12} = 2\left(v_1v_2 - uv\right) & R_{13} = 2\left(v_1v + uv_2\right) \\ R_{21} = 2\left(uv + v_1v_2\right) & R_{22} = u^2 - v_1^2 + v_2^2 - v^2 & R_{23} = 2\left(v_2v - uv_1\right) \\ R_{31} = 2\left(v_1v - uv_2\right) & R_{32} = 2\left(v_2v + uv_1\right) & R_{33} = u^2 - v_1^2 - v_2^2 + v^2 \end{pmatrix}$$

$$(7.40)$$

The three phase functions are defined as follows

$$\phi_1(x_1,x_2) = \frac{1}{2}\,\mathrm{a\,tan}2\big(R_{32},R_{22}\big) = \frac{1}{2}\,\mathrm{a\,tan}2\Big[2\big(uv_1 + vv_2\big),u^2 - v_1^2 + v_2^2 - v^2\Big]$$

$$(7.41)$$

$$\phi_2(x_1,x_2) = 0.5\,\mathrm{a\,tan}2\big(R_{13},R_{11}\big) = 0.5\,\mathrm{a\,tan}2\Big[2\big(uv_2 + vv_1\big),u^2 + v_1^2 - v_2^2 - v^2\Big]$$

$$(7.42)$$

$$\phi_3(x_1,x_2) = \frac{1}{2}\arcsin\left(\frac{-R_{12}}{A_q^2}\right) = \frac{1}{2}\arcsin\left(\frac{A_1^2 - A_2^2}{A_1^2 + A_2^2}\right) \qquad (7.43)$$

where A_1 and A_2 are two amplitudes defined by (7.30) and (7.31).

The reconstruction formula of a 2-D real signal from its polar quaternion representation is

$$u = A_q\Big[\cos(\phi_1)\cos(\phi_2)\cos(\phi_3) - \sin(\phi_1)\sin(\phi_2)\sin(\phi_3)\Big] \qquad (7.44)$$

7.4.3 Relations between the Analytic and Quaternion 2-D Phase Functions

The polar form of the analytic 2-D signals with single-quadrant spectra is represented by two amplitudes and two phase functions. The 2-D quaternion signal is represented by a single amplitude and three phase functions. The following formulae enable calculation of the quaternion representation starting with the complex one [4]. We have

$$\phi_1(x_1,x_2) = \frac{\varphi_1 + \varphi_2}{2}, \ \ \phi_2(x_1,x_2) = \frac{\varphi_1 - \varphi_2}{2}, \ \ \sin(2\phi_3) = \frac{A_1^2 - A_2^2}{A_1^2 + A_2^2} \qquad (7.45)$$

7.4.3.1 The Case of Separable Real Signals

For separable real signals, the polar representations of the complex and quaternion signals are both represented by a single amplitude A_q and two phase functions. The insertion of (7.35) into (7.45) yields

$$\phi_1(x_1,x_2) = \alpha_1(x_1), \ \ \phi_2(x_1,x_2) = \alpha_2(x_2) \qquad (7.46)$$

Note that the two amplitudes (7.35) can be written in the form

$$A_1\left(x_1,x_2\right)= A_q\sqrt{1+\sin\left(2\phi_3\right)} \; ; \; A_2\left(x_1,x_2\right)= A_q\sqrt{1-\sin\left(2\phi_3\right)} \quad (7.47)$$

and

$$A_q\left(x_1,x_2\right)=\sqrt{\frac{A_1^2+A_2^2}{2}} \tag{7.48}$$

The above reasoning shows that the complex and quaternion representation is a matter of choice of the researchers, since both methods are equivalent (see discussion in Chapter 11).

### 7.4.4	Polar Representation of the Monogenic 2-D Signal

The monogenic 2-D hypercomplex signal has the form (see Chapter 5, (5.84))

$$\psi_M\left(x_1,x_2\right)= u\left(x_1,x_2\right)+ v_{r1}\left(x_1,x_2\right)\cdot e_1 + v_{r2}\left(x_1,x_2\right)\cdot e_2 \tag{7.49}$$

The spherical representation (Figure 7.2) defines the amplitude

$$A_M\left(x_1,x_2\right)=\sqrt{u^2 + v_{r1}^2 + v_{r2}^2} \tag{7.50}$$

and two angles Φ and Θ. Cohen, Loughlin, and Vakman [3] called the first one the *orientation angle* given by

$$\Phi\left(x_1,x_2\right)= a\tan2\left(v_{r2},v_{r1}\right) \tag{7.51}$$

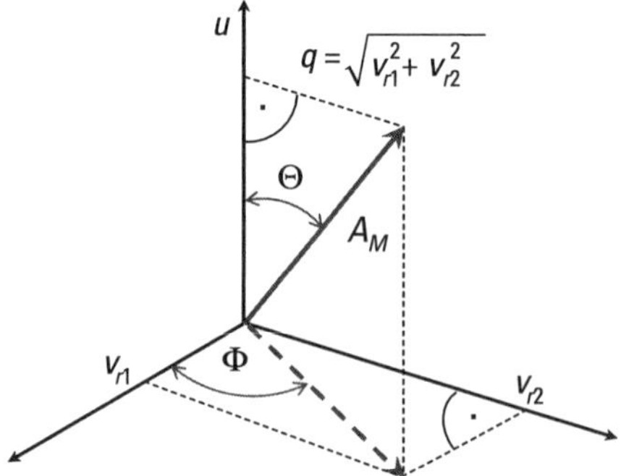

Figure 7.2	Polar representation of the monogenic signal.

and the second one is the *phase* given by

$$\Theta(x_1,x_2) = \arccos\left(\frac{u}{A_M}\right) = \operatorname{a\,tan2}\left(\sqrt{v_{r1}^2 + v_{r2}^2}\,,u\right) \tag{7.52}$$

In terms of the polar representation, the monogenic signal has the form

$$\Psi_M(x_1,x_2) = A_M\left[\cos(\Theta) + \sin(\Theta)\cos(\Phi)e_1 + \sin(\Theta)\sin(\Phi)e_2\right] \tag{7.53}$$

Note that the reconstruction of u (first term in the above formula) requires knowledge of only the orientation angle since

$$u(x_1,x_2) = A_M\cos(\Theta) \tag{7.54}$$

Evidently, the polar representation of the monogenic signal differs from the polar representation of the quaternion signal. The relations between both representations are actually unknown. Differently, the relations between complex and quaternion polar representations are described in Section 7.4.3.

7.4.5 Common Examples for 2-D Polar Representations of Analytic, Quaternion, and Monogenic Signals

Example 7.4

Consider the real signal $u(x_1,x_2) = \cos(\omega_{10}x_1 + \omega_{20}x_2)$ (see Chapter 5, Example 5.3). The total Hilbert transform is $v = -u$ and the two partial HTs are equal $v_1 = v_2 = \sin(\omega_{10}x_1 + \omega_{20}x_2)$. The insertion into (7.26) and (7.27) yields

$$\Psi_1 = u - v + (v_1 + v_2)\cdot e_1 = 2\cos(\omega_{10}x_1 + \omega_{20}x_2) + 2\sin(\omega_{10}x_1 + \omega_{20}x_2)\cdot e_1$$

$$= 2\exp\left[e_1(\omega_{10}x_1 + \omega_{20}x_2)\right]$$

$$\Psi_3 = u + v + (v_1 - v_2)\cdot e_1 = 0$$

Evidently, the amplitude $A_1 = 2$ and $A_2 = 0$. The complex and quaternion phase functions are the same. The monogenic signal (see Chapter 5, Example 5.8) is

$$\Psi_M(x_1,x_2) = \cos\left[2\pi\left(f_{10}x_1 + f_{20}x_2\right)\right] + \frac{1}{\sqrt{1+k^2}}\sin\left[2\pi\left(f_{10}x_1 + f_{20}x_2\right)\right]\cdot e_1$$

$$+ \frac{k}{\sqrt{1+k^2}}\sin\left[2\pi\left(f_{10}x_1 + f_{20}x_2\right)\right]\cdot e_2$$

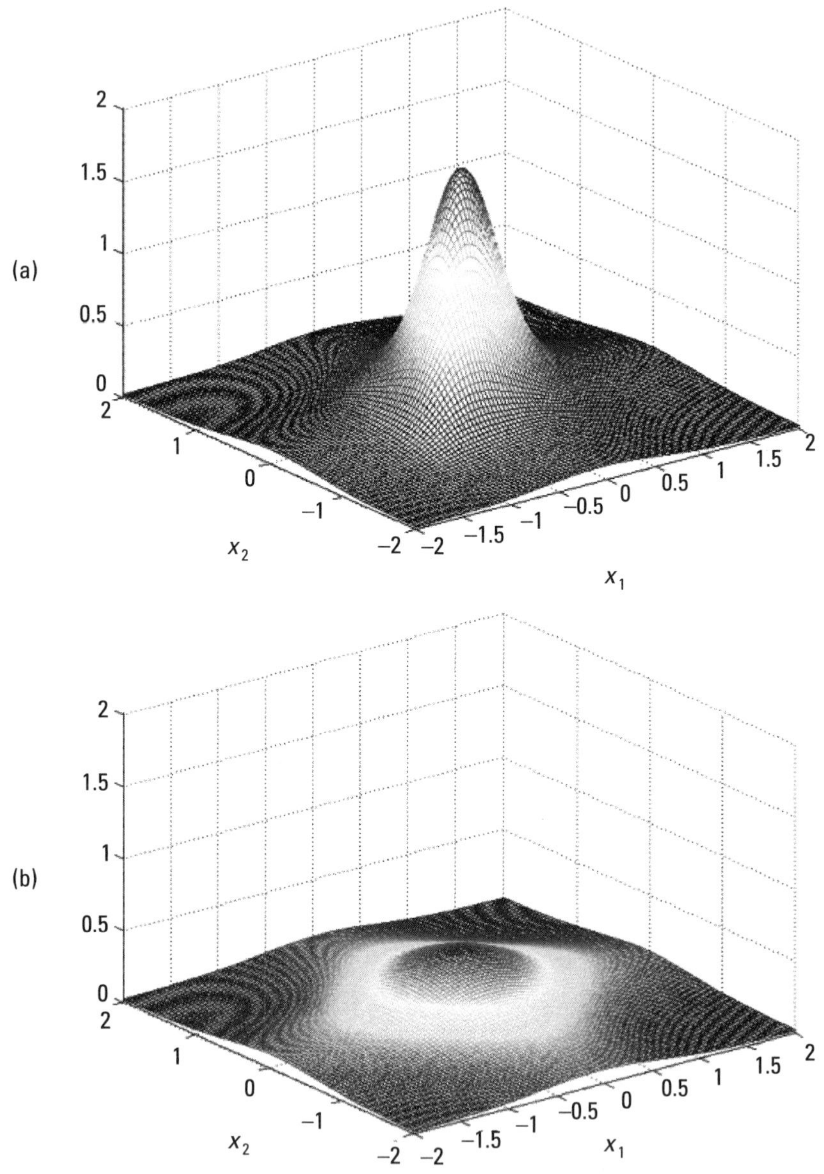

Figure 7.3 Amplitudes (a) $A_1 = \sqrt{(u-v)^2 + \left(v_1 + v_2\right)^2}$ and (b) $A_2 = \sqrt{(u+v)^2 + \left(v_1 - v_2\right)^2}$ of 2-D complex analytic signals.

The amplitude $A_M = 1$, the orientation angle $\Phi(x_1,x_2) = \text{atan}(v_{r2},v_{r1}) = \text{atan2}(k,1)$ is constant and the phase $\Theta(x_1,x_2) = \text{atan}(v_{r2}^2,v_{r1}^2,u) = \varphi_1 = \omega_{10}x_1 + \omega_{20}x_2$ equals the complex and quaternion phase.

Example 7.5

Consider the 2-D nonseparable Gaussian signal (see Chapter 5, Example 5.4):

$$u\left(x_1 - a, x_2 - b\right) = \frac{1}{\sqrt{1-\rho^2}} e^{\frac{-\pi}{1-\rho^2}\left[(x_1-a)^2 + (x_2-b)^2 - 2\rho(x_1-a)(x_2-b)\right]}, \quad 0 \leq \rho < 1$$

This example is illustrated by mesh plots of amplitude and phase functions of complex, quaternion and monogenic signals (Figures 7.3, 7.4, 7.6, and 7.8) as well as reconstructed 2-D Gaussian signals from the polar representations (see (7.34), (7.44), and (7.54); see also Figures 7.5, 7.7, and 7.9).

Example 7.6

Let us define a cylinder function of unit length and a support in the form of a difference of two circles of radiuses r and $r + d$, $d \ll r$. The circular support is defined

$$\text{circ}\left(\frac{\sqrt{x_1^2 + x_2^2}}{r}\right) = \begin{cases} 1, & \sqrt{x_1^2 + x_2^2} < r \\ 0.5, & \sqrt{x_1^2 + x_2^2} = r \\ 0. & \sqrt{x_1^2 + x_2^2} > r \end{cases}$$

The Fourier transform of the circ function is (J_1 − Bessel function)

$$U\left(f_1,f_2\right) = \frac{r}{\rho} J_1(2\pi r\rho); \quad \rho = \sqrt{f_1^2 + f_2^2}; \quad J_1(z) = \frac{1}{\pi} \int_0^\pi \cos(\varphi - z\sin(\varphi)) d\varphi$$

The cylindrical support is given by the difference of two circular supports. The spectrum of the difference is

$$U\left(f_1,f_2\right) = \frac{r_2 J_1\left(2\pi r_2\rho\right) - r_1 J_1\left(2\pi r_1\rho\right)}{\rho}, \quad r_2 = r_1 + d$$

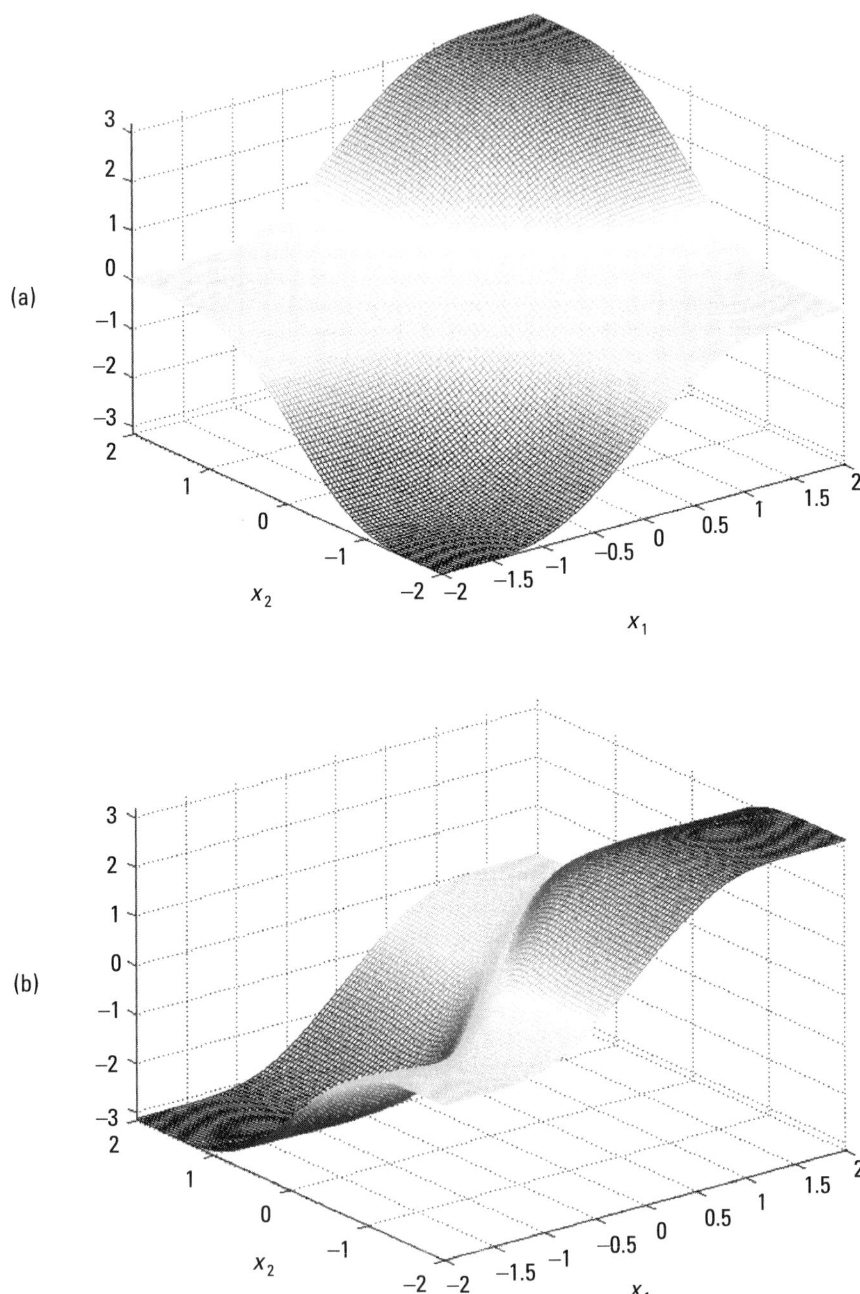

Figure 7.4 Phase functions (a) $\varphi_1 = \mathrm{atan2}(v_1 + v_2, u - v)$ and (b) $\varphi_2 = \mathrm{atan2}(v_1 - v_2, u + v)$ of 2-D complex analytic signals.

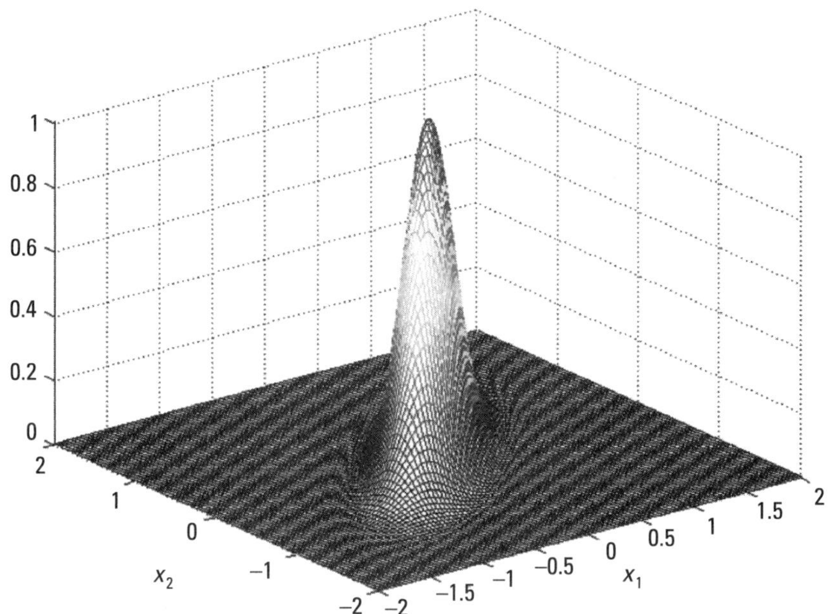

Figure 7.5 Reconstruction of a 2-D Gaussian signal from the complex polar representation using (7.34).

The 2-D analytic signal with the spectrum in the first quadrant is given by the inverse Fourier transform

$$\psi_1\!\left(x_1,x_2\right)=\int\limits_{0}^{\infty}\int\limits_{0}^{\infty}U\!\left(f_1,f_2\right)\exp\!\left[2\pi\!\left(f_1 x_1 + f_2 x_2\right)\right]df_1\,df_2$$

$$\psi_1\!\left(x_1,x_2\right)=X_1\!\left(x_1,x_2\right)+e_1 Y_1\!\left(x_1,x_2\right)=A_1\!\left(x_1,x_2\right)\exp\!\left[e_1\Phi_1\!\left(x_1,x_2\right)\right]$$

Because the cylinder function is nonseparable, we have to calculate the second signal with a single quadrant spectrum in the third quadrant. Its polar form defines the amplitude A_2 and the phase function ϕ_2. The amplitude and phase functions are displayed in Figure 7.10.

$$\psi_3\!\left(x_1,x_2\right)=\int\limits_{0}^{\infty}\int\limits_{-\infty}^{0}U\!\left(f_1,f_2\right)\exp\!\left[2\pi\!\left(f_1 x_1 + f_2 x_2\right)\right]df_1\,df_2$$

$$\psi_3\!\left(x_1,x_2\right)=X_2\!\left(x_1,x_2\right)+e_1 Y_2\!\left(x_1,x_2\right)=A_2\!\left(x_1,x_2\right)\exp\!\left[e_1\Phi_2\!\left(x_1,x_2\right)\right]$$

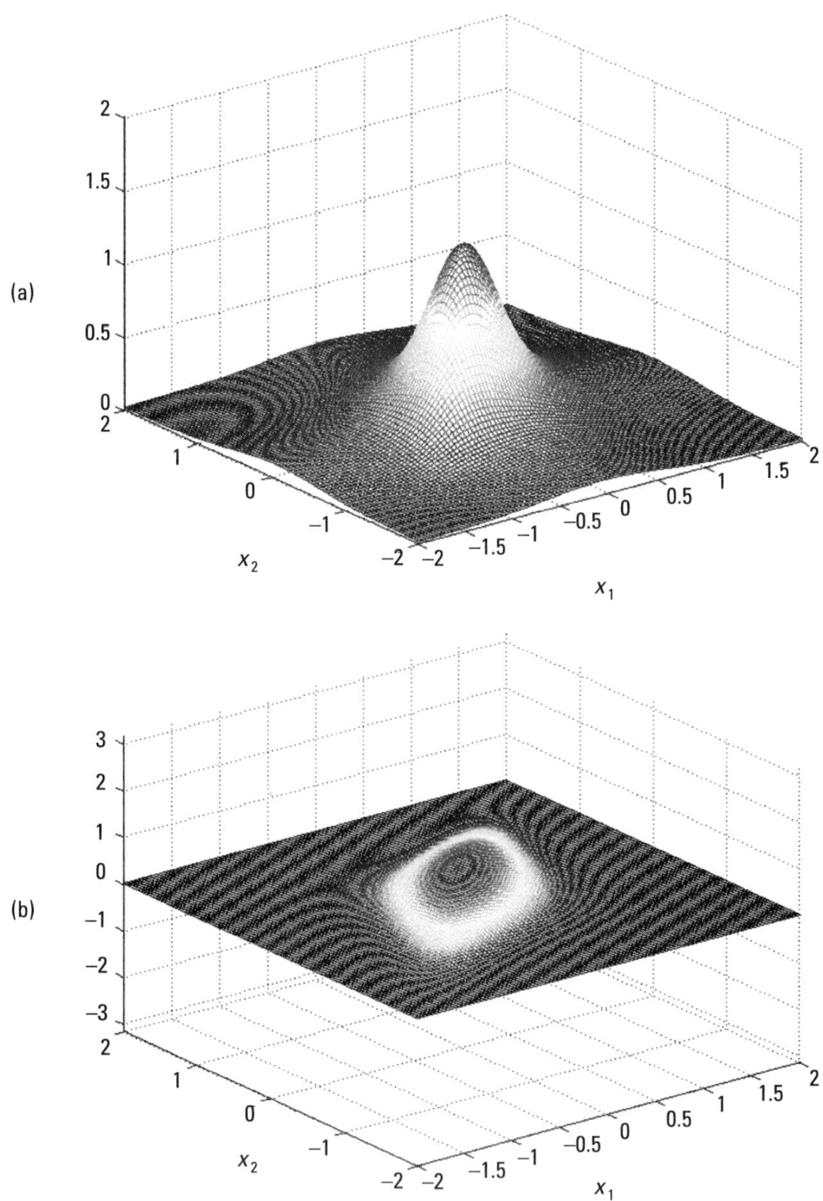

Figure 7.6　Amplitude (a) $A_q(x_1, x_2) = \sqrt{(A_1^2 + A_2^2)/2}$　and phase functions of quaternion analytic signals:　(b) $\phi_3 = 0.5 \mathrm{A}\sin\left(\dfrac{A_1^2 - A_2^2}{A_1^2 + A_2^2}\right)$, (c) $\phi_1 = 0.5 \cdot \mathrm{atan2}\left(2(uv_1 + vv_2), u^2 - v_1^2 + v_2^2 - v^2\right)$, (d) $\phi_2 = 0.5 \cdot \mathrm{atan2}\left(2(uv_2 + vv_1), u^2 + v_1^2 - v_2^2 - v^2\right)$.

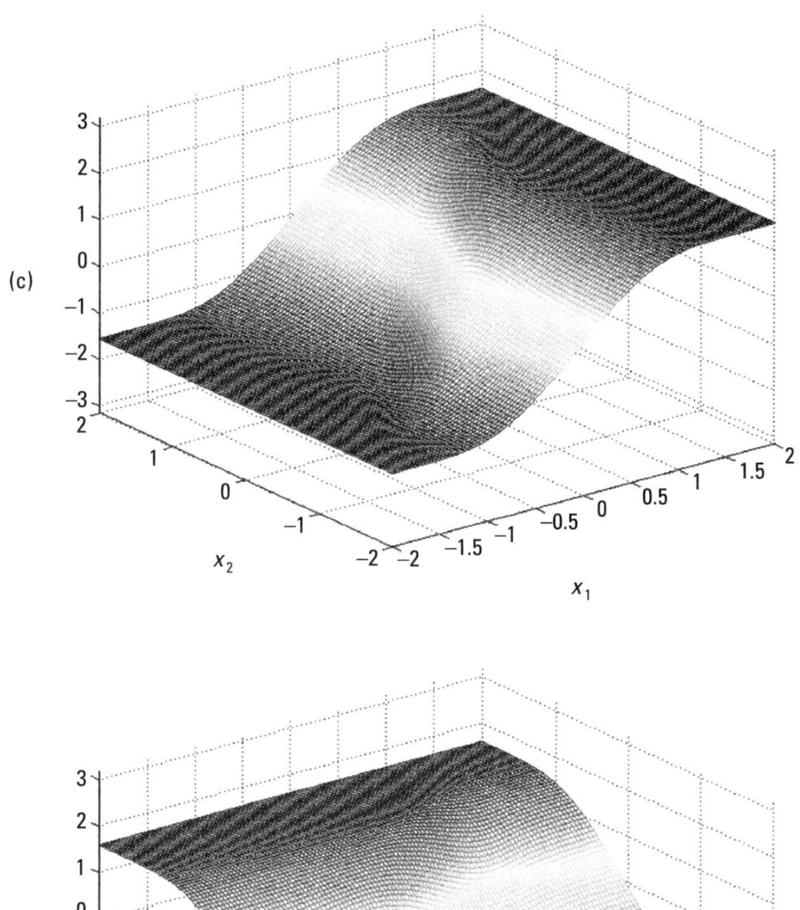

Figure 7.6 Continued

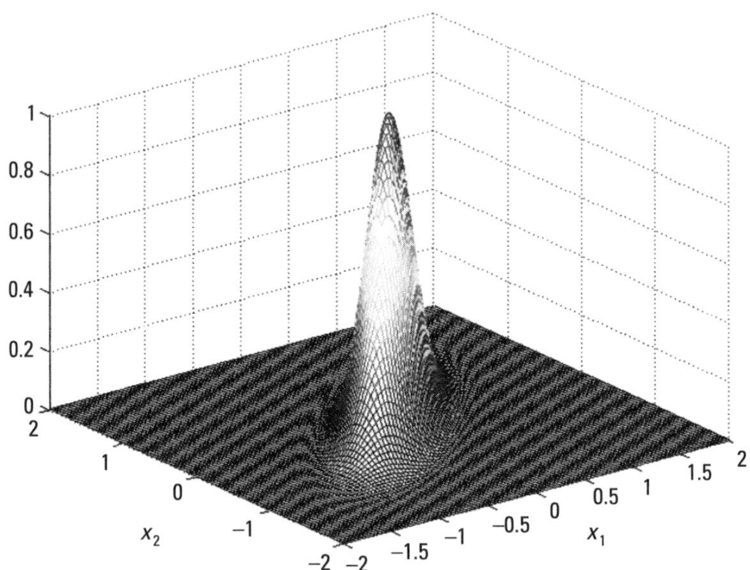

Figure 7.7 Reconstruction of a 2-D Gaussian signal from the quaternion polar representation using (7.44) (see Figure 7.5 for the complex case).

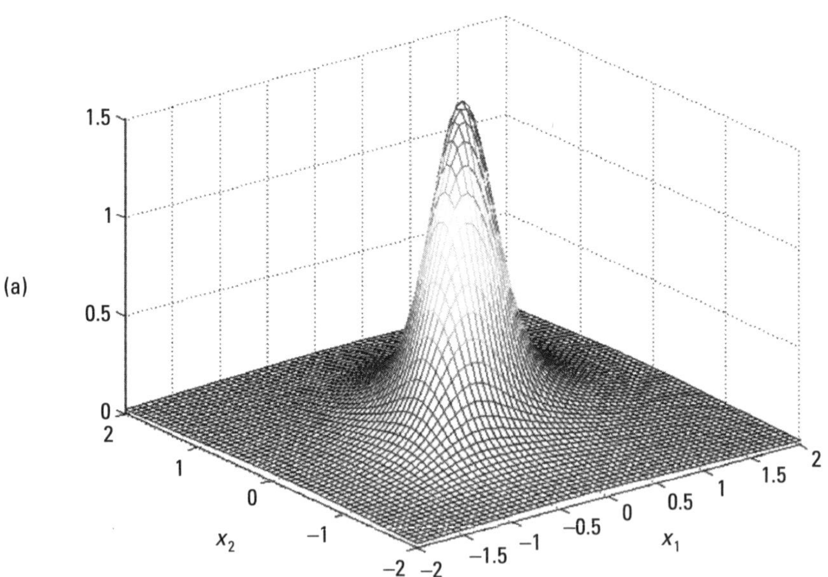

(a)

Figure 7.8 Amplitude (a) $A_M = \sqrt{u^2 + v_{r1}^2 + v_{r2}^2}$ and two phase functions of a monogenic signal: (b) $\Theta = \mathrm{atan2}\left(\sqrt{v_{r1}^2 + v_{r2}^2}, u\right)$, (c) $\Phi = \mathrm{atan2}(v_{r2}, v_{r1})$.

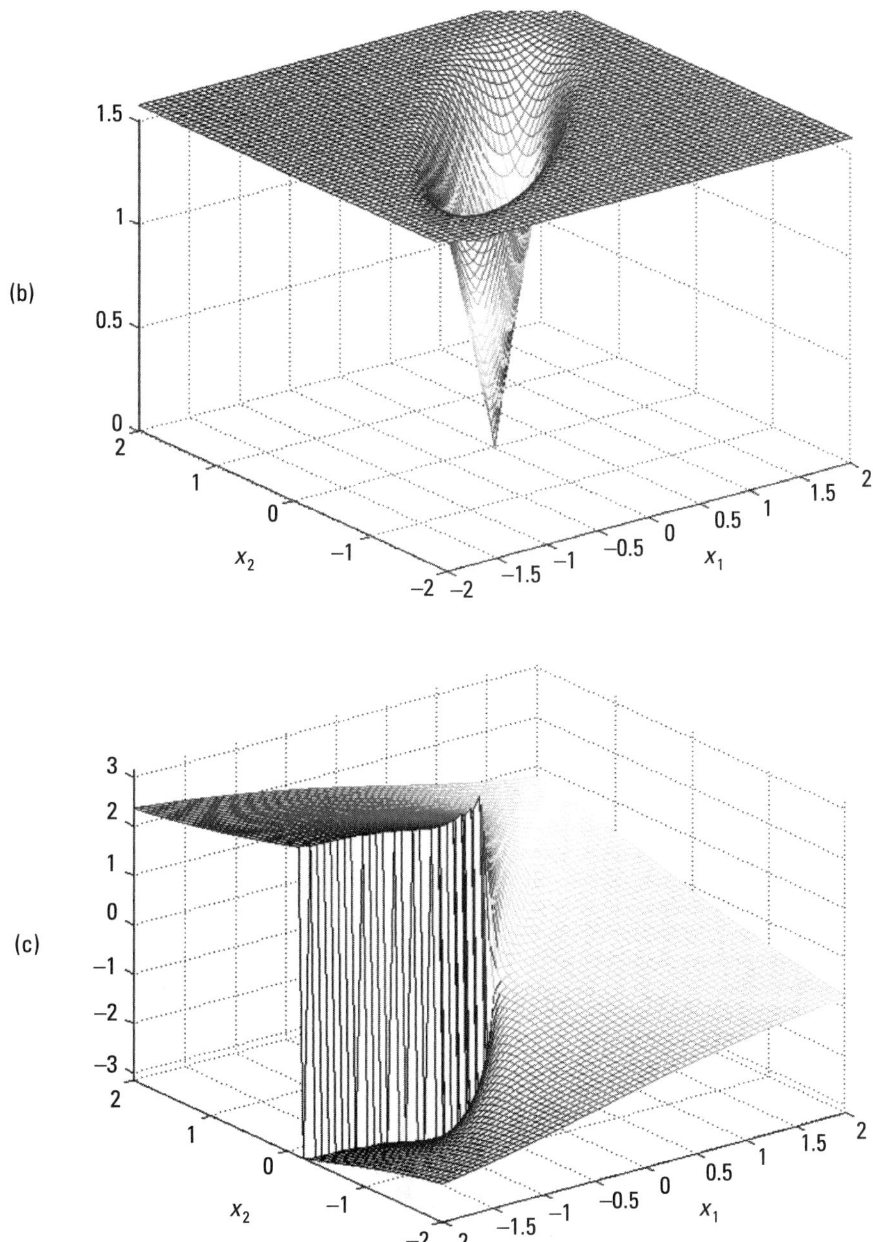

Figure 7.8　Continued

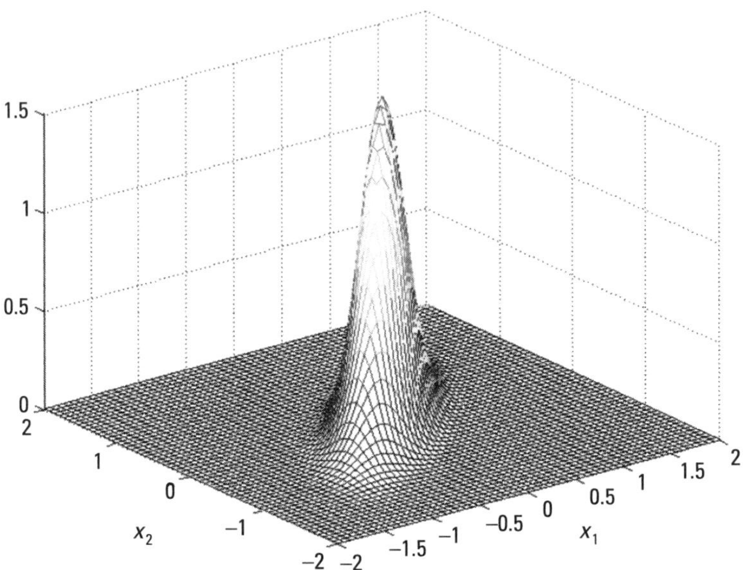

Figure 7.9   Reconstruction of a 2-D Gaussian signal from the monogenic polar representation using (7.54).

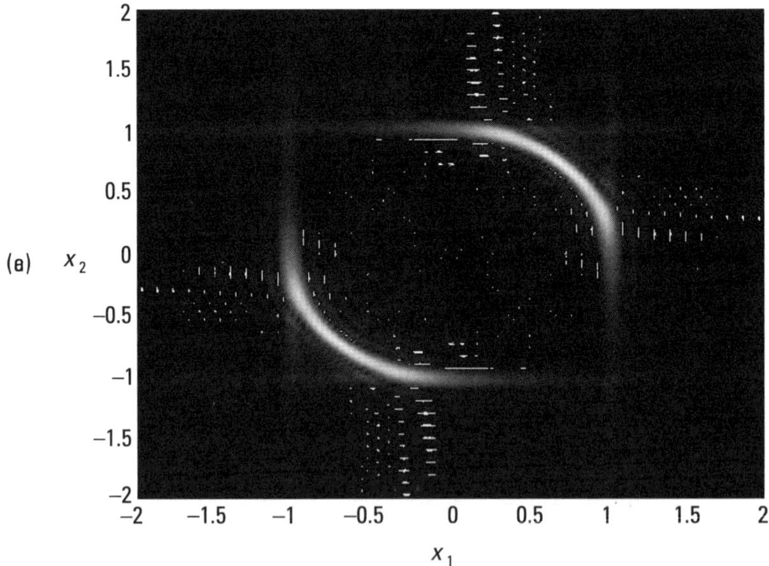

Figure 7.10   The amplitude and phase functions of the cylinder 2-D analytic signal. (a) Amplitude A_1. (b) Amplitude A_2. (c) The amplitude of the quaternion. (d) The reconstructed amplitude using (7.48). (e) The phase function ϕ_1. (f) The phase function ϕ_2.

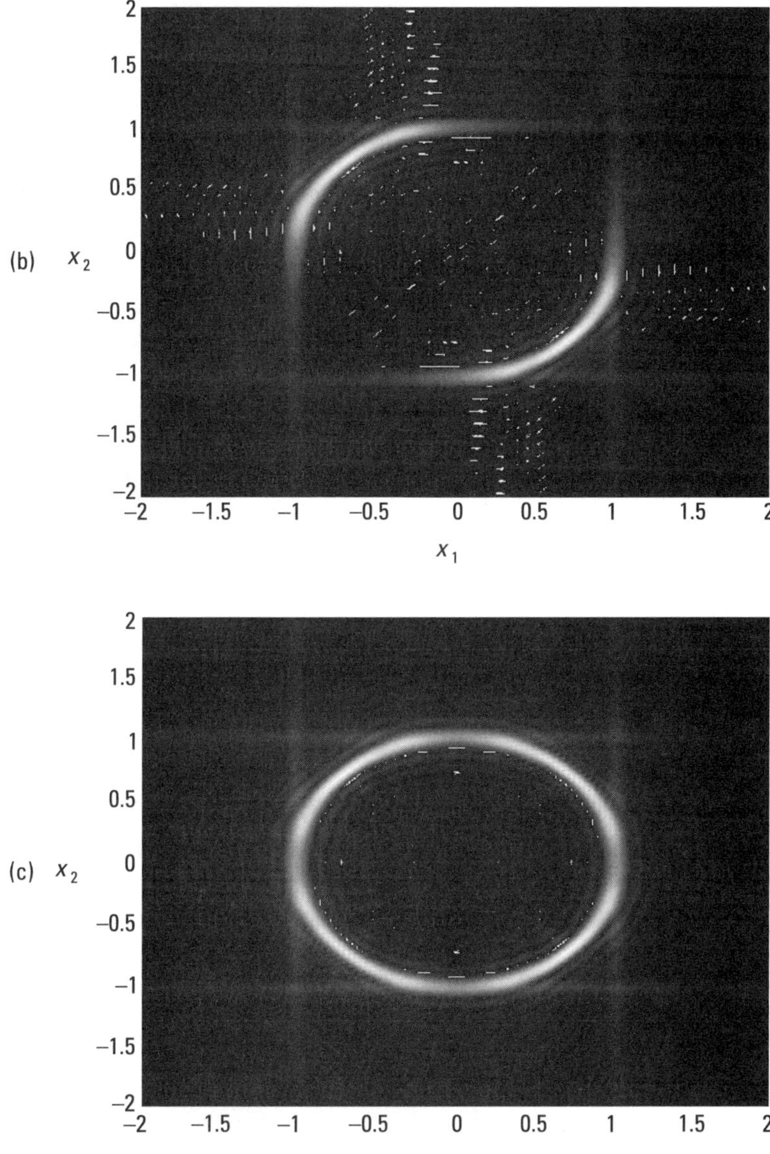

Figure 7.10 Continued

(d)

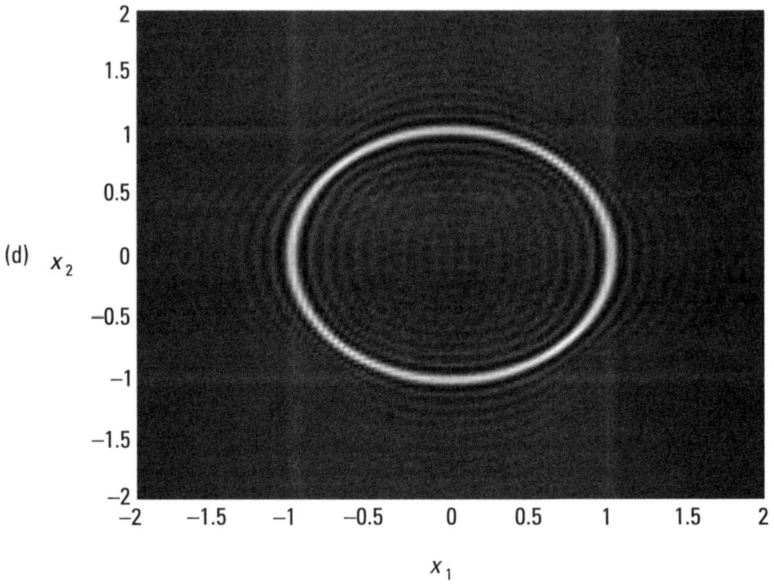

(e)

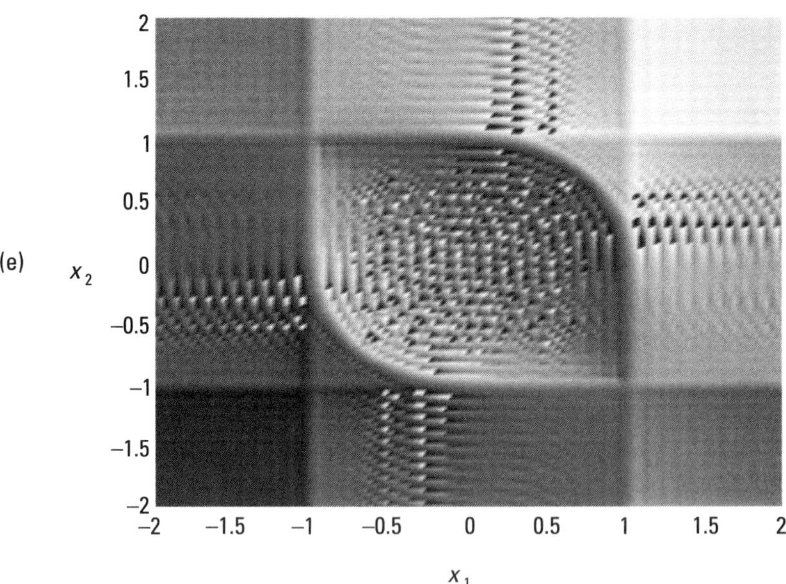

Figure 7.10   Continued

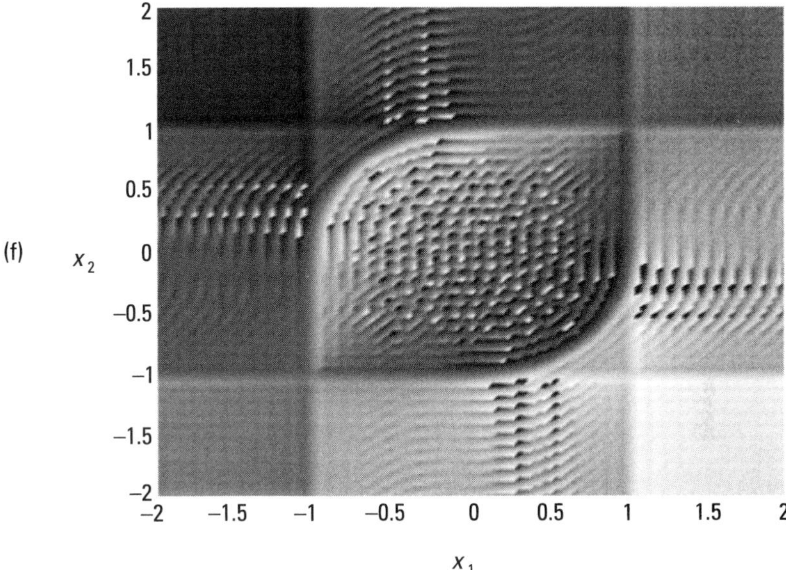

Figure 7.10 Continued

7.5 Polar Representation of 3-D Analytic Signals

7.5.1 3-D Complex Signals with Single Octant Spectra

Due to the Hermitian symmetry of the complex Fourier transformation, 3-D signals are represented by four pairs of complex analytic signals. Their spectra in the half space $f_1 > 0$ are located in octants 1, 3, 5, and 7 (see Chapter 4, Figure 4.8). The analytic signals are displayed in Table 5.4 (Chapter 5). Their polar representation defines four amplitudes and four phase functions of the form

$$\psi_1(x_1,x_2,x_3) = \mathrm{Re}_1 + e_1\,\mathrm{Im}_1 = A_1(x_1,x_2,x_3)\exp\left[e_1\varphi_1(x_1,x_2,x_3)\right] \tag{7.55}$$

$$\psi_3(x_1,x_2,x_3) = \mathrm{Re}_2 + e_1\,\mathrm{Im}_2 = A_2(x_1,x_2,x_3)\exp\left[e_1\varphi_2(x_1,x_2,x_3)\right] \tag{7.56}$$

$$\psi_5(x_1,x_2,x_3) = \mathrm{Re}_3 + e_1\,\mathrm{Im}_3 = A_3(x_1,x_2,x_3)\exp\left[e_1\varphi_3(x_1,x_2,x_3)\right] \tag{7.57}$$

$$\psi_7(x_1,x_2,x_3) = \mathrm{Re}_4 + e_1\,\mathrm{Im}_4 = A_4(x_1,x_2,x_3)\exp\left[e_1\varphi_4(x_1,x_2,x_3)\right] \tag{7.58}$$

Note that subscripts of ψ denote the corresponding octant while the amplitudes and phases are numbered with 1, 2, 3, and 4. The amplitudes are defined as $A_i = \sqrt{\mathrm{Re}_i^2 + \mathrm{Im}_i^2}$ and the phases are $\varphi_i = \mathrm{atan2}(\mathrm{Im}_i, \mathrm{Re}_i)$. The insertion of data from Table 5.4 yields the following four amplitudes:

$$
\begin{aligned}
A_1 &= \sqrt{\left(u - v_{12} - v_{13} - v_{23}\right)^2 + \left(v_1 + v_2 + v_3 - v\right)^2} \\
A_2 &= \sqrt{\left(u + v_{12} - v_{13} + v_{23}\right)^2 + \left(v_1 - v_2 + v_3 + v\right)^2} \\
A_3 &= \sqrt{\left(u - v_{12} + v_{13} + v_{23}\right)^2 + \left(-v_1 - v_2 + v_3 - v\right)^2} \\
A_4 &= \sqrt{\left(u + v_{12} + v_{13} - v_{23}\right)^2 + \left(-v_1 + v_2 + v_3 + v\right)^2}
\end{aligned}
\tag{7.59}
$$

and four phases

$$
\begin{aligned}
\varphi_1 &= \mathrm{atan2}\left(v_1 + v_2 + v_3 - v,\, u - v_{12} - v_{13} - v_{23}\right) \\
\varphi_2 &= \mathrm{atan2}\left(v_1 - v_2 + v_3 + v,\, u + v_{12} - v_{13} + v_{23}\right) \\
\varphi_3 &= \mathrm{atan2}\left(-v_1 - v_2 + v_3 - v,\, u - v_{12} + v_{13} + v_{23}\right) \\
\varphi_4 &= \mathrm{atan2}\left(-v_1 + v_2 + v_3 + v,\, u + v_{12} + v_{13} - v_{23}\right)
\end{aligned}
\tag{7.60}
$$

The real signal can be recovered from the amplitude-phase representation using the formula

$$
u\left(x_1, x_2, x_3\right) = \frac{\psi_1 + \psi_1^* + \psi_3 + \psi_3^* + \psi_5 + \psi_5^* + \psi_7 + \psi_7^*}{8}
$$

that is,

$$
u\left(x_1, x_2, x_3\right) = \frac{A_1 \cos\left(\varphi_1\right) + A_2 \cos\left(\varphi_2\right) + A_3 \cos\left(\varphi_3\right) + A_4 \cos\left(\varphi_4\right)}{4}
\tag{7.61}
$$

7.5.1.1 Simplifications for Separable Real 3-D Signals

A separable 3-D signal is a product of three 1-D signals: $u(x_1, x_2, x_3) = g_1(x_1) g_2(x_2) g_3(x_3)$. The corresponding 1-D analytic signals are (h_i denotes the 1-D Hilbert transform of g_i): $\psi_{g1}(x_1) = g_1(x_1) + e_1 h_1(x_1) = a_1 e^{e_1 \alpha_1}$, $\psi_{g2}(x_2) = g_2(x_2) + e_1 h_2(x_2) = a_2 e^{e_1 \alpha_2}$, and $\psi_{g3}(x_3) = g_3(x_3) + e_1 h_3(x_3) = a_3 e^{e_1 \alpha_3}$. Elementary multiplications show that all four amplitudes are equal

$$A_0 = a_1 a_2 a_3 = \sqrt{u^2 + v_1^2 + v_2^2 + v_3^2 + v_{12}^2 + v_{13}^2 + v_{23}^2 + v^2} \tag{7.62}$$

and the four phase functions are given by a linear summation of angles α_i.

$$\begin{aligned}
\varphi_1\left(x_1, x_2, x_3\right) &= \alpha_1\left(x_1\right) + \alpha_2\left(x_2\right) + \alpha_3\left(x_3\right) \\
\varphi_2\left(x_1, x_2, x_3\right) &= \alpha_1\left(x_1\right) - \alpha_2\left(x_2\right) + \alpha_3\left(x_3\right) \\
\varphi_3\left(x_1, x_2, x_3\right) &= \alpha_1\left(x_1\right) + \alpha_2\left(x_2\right) - \alpha_3\left(x_3\right) \\
\varphi_4\left(x_1, x_2, x_3\right) &= \alpha_1\left(x_1\right) - \alpha_2\left(x_2\right) - \alpha_3\left(x_3\right)
\end{aligned} \tag{7.63}$$

The real signal can be reconstructed in terms of the polar representation

$$u = A_0 \cos\left(\alpha_1\right)\cos\left(\alpha_2\right)\cos\left(\alpha_3\right) \tag{7.64}$$

that is, using a single amplitude A_0 and three phase functions α_1, α_2, α_3.

7.5.2 3-D Octonion Signals with Single Octant Spectra

Let us recall the form of the octonion signal (Cayley-Dickson algebra) with the spectrum in the first octant (see Chapter 5, Table 5.7).

$$\psi_1^{CD}\left(x_1, x_2, x_3\right) = u + v_1 \cdot e_1 + v_2 \cdot e_2 + v_{12} \cdot e_3 + v_3 \cdot e_4 + v_{13} \cdot e_5 + v_{23} \cdot e_6 + v \cdot e_7 \tag{7.65}$$

The exact polar representation of this signal is unknown. Let us recall that in the 2-D case, there are exact formulae enabling the calculation of a single amplitude and three phase function of a quaternion starting with two amplitudes and two phase functions of the 2-D complex representation. The derivation is using a formula for the tangent of a sum of two angles. A similar formula of the tangent of a sum of four angles contains a formidable number of terms. The quaternion amplitudes and phases are defined using the Rodriquez matrix (7.40). No such matrix is known for octonions. Hahn and Snopek [12] proposed an approximate solution of the problem using similarities with the 2-D case and some arbitrary assumptions. The proposed polar form of (7.65) is

$$\psi_1^{CD}\left(x_1, x_2, x_3\right) = A_0\left(x_1, x_2, x_3\right) e^{e_1 \phi_1} e^{e_2 \phi_2} e^{e_3 \phi_3} e^{e_7 \phi_7} e^{\phi_4 \phi_4} e^{e_5 \phi_5} e^{e_6 \phi_6} \tag{7.66}$$

We have a single amplitude and seven phase functions. The term with the angle ϕ_7 is located arbitrarily at the center. The amplitude is the same as given by (7.62)

$$A_0 = \sqrt{u^2 + v_1^2 + v_2^2 + v_3^2 + v_{12}^2 + v_{13}^2 + v_{23}^2 + v^2} \tag{7.67}$$

Recapitulating, our goal is to deduce the seven phase functions in (7.66) denoted ϕ_i starting with four amplitudes A_i and four phase functions φ_i defined by (7.59) and (7.60). Using analogies to the 2-D case, the seven phase functions form two groups. The members of the first group are unions of four phase functions as follows

$$\phi_1 = \frac{1}{4}\left(\underbrace{\varphi_1 + \varphi_3}_{a} + \underbrace{\varphi_5 + \varphi_7}_{b} \right), \quad \phi_2 = \frac{1}{4}\left(\underbrace{\varphi_1 + \varphi_3}_{a} - \underbrace{\varphi_5 + \varphi_7}_{b} \right)$$

$$\phi_4 = \frac{1}{4}\left(\underbrace{\varphi_1 - \varphi_3}_{c} + \underbrace{\varphi_5 - \varphi_7}_{d} \right), \quad \phi_6 = \frac{1}{4}\left(\underbrace{\varphi_1 - \varphi_3}_{c} - \underbrace{\varphi_5 + \varphi_7}_{-d} \right) \tag{7.68}$$

The second group is defined by the four amplitudes

$$\sin\left(4\phi_3\right) = \frac{A_1^2 - A_3^2}{A_1^2 + A_3^2}, \quad \sin\left(4\phi_6\right) = \frac{A_5^2 - A_7^2}{A_5^2 + A_7^2},$$

$$\sin\left(4\phi_7\right) = \frac{A_1^2 + A_3^2 - A_5^2 - A_7^2}{A_1^2 + A_3^2 + A_5^2 + A_7^2} \tag{7.69}$$

The real signal can be reconstructed using the formula ($c_i = \cos\phi_i$; $s_i = \sin\phi_i$):

$$
\begin{aligned}
u_{\text{rec}}\left(x_1, x_2, x_3\right) = A_0\Big[& c_1 c_2 c_3 c_4 c_5 c_6 c_7 + s_1 s_2 s_3 c_4 c_5 c_6 c_7 - s_1 c_2 c_3 s_4 s_5 c_6 c_7 \\
& + c_1 s_2 s_3 s_4 s_5 c_6 c_7 - s_1 s_2 c_3 s_4 c_5 s_6 c_7 + s_1 c_2 s_3 s_4 c_5 s_6 c_7 \\
& - c_1 c_2 s_3 c_4 s_5 s_6 c_7 - s_1 s_2 c_3 c_4 s_5 s_6 c_7 + c_1 c_2 s_3 s_4 c_5 c_6 s_7 \\
& + s_1 s_2 c_3 s_4 c_5 c_6 s_7 + c_1 c_2 c_3 s_4 s_5 c_6 s_7 - s_1 c_2 s_3 c_4 s_5 c_6 s_7 \\
& - s_1 c_2 c_3 c_4 c_5 s_6 s_7 + c_1 s_2 s_3 c_4 c_5 s_6 s_7 - c_1 c_2 c_3 s_4 s_5 s_6 s_7 - s_1 s_2 s_3 s_4 s_5 s_6 s_7 \Big]
\end{aligned}
\tag{7.70}
$$

Computer calculations have shown that only the first term is significant and the calculation using all terms in (7.70) almost do not change the result.

7.5.2.1 The Case of Separable Real Signals (see Section 7.4.1.1)

In the case of 3-D separable real signals, their octonion polar notation is considerably simplified. Because all four amplitudes are equal, the phase angles of the second group are all equal to zero. The four phase angles of the first group are given by insertion of (7.63) into (7.68) yielding $\phi_1 = \alpha_1$, $\phi_2 = \alpha_3$, $\phi_4 = \alpha_2$, and $\phi_5 = 0$. In consequence, we get a polar representation exactly of the same form as in the 3-D complex case. They differ only by the imaginary units.

$$\psi_1^{CD}\left(x_1,x_2,x_3\right) = A_0 e^{e_1\alpha_1} e^{e_2\alpha_2} e^{e_4\alpha_3} \tag{7.71}$$

This fact validates strongly the deduced (7.68) and (7.69). Of course, (7.71) is not an approximation. The real signal can be recovered in terms of the above polar representation using $u_{rec} = A_0\cos(\alpha_1)\cos(\alpha_2)\cos(\alpha_3)$.

Example 7.7

The 3-D Gaussian signal is defined by

$$u\left(x_1,x_2,x_3\right) = (2\pi)^{-3/2}|M|^{-1/2}\exp\left\{\frac{-1}{2|M|}\sum_{i,j=1}^{3}\left|M_{ij}\right|x_i x_j\right\}$$

where

$$|M| = \sigma_1^2\sigma_2^2\sigma_3^2\left(1 + 2\rho_{12}\rho_{23}\rho_{13} - \rho_{12}^2 - \rho_{13}^2 - \rho_{23}^2\right)$$

$$\left|M_{11}\right| = \left(1-\rho_{23}^2\right)\sigma_2^2\sigma_3^2, \left|M_{22}\right| = \left(1-\rho_{13}^2\right)\sigma_1^2\sigma_3^2, \left|M_{33}\right| = \left(1-\rho_{12}^2\right)\sigma_1^2\sigma_2^2$$

$$\left|M_{12}\right| = \left|M_{21}\right| = \sigma_1\sigma_2\sigma_3^2\left(\rho_{13}\rho_{23} - \rho_{12}\right), \left|M_{12}\right| = \left|M_{21}\right| = \sigma_1\sigma_2\sigma_3^2\left(\rho_{13}\rho_{23} - \rho_{12}\right)$$

$$\left|M_{12}\right| = \left|M_{21}\right| = \sigma_1\sigma_2\sigma_3^2\left(\rho_{13}\rho_{23} - \rho_{12}\right)$$

If all correlation coefficients $\rho_{ij} = 0$, we deal with a 3-D separable Gaussian signal. The Fourier spectrum is

$$U\left(\omega_1,\omega_2,\omega_3\right) = e^{-\frac{1}{2}\left(\omega_1^2\sigma_1^2 + \omega_2^2\sigma_2^2 + \omega_3^2\sigma_3^2\right)} e^{-\omega_1\omega_2\rho_{12}\sigma_1\sigma_2 - \omega_1\omega_3\rho_{13}\sigma_1\sigma_3 - \omega_2\omega_3\rho_{23}\sigma_2\sigma_3}$$

There are no means to display a 3-D function on a 2-D page or diagram. Figure 7.11 shows 2-D cross sections of 3-D signals assuming $u(x_1,x_2,x_3) = 0$.

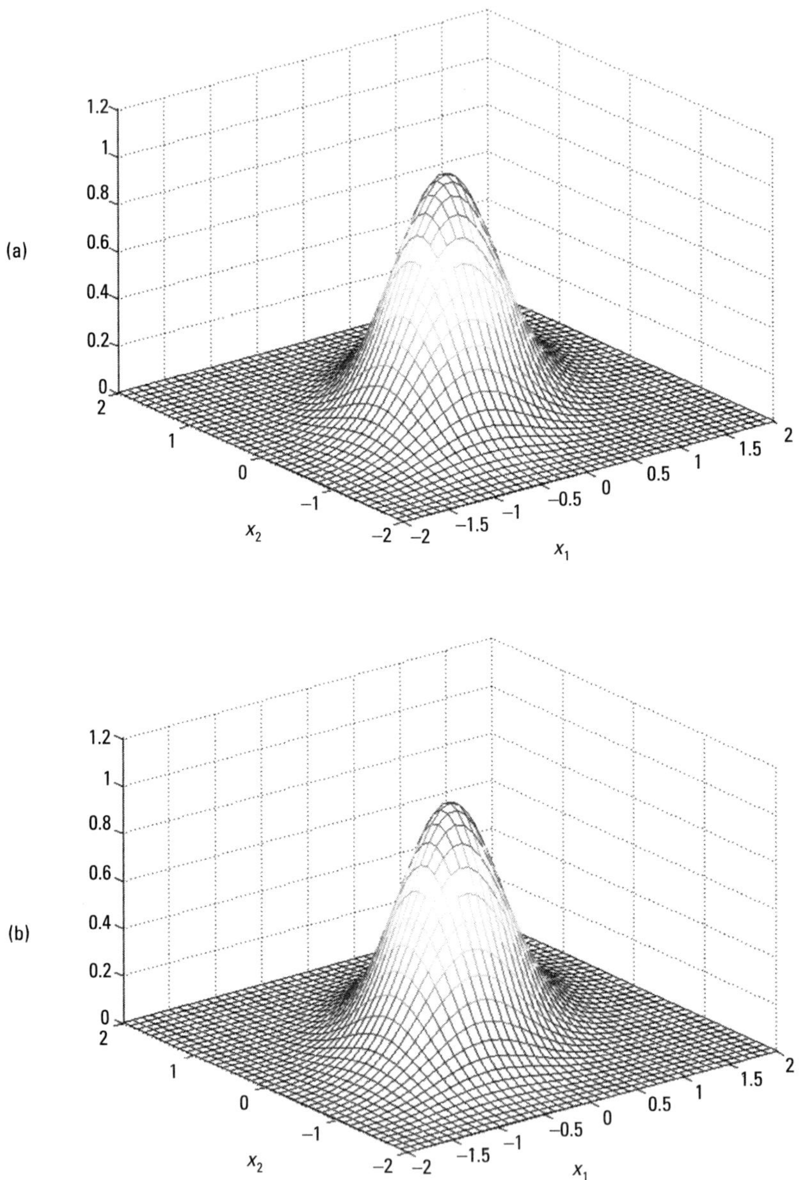

Figure 7.11 (a) The cross section of the real signal u. (b): The reconstructed signal using (7.61) (3-D complex case, the difference (a)–(b) = 0). (c) The reconstructed signal using (7.64). (d). The difference (a)–(c).

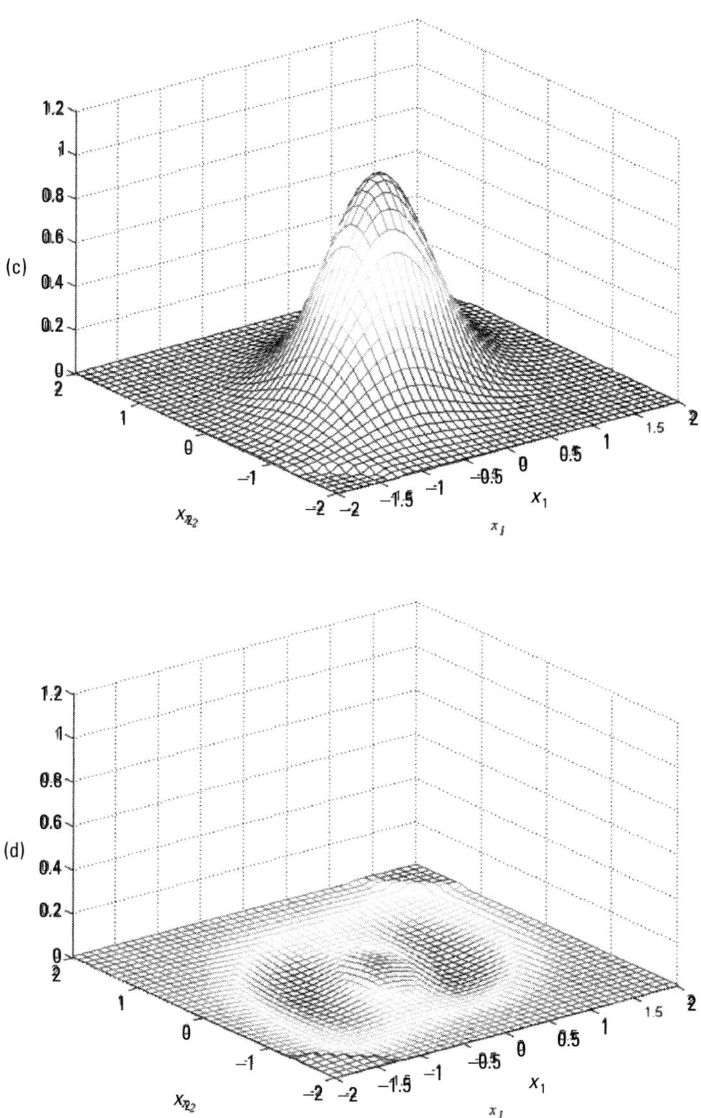

Figure 7.11 Continued

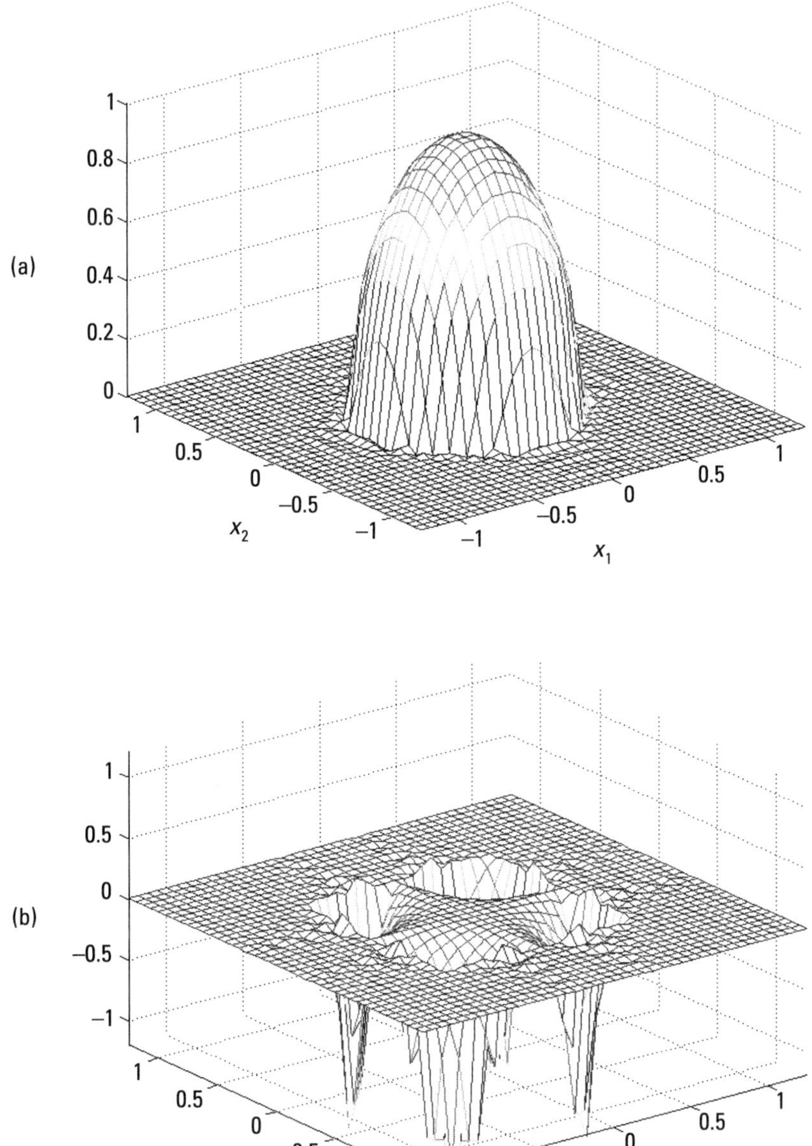

Figure 7.12   (a) The reconstructed signal. No practical difference between the complex and octonion approach. (b) Ten times the difference between the complex and octonion reconstruction.

We applied all variances $\sigma_i = 1$ and all correlation coefficients $\rho_{ij} = 0.8$. We will display the original real signal and its reconstructed versions using in the complex case (7.61) and in the octonion case (7.70). The amplitude and phase functions are obtained by using the real and imaginary parts of the complex signals calculated by the inverse Fourier transform of their spectra. The reconstruction in the case of the complex signal is perfect and in the case of the octonion signal nearly perfect. This gives evidence, that the equations for the octonion phase functions have been correctly deduced.

Example 7.8

Consider a 3-D spherical signal $u(r) = \begin{cases} 1 & r < r_1 \\ 0 & r > r_1 \end{cases}$. Its Fourier transform is

$$U\left(\omega_1,\omega_2,\omega_3\right) = \frac{4\pi}{\rho^3}\left[\sin\left(r_1\rho - r_1\rho\cos\left(r_1\rho\right)\right)\right],$$

$$\rho = \sqrt{\omega_1^2 + \omega_2^2 + \omega_3^2}, \ \omega_i = 2\pi f_i$$

Figure 7.12 shows the cross sections of reconstructed signals. They again validate the accuracy of the deduced seven phase functions of the octonion analytic signal.

References

[1]   Bülow, T., "Hypercomplex spectral signal representation for the processing and analysis of images," Bericht Nr. 99–3, Institut für Informatik und Praktische Mathematik, Christian-Albrechts-Universität Kiel, Aug. 1999.

[2]   Bülow, T., and G. Sommer, "The Hypercomplex Signal—A Novel Extension of the Analytic Signal to the Multidimensional Case," *IEEE Trans. Signal Processing*, Vol. 49, No. 11, Nov. 2001, pp. 2844–2852.

[3]   Cohen, L., P. Loughlin, and D. Vakman, "On the ambiguity in the definition of the amplitude and phase of a signal," *Signal Processing*, Vol. 79, 1999, pp.301–307.

[4]   Hahn, S. L, and K. M. Snopek, "Comparison of Properties of Analytic, Quaternionic and Monogenic 2D Signals," *WSEAS Transactions on Computers*, Issue 3, Vol. 3, July 2004, pp. 602–611.

[5]   Sommer, G. (ed.), *Geometric Computing with Clifford Algebras*, Berlin: Springer-Verlag, 2001.

[6]   Bülow, T., M. Felsberg, and G. Sommer, "Non-Commutative Hypercomplex Fourier Transforms of Multidimensional Signals" in *Geometric Computing with Clifford Algebra*, G. Sommer (Ed.), Berlin: Springer-Verlag, 2001.

[7]   Hahn, S. L., "The instantaneous complex frequency concept and its application to the analysis of building-up oscillations in oscillators," *Proceedings of Vibration Problems*, No. 1, 1959, pp. 29–47.

[8]   Hahn, S. L., "Complex variable frequency electric circuit theory," *Proc. IEEE*, Vol. 52, No. 6, June 1064, pp. 735–736.

[9]   Hahn, S. L., "Multidimensional Complex Signals with Single-Orthant Spectra," *Proc. IEEE*, Vol. 80, No. 8, August 1992, pp. 1287–1300.

[10]   Hahn, S. L., *Hilbert Transforms in Signal Processing*, Norwood, MA: Artech House, 1996.

[11]   Hahn, S. L., "The instantaneous complex frequency of a sum of two harmonic signals," *Bull. Polish Ac. Sci.*, Technical Sciences, Bol. 48, No. 4, 2000.

[12]   Hahn, S. L., and K. M. Snopek, "The Unified Theory of n-Dimensional Complex and Hypercomplex Analytic Signals," *Bull. Polish Ac. Sci.*, Technical Sciences, Vol. 59, No. 2, 2001, pp. 167–181.

[13]   Hahn, S. L., "On the uniqueness of the definition of the amplitude and phase of analytic signals," *Signal Processing*, Vol. 83, No. 8, 2003, pp. 1815–1820.

[14]   Oppenheim, A.V., and J. S. Lim, "The importance of phase in signals," *Proc. IEEE*, Vol. 69, May 1981, pp.529–541.

[15]   Vakman, D., "On the analytic signal, the Teager-Kaiser energy algorithm, and other methods for defining amplitude and frequency," *IEEE Trans. Signal Process.*, Vol. 44, No. 4, 1996, pp. 791–797.

[16]   Vakman, D., *Signals, Oscillations, and Waves: A Modern Approach*, Norwood, MA: Artech House, 1998.

8

Quasi-Analytic Signals

8.1 Definition of a Quasi-Analytic Signal

The theory of analytic signals with single orthant spectra [3] is presented in Chapter 5. The frequency domain space $\mathbb{R}^n$ is a union of 2^n orthants. However, due to the Hermitian symmetry of the complex single-orthant spectra, the real n-D signal $u(x)$ can be reconstructed using knowledge of 2^{n-1} analytic signals with single orthant spectra located in a half-frequency space $f_1 > 0$ (see Chapter 5). This statement does not apply for quasi-analytic signals. The support of the spectrum of a quasi-analytic signal is only approximately located in a single orthant, in 1-D—in a half-axis, in 2-D—in a single quadrant.

Consider a real n-D signal $u(x)$ and its complex Fourier spectrum $U(f)$. The multiplication of the real signal by the n-D carrier represented here by a the product of n exponentials, shifts the spectrum towards a single first orthant (i.e., $\mathbb{R}$)

$$
\begin{aligned}
g(x) &= u(x)e^{e_1 2\pi f_{10}x_1}e^{e_1 2\pi f_{20}x_2}\ldots e^{e_1 2\pi f_{n0}x_n} \overset{nF}{\Leftrightarrow} G\left(f - f_{i0}\right) \\
&= U\left(f_1 - f_{10}\right)U\left(f_2 - f_{20}\right)\ldots U\left(f_n - f_{n0}\right)
\end{aligned}
\tag{8.1}
$$

The quality of this shift is described by the leakage coefficient

$$\varepsilon = 1 - \frac{\text{Spectral energy in } \mathbb{R}^+}{\text{Total spectral energy}} = 1 - \frac{\int_{\mathbb{R}^+}\|G(f)\|^2 \, df}{\int_{\mathbb{R}}\|G(f)\|^2 df} \qquad (8.2)$$

If the value of $\varepsilon \ll 1$, the shifted signal is called *quasi-analytic*. If $\varepsilon = 0$, the shift is perfect and the signal is analytic. Note that if all frequencies of the carrier are positive, we get a single quasi-analytic (or analytic) signal. In principle, the carrier in (8.1) could be a hypercomplex function of the form $e^{e_1 2\pi f_{10} x_1} e^{e_2 2\pi f_{20} x_2} \ldots e^{e_n 2\pi f_{n0} x_n}$. However, because we define a signal with a single-orthant spectrum the use of the hypercomplex version seems to have no advantage. Bearing this in mind, in the remainder of this chapter, we will use the notation j instead of e_1.

8.2 The 1-D Quasi-Analytic Signals

The notion of a quasi-analytic signal applies for modulated harmonic carriers. The complex form of this carrier is: $\psi_c(t) = A_0 e^{j(2\pi f_0 t + \phi_0)}$. In the following, we assume a normalized amplitude $A_0 = 1$. Note that the carrier is itself an analytic signal. Any kind of modulation of this carrier—for example amplitude-, phase-, or frequency-modulation—can be represented by a product of a modulation function and a carrier [2–4]:

$$\psi(t) = \gamma(t)e^{j\phi_0}e^{j2\pi f_0 t} = \gamma_m(t)e^{j2\pi f_0 t} \qquad (8.3)$$

where $\gamma(t)$ is called the *modulation function* and $\gamma_m(t)$ is called the modified modulation function (including the carrier's phase). The carrier's phase plays a role by addition of two modulated signals and is relative with regard to the choice of the origin of the time scale. In the following we assume a carrier phase of zero, that is $\phi_0 = 0$.

Consider a real signal $x(t)$, which should be transmitted by modulation. We have a Fourier pair $x(t) \overset{F}{\Longleftrightarrow} X(f)$. The modulation function is a function of $x(t)$: $\gamma(t) = \gamma[x(t)]$. We have a new Fourier pair

$$\gamma_m(t) \overset{F}{\Longleftrightarrow} G(f) \qquad (8.4)$$

The above spectrum may be called *the spectrum of the modulated signal at zero carrier frequency*. This is a two-sided spectrum with one part at positive frequencies and another part at negative frequencies. The part at negative frequencies represents the lower sideband of the modulated signal. The part at positive frequencies represents the upper sideband The multiplication with the analytic carrier shifts the spectrum of the modulation function towards positive frequencies. The analytic carrier and its spectrum are:

$$\psi_c(t) = e^{j2\pi f_0 t} \overset{F}{\Leftrightarrow} \Gamma(f) = \delta\left(f - f_0\right) \tag{8.5}$$

The spectrum of the modulated signal (8.3) is given by the convolution

$$G_s(f) = G(f) * \delta\left(f - f_0\right) = G\left(f - f_0\right) \tag{8.6}$$

The shifted signal may be quasi-analytic if the shift is imperfect (still some energy at negative frequencies). Otherwise it is analytic. Equation (8.2) under these conditions takes the form

$$\varepsilon = 1 - \frac{\text{Spectral energy for } f > 0}{\text{Total spectral energy}} = 1 - \frac{\int\limits_{0}^{\infty} \left\| G\left(f - f_0\right) \right\|^2 df}{\int\limits_{-\infty}^{\infty} \left\| G(f) \right\| df} \tag{8.7}$$

Example 8.1 Modulated Gaussian Pulse

The modulated Gaussian pulse and its spectrum are:

$$g(t) = \frac{1}{\sqrt{2\pi}\sigma} e^{-\frac{t^2}{2\sigma^2}} e^{j2\pi f_0 t} \overset{F}{\Leftrightarrow} G(f) = e^{-2\pi^2\left(f - f_0\right)^2 \sigma^2} \tag{8.8}$$

Figure 8.1 displays the waveforms for two values of the carrier frequency and the corresponding values of the leakage parameter ε, $\sigma = 1$.

Example 8.2 Modulated Rectangular Pulse

The modulated rectangular pulse and its spectrum are

$$g(t) = \Pi_T(t) e^{j2\pi f_0 t} \overset{F}{\Leftrightarrow} \frac{\sin\left[2\pi\left(f - f_0\right)T\right]}{2\pi\left(f - f_0\right)}, \quad \Pi_T(t) = 1 \text{ in the interval } -1 < t < 1$$

Figure 8.2 displays the waveforms for two values of the carrier frequency and the corresponding values of the leakage parameter ε.

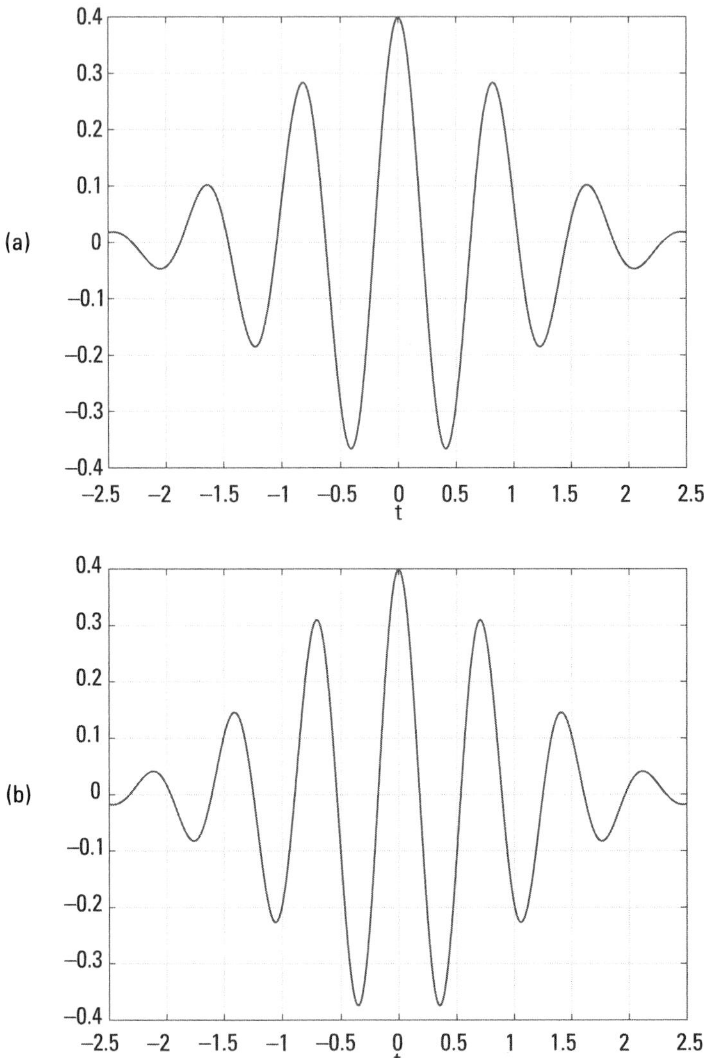

Figure 8.1   The waveforms of a modulated Gaussian pulse, $g(t) = (1/\sqrt{2\pi})e^{-t^2/2}\cos(2\pi f_0 t)$. (a) $f_0 = 1.2$, $\varepsilon = 0.0415$; (b) $f_0 = 1.4$, $\varepsilon = 0.00023$. It should be noted that a small increase in frequency induces a large decrease of the leakage parameter.

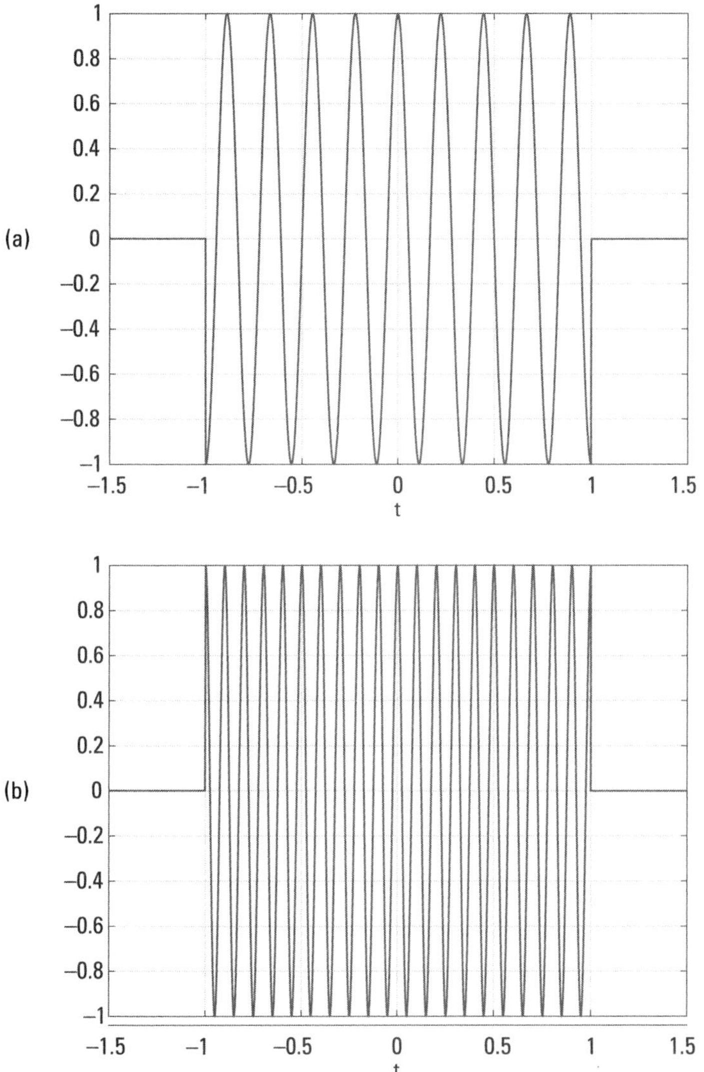

Figure 8.2 The waveforms of a modulated rectangular pulse. The comparison with Figure 8.1 shows that much larger carrier frequencies should be applied in order to get small values of the leakage. (a) $g(t) = \Pi(t)\cos(2\pi f_0 t)$; $f_0 = 4.5$; $\varepsilon = 0.0456$. (b) $g(t) = \Pi(t)\cos(2\pi f_0 t)$; $f_0 = 10.0$; $\varepsilon = 0.002$.

Example 8.3 Phase and Frequency Modulation

The modulation function in case of frequency modulation of a harmonic carrier $\psi_c(t) = A_0 e^{j(2\pi f_0 t + \phi_0)}$ and the low-frequency sine modulating signal $x(t) = \sin(2\pi a t)$ is a periodic function. Its expansion into a Fourier series is (assuming $A_0 = 1$ and $\phi_0 = 0$)

$$g(t) = e^{j\beta \sin(2\pi a t)} = \sum_{n=-\infty}^{\infty} J_n(\beta) e^{j2\pi n a t} \tag{8.9}$$

The coefficient β equals the phase deviation or the ratio $\Delta f / a$, where Δf is the frequency deviation, and $J_n(\beta)$ are Bessel functions. The spectrum of the modulation function representing the spectrum of the modulated signal at zero carrier frequency is

$$\begin{aligned}
G(f) = {}& J_0(\beta)\delta(f) + J_1(\beta)\big[\delta(f+a) - \delta(f-a)\big] \\
& + J_2(\beta)\big[\delta(f+2a) + \delta(f-2a)\big] \\
& + \ldots\ldots\ldots\ldots\ldots\ldots\ldots\ldots\ldots \\
& + J_n(\beta)\big[\delta(f+na) + (-1)^n \delta(f-na)\big] \\
& + \ldots\ldots\ldots\ldots\ldots\ldots\ldots\ldots\ldots
\end{aligned} \tag{8.10}$$

The average power of the modulated signal equals the carrier power

$$\overline{P} = \frac{1}{2} \sum_{n=-\infty}^{\infty} J_n^2(\beta) \tag{8.11}$$

The modulation shifts the zero carrier frequency spectrum towards positive frequencies. Let us assume that the carrier frequency is a multiple of the frequency of modulation $f_0 = ka$. The leakage parameter (8.2) equals the sum of the powers of the terms of the shifted spectrum (8.10) having support at negative frequencies

$$\varepsilon = \sum_{n=-k}^{n=-\infty} J_n^2(\beta) \tag{8.12}$$

When, for example, $\beta = 5$: if $k = 5$, then $\varepsilon = 0.0446$ and if $k = 7$, then $\varepsilon = 0.00169$. For $\beta = 10$, if $k = 10$, then $\varepsilon = 0.032$ and if $k = 12$, then $\varepsilon = 0.00267$.

8.3 Phase Signals

The polar form of the analytic signal is $\psi(t) = A(t)e^{j\varphi(t)}$ (see Chapter 7). The phase signal has the form $\psi(t) = A_0 e^{j\varphi(t)}$, that is, the amplitude is a time-independent constant. However, not every phase signal is analytic. This is because, in general, the Hilbert transform of the real part $\mathrm{H}[\cos\varphi(t)]$ is not equal to $\sin\varphi(t)$. A nonanalytic phase signal multiplied by the carrier $e^{j\Omega_0 t}$ is converted to a quasi-analytic signal with a negligible leakage of its spectrum into the negative frequency range. A good example is the familiar FM signal shown in Example 8.3.

Let us derive a special class of analytic phase signals starting with a complex function of a complex variable $z = t + j\tau$ of the form

$$\psi(z) = \prod_{k=1}^{N} \frac{z - z_k}{z - z_k^*} = \psi_1(z)\psi_2(z)\cdots\psi_N(z) \tag{8.13}$$

where the coefficients $z_k = \alpha_k + j\beta_k$ are complex constants. In the following, it is convenient to write the kth term of the product given by (8.13) in the form

$$\psi_k(t) = \frac{z - \left(\alpha_k + j\beta_k\right)}{z - \left(\alpha_k - j\beta_k\right)} = u_k(t,\tau) + jv_k(t,\tau) \tag{8.14}$$

of an analytic function satisfying the Cauchy-Riemann equations (see Chapter 3). The function $\psi(z)$ given by (8.14) is analytic as a product of analytic functions. In the limit for $\tau \to 0_+$, we get the following analytic phase signal

$$\psi_k(t) = u_k(t) + jv_k(t) = 1 - \tilde{u}_k(t) + jv_k(t) \tag{8.15}$$

The signal $\tilde{u}_k(t)$ is called the finite energy part of the signal $u_k(t) = 1 - \tilde{u}_k(t)$. Notice that the Hilbert transforms $\mathrm{H}[u_k(t)] = \mathrm{H}[1 - \tilde{u}_k(t)] = \mathrm{H}[-\tilde{u}_k(t)]$ are equal because $\mathrm{H}[1] = 0$. The evaluation of the real and imaginary parts of (8.15) yields

$$-\tilde{u}_k(t) = -2\beta_k \frac{\beta_k}{\left(t - \alpha_k\right)^2 + \beta_k^2}, \quad -v_k(t) = -2\beta_k \frac{t - \alpha_k}{\left(t - \alpha_k\right)^2 + \beta_k^2} \tag{8.16}$$

We recognize a pair of negated Cauchy Hilbert transforms (see Appendix G) multiplied by $2\beta_k$ and shifted in time by α_k. The calculation of the polar form shows that the amplitude of $\psi_k(t)$ equals 1 and the calculation of the phase functions shows that

$$\varphi_k(t) = \tan^{-1}\left(\frac{v_k(t)}{u_k(t)}\right) = 2\tilde{\varphi}(t) = 2\tan^{-1}\left(\frac{v_k}{\tilde{u}(t)}\right) \tag{8.17}$$

It should also be noted that the instantaneous angular frequency of the phase signal $\omega(t) = d\varphi_k(t)/dt$ equals twice the instantaneous frequency of the Cauchy signal $\tilde{\omega}(t) = d\tilde{\varphi}_k(t)/dt$. The phase signals $\psi_k(t)$ are noncausal and can be made nearly causal using a large value of the time shift constant α_k, if $\alpha_k > 0$. The total analytic phase signal defined by (8.14) can be written in the form

$$\psi(t) = 1 \; e^{j\left[\varphi_1(t)+\varphi_2(t)+\cdots+\varphi_N(t)\right]} \tag{8.18}$$

with the amplitude $A(t) = 1$ for all t. Let $\Gamma(f) = F\{\psi(t)\}$ be the Fourier transform of the phase signal and $\Gamma_k(f) = F\{\psi_k(t)\}$ the transform of the kth term. Due to the multiplication-to-convolution theorem (see Appendix A) we get

$$\Gamma(f) = \Gamma_1(f) * \Gamma_2(f) * \cdots * \Gamma_N(f) \tag{8.19}$$

All terms $\Gamma_k(f)$ and the spectrum $\Gamma(f)$ are one-sided being the spectra of analytic signals. Since $\psi(t)$ is a rational function, it can be expanded in partial fractions

$$\psi(t) = 1 + \frac{c_1}{t - z_1^*} + \frac{c_2}{t - z_2^*} + \cdots + \frac{c_N}{t - z_N^*} \tag{8.20}$$

with the complex coefficients $c_k = c_{kr} + jc_{ki} = \lim\limits_{z \to z_k^*}\left[\left(z - z_k^*\right)\psi(z)\right]$. For $N = 1$, we get $c_1 = z_1^* - z_1 = -2j\beta_1$ and for $N > 1$ we have

$$c_k = \left(\prod_{n=1}^{N}\left(z_k^* - z_n\right)\right)\left(\prod_{\substack{n=1 \text{ except } n=k}}^{N}\left(z_k^* - z_n^*\right)\right)^{-1} \tag{8.21}$$

It follows from (8.20) that the multiple convolution (8.19) can be replaced by the summation of spectra; that is,

$$\Gamma(f) = \delta(f) + c_1 F\left[\left(t - z_1^*\right)^{-1}\right] + c_2 F\left[\left(t - z_2^*\right)^{-1}\right] + \cdots + c_N F\left[\left(t - z_N^*\right)^{-1}\right] \tag{8.22}$$

The individual Fourier transform in this equation is

$$\Gamma_k(f) = c_k \mathrm{F}\left[\left(t - z_k^*\right)^{-1}\right] = -jc_k 2\pi 1(f)e^{-2\pi f \beta_k}e^{-j2\pi f \alpha_k} \qquad (8.23)$$

where $1(f)$ is a unit step function (distribution) and $\exp(-j2\pi f \alpha_k)$ is a well-known time shift operator.

8.4   The *n*-D Quasi-Analytic Signals

The *n*-D quasi-analytic signals are defined in exactly the same way as in the 1-D case. Consider the *n*-D low-pass signal and its spectrum

$$u(\boldsymbol{x}) \overset{n-F}{\Leftrightarrow} U(f), \ \boldsymbol{x} = \left(x_1, x_2, \ldots, x_n\right), \ f = \left(f_1, f_2, \ldots, f_n\right) \qquad (8.24)$$

The quasi-analytic signal is defined by multiplication of *u* by the *n*-D harmonic complex (or hypercomplex) carrier. The complex carrier is

$$\psi_c(\boldsymbol{x}) = e^{j2\pi f_{10}x_1}e^{j2\pi f_{20}x_2} \times \ldots \times e^{j2\pi f_{n0}x_n} = \prod_{i=1}^{n} e^{j2\pi f_{i0}x_i} \qquad (8.25)$$

and the hypercomplex carrier is

$$\psi_{hc}(\boldsymbol{x}) = e^{e_1 2\pi f_{10}x_1}e^{e_2 2\pi f_{20}x_2} \times \ldots \times e^{e_n 2\pi f_{n0}x_n} = \prod_{i=1}^{n} e^{e_i 2\pi f_{i0}x_i} \qquad (8.26)$$

The algebra of the imaginary units e_i should be defined. It should be noted that the spectrum of the complex and hypercomplex carriers is the same. Of course, the spectrum of the complex carrier is given by the complex *n*-D Fourier transform and the hypercomplex one by a hypercomplex Fourier transform. In both cases, we have

$$F_c\{\psi_c(\boldsymbol{x})\} = F_{hc}\{\psi_{hc}(\boldsymbol{x})\} = \delta\left(f_1 - f_{10}\right)\delta\left(f_2 - f_{20}\right)\ldots\delta\left(f_n - f_{n0}\right)$$

$$= \prod_{i=1}^{n}\delta\left(f_i - f_{i0}\right) \qquad (8.27)$$

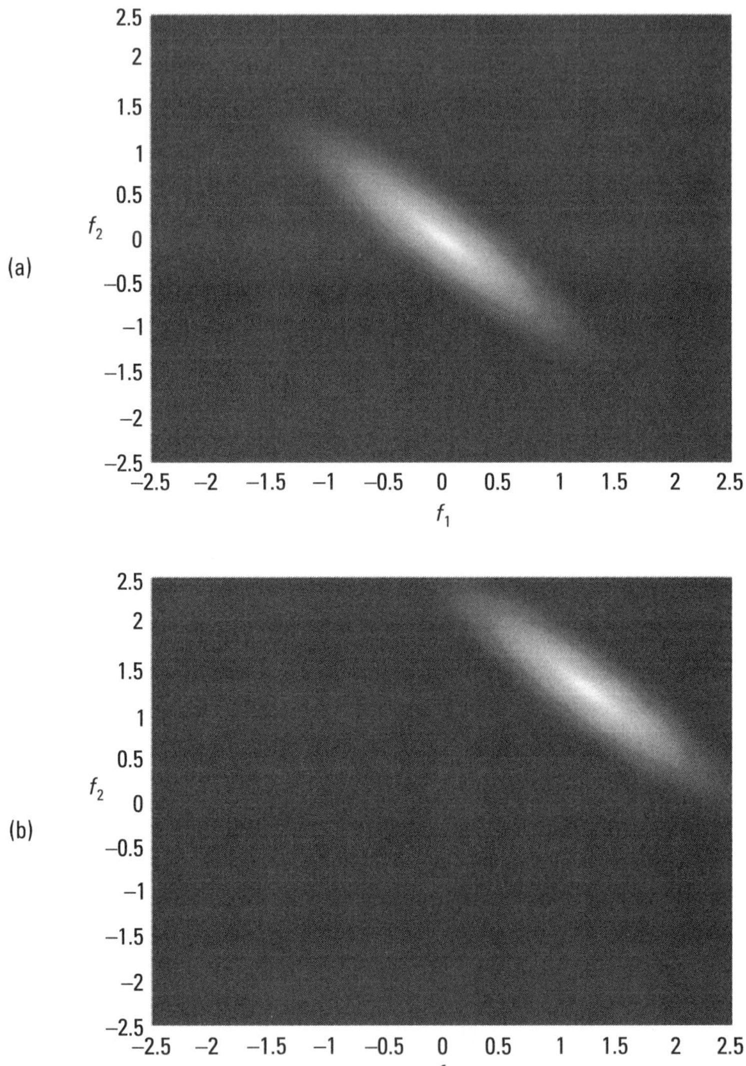

Figure 8.3   (a) The low-pass spectrum of a 2-D Gaussian signal, $\sigma_1 = \sigma_2 = 0.7$, $\rho = 0.9$.
(b) The shifted spectrum $G(f_1 - 1.25, f_2 - 1.25)$.

The shifted spectrum is defined by the n-fold convolution

$$G(f) = U(f) ** \ldots * \prod_{i=1}^{n} \delta\left(f_i - f_{i0}\right) \tag{8.28}$$

Equations (8.25) and (8.26) define quasi-analytic signals if the value of the shift frequencies is sufficiently high enough to keep the leakage parameter (8.2) small. Observe that in both cases, we have a single quasi-analytic signal with single orthant spectrum. For this reason, there is no advantage for applying the hypercomplex version with regard to the complex one.

Example 8.4

The 2-D Gaussian pulse is given by the formula

$$g(x_1,x_2) = \frac{1}{2\pi\sigma_1\sigma_2\sqrt{1-\rho^2}}\exp\left[\frac{\left(x_1/\sigma_1\right)^2 + \left(x_2/\sigma_2\right)^2 + 2\rho\left(x_1/\sigma_1\right)\left(x_2/\sigma_2\right)}{2\left(1-\rho^2\right)}\right]$$

(8.29)

and its low-pass spectrum displayed in Figure 8.3(a) is

$$G(f_1,f_2) = \exp\left\{-0.5\left[\left(2\pi f_1\sigma_1\right)^2 + \left(2\pi f_2\sigma_2\right)^2 + 8\pi^2\rho f_1 f_2\sigma_1\sigma_2\right]\right\} \qquad (8.30)$$

Figure 8.3(a) shows the low-pass spectrum of a Gaussian signal and Figure 8.3(b) the shifted spectrum $G(f_1 - 1.25, f_2 - 1.25)$. The presented Gaussian signal and its spectrum have only even-even and odd-odd terms (see Chapter 3). Therefore, the Fourier spectrum is a real function. It should be noted that the corresponding quaternion Fourier spectrum is a complex function.

References

[1]   Hahn, S. L., and K. M. Snopek, "Quasi-analytic multidimensional signals," *Bulletin of the Polish Academy Sciences*, Vol. 61, No.4, 2013, pp. 1017–1024.

[2]   Hahn, S. L., *Hilbert Transforms in Signal Processing*, Norwood, MA: Artech House, 1996.

[3]   Hahn, S. L., "Multidimensional complex signals with single-orthant spectra," *Proc. IEEE*, Vol. 80, No. 8, 1992, pp. 1287–1300.

[4]   Voelcker, H.B., "Toward a unified theory of modulation, part I: Phase-envelope relationships," *Proc. IEEE*, Vol. 54, No. 3, March 1966, pp. 340–353.

9

Space-Frequency Representations of n-D Complex and Hypercomplex Analytic Signals

In this chapter, we recall notions of the *Wigner distribution* (WD) and the corresponding *Woodward's ambiguity function* (AF) called *space-frequency representations* of n-D signals. Section 9.1 is devoted to WDs and AFs of complex signals (including analytic signals). In Section 9.2, the corresponding space-frequency distributions of quaternion and monogenic analytic signals are described. In Section 9.3 we present the idea of *double-dimensional distributions* introduced by Hahn and Snopek in 2002. Section 9.4 presents a short description of some practical applications of WDs and AFs in the domain of signal processing.

9.1 Wigner Distributions and Woodward Ambiguity Functions of Complex Analytic Signals

The Wigner distribution of n-D complex functions was introduced into quantum mechanics in 1932 by the American physicist of Hungarian origin, Eugene

Paul Wigner [1]. Originally, the Wigner distribution was defined as a $2n$-D function of positions and velocities of elementary particles in n-D space. In 1948, Jean Ville redefined the WD for the 1-D Gabor's analytic signal, and this distribution is often referred to as the *Wigner-Ville distribution* [2]. Some important works concerning the study of properties of WDs of continuous- and discrete-time complex signals appeared in the 1980s [3–5]. The detailed description of different time-frequency distributions belonging to the Cohen's class is presented in [6]. Research concerning space-frequency representation of n-D complex analytic signals with single-orthant spectra [7] is the subject of [8–12].

Let us consider the n-D complex signal $\psi(\boldsymbol{x})$, $\boldsymbol{x} = (x_1,x_2,\ldots,x_n)$ and its complex FT $\Psi(\boldsymbol{f})$, $\boldsymbol{f} = (f_1,f_2,\ldots,f_n)$. We introduce the n-D shift variable $\boldsymbol{\chi} = (\chi_1,\chi_2,\ldots,\chi_n)$ in the $\boldsymbol{x}$-domain (signal domain) and define the $2n$-D signal correlation product as a function

$$r(\boldsymbol{x},\boldsymbol{\chi}) = \psi(\boldsymbol{x} + \boldsymbol{\chi}/2)\psi^*(\boldsymbol{x} - \boldsymbol{\chi}/2) \tag{9.1}$$

The $2n$-D Wigner distribution $W_\psi(\boldsymbol{x},\boldsymbol{f})$ of $\psi(\boldsymbol{x})$ is a n-D Fourier transform of (9.1) with regard to $\boldsymbol{\chi}$:

$$\text{direct } 2n\text{-D FT:} \quad W_\psi(\boldsymbol{x},\boldsymbol{f}) = \int\ldots\int r(\boldsymbol{x},\boldsymbol{\chi})\, e^{-j2\pi(\chi_1 f_1+\ldots+\chi_n f_n)} d\boldsymbol{\chi}$$
$$\text{inverse } 2n\text{-D FT:} \quad r(\boldsymbol{x},\boldsymbol{\chi}) = \int\ldots\int W_\psi(\boldsymbol{x},\boldsymbol{f})\, e^{j2\pi(\chi_1 f_1+\ldots+\chi_n f_n)} d\boldsymbol{f} \tag{9.2}$$

The WD can be calculated equivalently in the frequency domain as the inverse n-D Fourier transform of the $2n$-D spectrum correlation product

$$R(\boldsymbol{\mu},\boldsymbol{f}) = \Psi(\boldsymbol{f} + \boldsymbol{\mu}/2)\Psi^*(\boldsymbol{f} - \boldsymbol{\mu}/2) \tag{9.3}$$

$$r(\boldsymbol{x},\boldsymbol{\chi}) \overset{2nF}{\Leftrightarrow} R(\boldsymbol{\chi},\boldsymbol{f}) \tag{9.4}$$

and $\Psi(\boldsymbol{f})$ is the n-D Fourier spectrum of $\psi(\boldsymbol{x})$ and $\boldsymbol{\mu} = (\mu_1,\mu_2,\ldots,\mu_n)$ is the frequency-domain shift variable, that is,

$$\text{inverse } 2n\text{-D FT:} \quad W_\psi(\boldsymbol{x},\boldsymbol{f}) = \int\ldots\int R(\boldsymbol{\mu},\boldsymbol{f})\, e^{j2\pi(\mu_1 x_1+\ldots+\mu_n x_n)} d\boldsymbol{\mu}$$
$$\text{direct } 2n\text{-D FT:} \quad R(\boldsymbol{\mu},\boldsymbol{f}) = \int\ldots\int W_\psi(\boldsymbol{x},\boldsymbol{f})\, e^{-j2\pi(\mu_1 x_1+\ldots+\mu_n x_n)} d\boldsymbol{x} \tag{9.5}$$

Let us recall the definitions of 2n-D AF of n-D complex signals introduced by Snopek in 1999 [13] and further exploited in [14, 15]. The 2n-D AF is defined as the n-D inverse Fourier transform of the correlation product (9.1) calculated with regard to the signal-domain variable $\mathbf{x} = (x_1,\ldots,x_n)$:

inverse 2n-D FT:
$$A_\psi(\boldsymbol{\mu},\boldsymbol{\chi}) = \int\ldots\int r(\mathbf{x},\boldsymbol{\chi})\, e^{j2\pi(\mu_1 x_1+\ldots+\mu_n x_n)}d\mathbf{x}$$

direct 2n-D FT:
$$r(\mathbf{x},\boldsymbol{\chi}) = \int\ldots\int A_\psi(\boldsymbol{\mu},\boldsymbol{\chi})\, e^{-j2\pi(\mu_1 x_1+\ldots+\mu_n x_n)}d\boldsymbol{\mu}$$

$$(9.6)$$

It is a function of the n-D frequency shift variable (frequency lag) $\boldsymbol{\mu}$ and the n-D signal shift variable (signal-domain lag) $\boldsymbol{\chi}$.

Similar to the 2n-D WD, the 2n-D AF can be defined equivalently in the frequency domain as the n-D Fourier transform of (9.3) with regard to $\boldsymbol{f}$:

direct 2n-D FT:
$$A_\psi(\boldsymbol{\mu},\boldsymbol{\chi}) = \int\ldots\int R(\boldsymbol{\mu},\boldsymbol{f})\, e^{-j2\pi(\chi_1 f_1+\ldots+\chi_n f_n)}d\boldsymbol{f}$$

inverse 2n-D FT:
$$R(\boldsymbol{\mu},\boldsymbol{f}) = \int\ldots\int A_\psi(\boldsymbol{\mu},\boldsymbol{\chi})\, e^{j2\pi(\chi_1 f_1+\ldots+\chi_n f_n)}d\boldsymbol{\chi}$$

$$(9.7)$$

It should be noted that there is a closed relation between the 2n-D WDs and AFs. Let us introduce into (9.5) the second relation of (9.6) as follows

$$W_\psi(\mathbf{x},\mathbf{f}) = \int\ldots\int\left[\int\ldots\int A_\psi(\boldsymbol{\mu},\boldsymbol{\chi})e^{-j2\pi(\mu_1 x_1+\ldots+\mu_n x_n)}\,d\boldsymbol{\mu}\right]e^{-j2\pi(\chi_1 f_1+\ldots+\chi_n f_n)}d\boldsymbol{\chi}$$

$$(9.8)$$

Changing the order of integration in (9.8) we obtain

$$W_\psi(\mathbf{x},\mathbf{f}) = \iint\ldots\iint A_\psi(\boldsymbol{\mu},\boldsymbol{\chi})\, e^{-j2\pi(\mu_1 x_1+\ldots+\mu_n x_n+\chi_1 f_1+\ldots+\chi_n f_n)}d\boldsymbol{\chi}d\boldsymbol{\mu}$$

$$A_\psi(\boldsymbol{\mu},\boldsymbol{\chi}) = \iint\ldots\iint W_\psi(\mathbf{x},\mathbf{f})\, e^{j2\pi(\mu_1 x_1+\ldots+\mu_n x_n+\chi_1 f_1+\ldots+\chi_n f_n)}d\boldsymbol{f}d\mathbf{x}$$

$$(9.9)$$

The formulas (9.9) show that the 2n-D WD and 2n-D AF form a pair of Fourier transforms are presented schematically in Figure 9.1. Solid arrows represent direct and inverse n-D Fourier transforms $F_{(\cdot)}$ with regard to the n-D variable in the subscript. So, F_χ is the Fourier relation (9.2) between the signal-domain correlation product (9.1) and the 2n-D WD. Then, F_μ defined the n-D Fourier relation (9.5) between the frequency-domain correlation product (9.3) and the WD given by (9.5). F_x and F_f represent the Fourier relations (9.6) and (9.7), respectively. The dotted arrows represent 2n-D Fourier relations defined by (9.4) and (9.9), respectively.

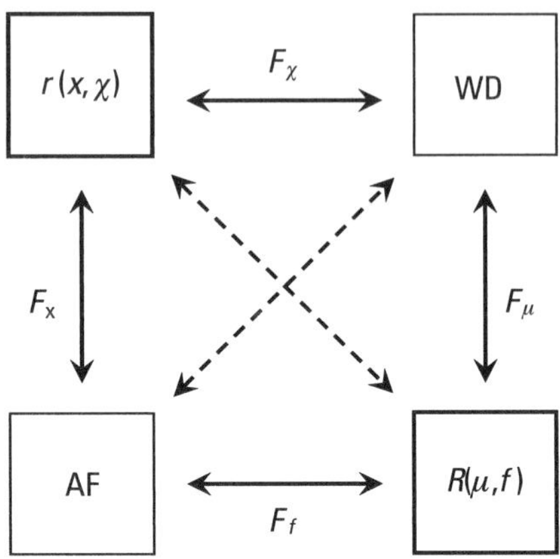

Figure 9.1 Fourier relations between correlation products $r(x,\chi)$ and $R(\mu,f)$, $2n$-D WDs and AFs. F_χ is the n-D direct FT of (9. 2), F_μ is the n-D inverse FT of (9.5), F_x is the n-D inverse FT of (9.6), F_f is the n-D direct FT of (9.7). Dotted arrows define $2n$-D Fourier relations (9.4) and (9.9).

9.1.1 WDs and AFs of 1-D Signals

Consider a 1-D real or complex signal $\psi(t)$, $t \in \mathbb{R}$, and its 1-D complex Fourier transform $\Psi(f)$, $f \in \mathbb{R}$. We introduce two shift variables: τ in the time domain (time lag), and μ in the frequency domain (frequency lag).

The *time-frequency Wigner distribution* of ψ, $W_\psi(t,f)$, is defined as the Fourier transform (with regard to τ) of the time-domain correlation product

$$r(t,\tau) = \psi(t + \tau/2)\psi^*(t - \tau/2) \tag{9.10}$$

Hence, we have

$$\begin{aligned}
\text{direct FT:} \quad & W_\psi(t,f) = \int r(t,\tau)e^{-j2\pi f\tau}\, d\tau \\
\text{inverse FT:} \quad & r(t,\tau) = \int W_\psi(t,f)e^{j2\pi f\tau}\, df
\end{aligned} \tag{9.11}$$

It should be noted that the WD of a real or complex signal $\psi(t)$ is always a *real valued* function. Equivalently, the WD is defined in the frequency domain as the inverse FT of the frequency-domain correlation product

$$R(\mu, f) = \Psi\left(f + \mu/2\right)\Psi^*\left(f - \mu/2\right) \tag{9.12}$$

where

$$r(t,\tau)\overset{2F}{\Longleftrightarrow}R(\mu,f) \tag{9.13}$$

We have next two Fourier relations

$$\begin{aligned}
\text{inverse FT:} \quad & W_\psi(t,f)=\int R(\mu,f)\,e^{j2\pi\mu t}\,d\mu \\[2mm]
\text{direct FT:} \quad & R(\mu,f)=\int W_\psi(t,f)e^{-j2\pi\mu t}\,dt
\end{aligned} \tag{9.14}$$

Let us show the equivalence of definitions (9.11) and (9.14).

Proof: We substitute in (9.11): $t+\tau/2=\alpha$ and get

$$W_\psi(t,f)=2e^{j4\pi ft}\int\psi(\alpha)\psi^*(2t-\alpha)e^{-j4\pi f\alpha}\,d\alpha$$

Then, bearing in mind that a FT of a product of two signals is a convolution of the corresponding spectra and $\psi^*(2t-\alpha)\overset{F_\alpha}{\Longleftrightarrow}\Psi^*(2f)e^{-j4\pi(2f)t}$, we obtain

$$\begin{aligned}
W_\psi(t,f)&=2e^{j4\pi ft}\left[\Psi(2f)*\Psi^*(2f)e^{-j4\pi(2f)t}\right]\\[2mm]
&=2e^{j4\pi ft}\int\Psi(\beta)\Psi^*(2f-\beta)e^{-j4\pi(2f-\beta)t}\,d\beta
\end{aligned}$$

Substituting in the above equation $\beta=f+\mu/2$, we get the definition (9.14).

The Wigner distribution is closely related to the *Woodward's ambiguity function* $A_\psi(\mu,\tau)$, which is commonly used in radar and sonar imaging to track the distance and velocity of a moving object [16]. The AF is defined in two equivalent ways: as the inverse FT (with regard to t) of (9.10) and the FT (with regard to f) of (9.12), as follows

$$\begin{aligned}
\text{inverse FT:} \quad & A_\psi(\mu,\tau)=\int r(t,\tau)e^{j2\pi\mu t}\,dt \\[2mm]
\text{direct FT:} \quad & r(t,\tau)=\int A_\psi(\mu,\tau)e^{j2\pi\mu t}\,d\mu
\end{aligned} \tag{9.15}$$

$$\begin{aligned}
\text{direct FT:} \quad & A_\psi(\mu,\tau)=\int R(\mu,f)e^{-j2\pi f\tau}\,df \\[2mm]
\text{inverse FT:} \quad & R(\mu,f)=\int A_\psi(\mu,\tau)e^{j2\pi f\tau}\,d\tau
\end{aligned} \tag{9.16}$$

Note that the formula (9.15) defines the AF in *symmetrical form* [17] and that the original Woodward's definition applies a different sign in the exponent [16]. The AF of a real or complex signal $\psi(t)$ can be a *complex valued* function. Bearing in mind (9.11), (9.14), (9.15), and (9.16), we can derive the following 2-D Fourier relations between WD and AF

$$A_\psi(\mu,\tau) = \iint W_\psi(t,f)\, e^{-j2\pi(f\tau+\mu t)} dt df$$

$$W_\psi(t,f) = \iint A_\psi(\mu,\tau)\, e^{j2\pi(f\tau+\mu t)} d\mu d\tau$$

(9.17)

For even signals, $\psi(t) = \psi(-t)$, we get

$$A_\psi(\mu,\tau) = \tfrac{1}{2} W_\psi(\tau/2,\mu/2)$$

(9.18)

and for odd signals, $\psi(t) = -\psi(-t)$:

$$A_\psi(\mu,\tau) = -\tfrac{1}{2} W_\psi(\tau/2,\mu/2)$$

(9.19)

Figure 9.2 shows Fourier relations between time- and frequency-domain correlation products, the 2-D Wigner distribution, and the 2-D Woodward's ambiguity function. Solid arrows represent direct and inverse 1-D Fourier transforms with regard to the variable in the subscript. So, F_τ is the Fourier relation (9.11) between the time-domain correlation product (9.10) and the WD. Then, F_μ defines the Fourier relation between the frequency-domain

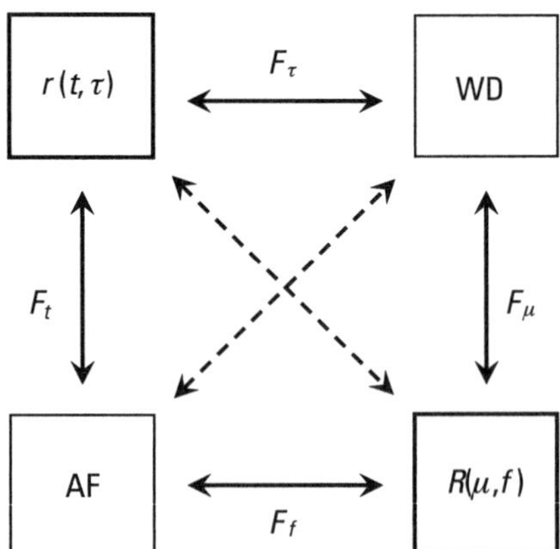

Figure 9.2   Fourier relations between correlation products, 2-D WDs and AFs. F_τ is the direct FT of (9.11), F_μ is the inverse FT of (9.14), F_t is the inverse FT of (9.15), and F_f is the direct FT of (9.16). Dotted arrows define 2-D Fourier relations (9.13) and (9.17).

correlation product (9.12) and the WD given by (9.14). F_t and F_f represent Fourier relations (9.15) and (9.16), respectively. The dotted arrows represent 2-D Fourier relations defined by (9.13) and (9.17), respectively.

In Tables 9.1 and 9.2, we collect basic properties of the 2-D WDs and AFs. They are found in many publications, such as [3–5, 18–24].

Table 9.1
Properties of the WD of 1-D Signals

	Property	Signal	Wigner Distribution
1	Time shift	$\psi(t - t_0)$	$W_\psi(t - t_0, f)$
2	Modulation	$\psi(t)e^{j2\pi f_0 t}$	$W_\psi(t, f - f_0)$
3	Time shift and modulation	$\psi(t - t_0)e^{j2\pi f_0 t}$	$W_\psi(t - t_0, f - f_0)$
4	Time scaling	$\psi(at)$	$\dfrac{1}{|a|}W_\psi\left(at, \dfrac{f}{a}\right)$
5	Time Marginal (zero moment)	$m_0(t) = \int W_\psi(t,f)\,df = |\psi(t)|^2$	
6	Frequency marginal	$\int W_\psi(t,f)\,dt = |\Psi(f)|^2$	
7	Parseval's equality	$\iint W_\psi(t,f)\,dt\,df = \int|\psi(t)|^2 dt = \int|\Gamma(f)|^2 df = E_\psi$	
8	Time support conservation	$\psi(t) \neq 0$ for $t \in \mathbb{D}_t,\ \mathbb{D}_t \subset \mathbb{R}$	$W_\psi(t,f) \neq 0$ for $t \in \mathbb{D}_t,\ \mathbb{D}_t \subset \mathbb{R},\ f \in \mathbb{R}$
9	Frequency support conservation	$\Psi(f) \neq 0$ for $f \in \mathbb{D}_f,\ \mathbb{D}_f \subset \mathbb{R}$	$W_\psi(t,f) \neq 0$ for $t \in \mathbb{R},\ f \in \mathbb{D}_f,\ \mathbb{D}_f \subset \mathbb{R}$
10	First moment	$m_1(t) = \int f W_\psi(t,f)\,df$	
11	Instantaneous frequency	$f(t) = \dfrac{m_1(t)}{m_0(t)} = \dfrac{\int f W_\psi(t,f)\,df}{\int W_\psi(t,f)\,df}$	
12	Instantaneous complex frequency	$s(t) = \dfrac{1}{2}\dfrac{\dot{m}_0(t)}{m_0(t)} + 2\pi\dfrac{m_1(t)}{m_0(t)} \cdot j$	

Table 9.2
Properties of the AF of 1-D Signals

	Property	Signal	Woodward's Ambiguity Function
1	Symmetry	$A_\psi^*(\mu,\tau) = A_\psi(-\mu,-\tau)$	
2	Time shift	$\psi(t-t_0)$	$A_\psi(\mu,\tau)\,e^{j2\pi\mu t_0}$
3	Modulation	$\psi(t)e^{j2\pi f_0 t}$	$A_\psi(\mu,\tau)\,e^{j2\pi f_0 \tau}$
4	Time shift and modulation	$\psi(t-t_0)e^{j2\pi f_0 t}$	$A_\psi(\mu,\tau)\,e^{j2\pi(f_0\tau+\mu t_0)}$
5	Time scaling	$\psi(at)$	$\dfrac{1}{|a|}A_\psi\left(\dfrac{\mu}{a},a\tau\right)$
6	Time marginal	$\int A_\psi(\mu,\tau)d\mu = \left|\psi\left(\dfrac{\tau}{2}\right)\right|^2$	
7	Frequency marginal	$\int A_\psi(\mu,\tau)d\tau = \left|\Psi\left(\dfrac{\mu}{2}\right)\right|^2$	
8	Price's formula [24]	$\iint |A(\mu,\tau)|^2 d\tau d\mu = |A(0,0)|^2 = \left[\int|\psi(t)|^2 dt\right]^2 = E_\psi^2$	
9	Self-transform (Siebert's theorem [22, 23])	$\iint |A_\psi(\mu,\tau)|^2 e^{j2\pi(\mu\chi_t+\tau\mu_t)}d\tau d\mu = \left|A_\psi\left[-\mu_t,\chi_t\right]\right|^2$	
10	Maximum value	$\max A_\psi(\mu,\tau) = A_\psi(0,0) = E_\psi$	

9.1.1.1   Auto- and Cross-Terms of WDs and AFs

The WDs and AFs are *quadratic* (nonlinear) transformations of a signal. These nonlinear functions also produce—besides a number of *autoterms*—unwanted *cross-terms*, which can make the analysis difficult.

Cross-Terms Reduction

There are a lot of different methods of reduction of unwanted cross-terms in Wigner distributions. They were introduced to signal processing in the 1990s and concerned only 1-D signals. The most important works in this field are [17, 25, 26]. The first step in all methods is to replace a real signal in (9.11) with the corresponding analytic signal [2, 35]. Because the analytic signal has a one-sided spectrum support, its WD is also one-sided, and there are no cross-terms resulting from the interaction between spectrum components at positive and negative frequencies. The next step is to introduce in (9.11) a specific kernel function. This idea was used in time-domain and frequency-domain pseudo-Wigner distributions and in smoothed pseudo-Wigner distributions [3, 17].

Number of Cross-Terms

Let us consider the WD of a sum of k components, $\psi(t) = \sum_{i=1}^{k} \psi_i(t)$. It is expressed as a sum:

$$W_\psi(t,f) = \sum_{i=1}^{k} W_i(t,f) + 2\sum_{i=1}^{k-1}\sum_{j=i+1}^{k} \mathrm{Re}\left\{W_{ij}(t,f)\right\} \tag{9.20}$$

in which $W_i(t,f)$ are WDs of ψ_i (*autoterms*) and $W_{ij}(t,f)$ are cross-Wigner distributions (*cross-terms*) defined as

$$W_i(t,f) = \int \psi_i\left(t+\frac{\tau}{2}\right)\psi_i^*\left(t-\frac{\tau}{2}\right)e^{-j2\pi f\tau}\,d\tau \tag{9.21}$$

$$W_{ij}(t,f) = \int \psi_i\left(t+\frac{\tau}{2}\right)\psi_j^*\left(t-\frac{\tau}{2}\right)e^{-j2\pi f\tau}\,d\tau \tag{9.22}$$

The number of autoterms is k and the number of cross-terms is $k(k-1)/2$. Let us illustrate this problem with some examples.

Consider a multicomponent signal $\psi(t)$ in the form of a sum of $\psi_i(t) = u(t-t_i)e^{j2\pi f_i t}$, $i = 1, \ldots, k$ (i.e., time-shifted and modulated signals $u(t)$) [20]. Its WD contains k autoterms centered at (t_i,f_i) and, for each pair of autoterms, a cross-term is situated midway between them (compare with Figure 9.3 for a bicomponent signal). Correspondingly, in the AF, all autoterms are focused in the center of the (μ,τ)-plane, while the cross-terms are mirrored with regard to the origin (0,0) (as in Figure 9.4 for a bicomponent signal).

We start with a simple cosine signal of frequency f_0, $\cos(2\pi f_0 t)$ and its analytic form $e^{j2\pi f_0 t}$.

Example 9.1　Cosine and Harmonic Exponential (Analytic) Signals

Consider the real signal $u(t) = \cos(2\pi f_0 t) = \frac{1}{2}\left(e^{j2\pi f_0 t} + e^{-j2\pi f_0 t}\right)$. Its WD is given by $W_u(t,f) = \frac{1}{4}\left[\mathbf{1}(t) \otimes \delta(f-f_0) + \mathbf{1}(t) \otimes \delta(f+f_0) + 2\cos(4\pi f_0 t)\delta(f)\right]$ (the symbol

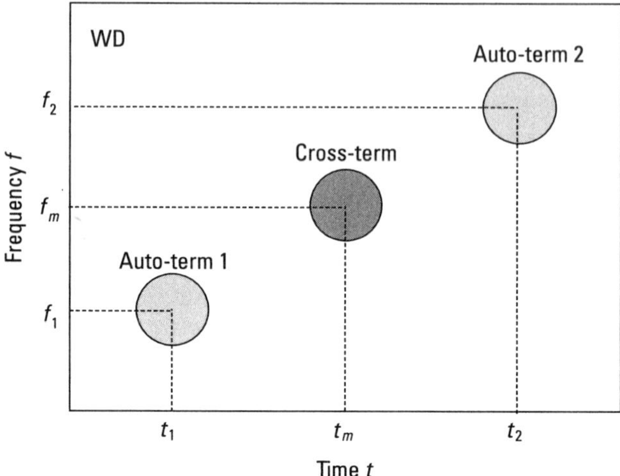

Figure 9.3 Geometry of auto- and cross-terms of the WD of bicomponent signals.

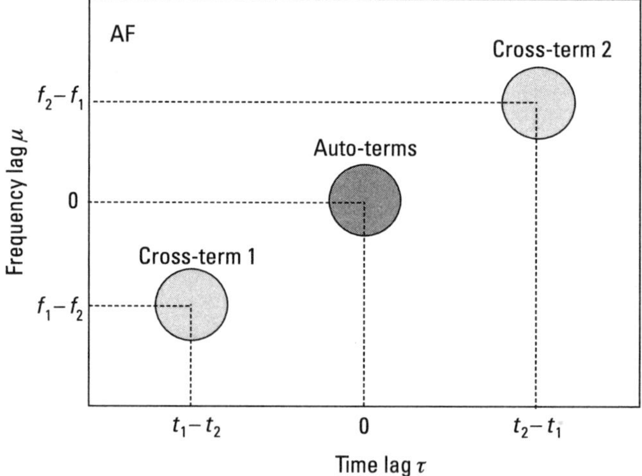

Figure 9.4 Geometry of auto- and cross-terms of the AF of bicomponent signals.

$\otimes$ denotes the multiplication of distributions). Figure 9.5(a) shows $W_u(t,f)$ for $f_0 = 10$. We have two autoterms in the form of *delta planes* at $f = \pm10$ and a cosine modulated cross-term at zero frequency. If we replace u with its analytic form $\psi(t) = e^{j2\pi f_0 t}$, we get the WD in the form of a single *delta plane* $W_u(t,f) = \mathbf{1}(t) \otimes \delta(f - f_0), f_0 = 10$ (Figure 9.5(b)). There are no cross-terms.

The next example is a generalized version of Example 9.1.

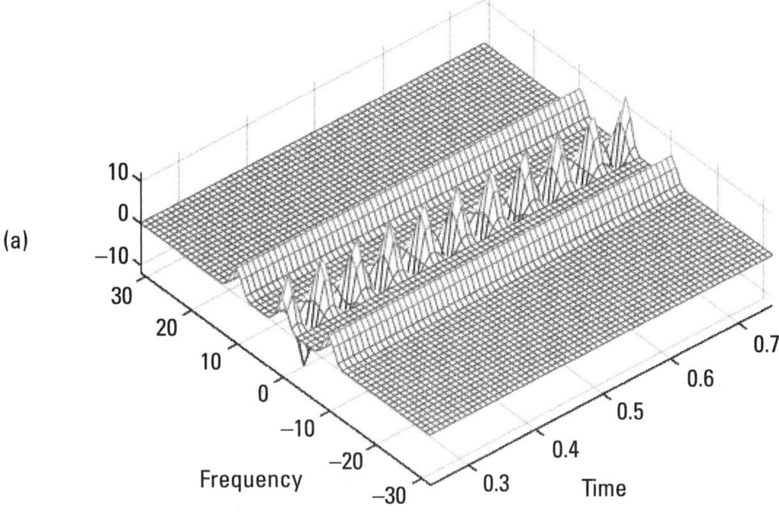

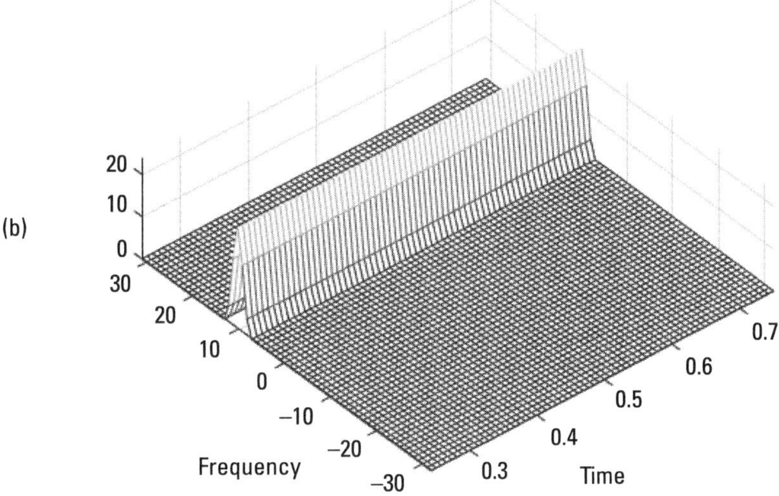

Figure 9.5 (a) The WD of the real cosine signal $u(t) = \cos(2\pi f_0 t)$, $f_0 = 10$. Autoterms: *delta planes* at ±10, cross-term: oscillating cosine-shaped *plane* at zero frequency. (b) The WD of the analytic signal $\psi(t) = \exp(j2\pi f_0 t)$, $f_0 = 10$. Autoterm: *delta plane* at $f = 10$. No cross-terms.

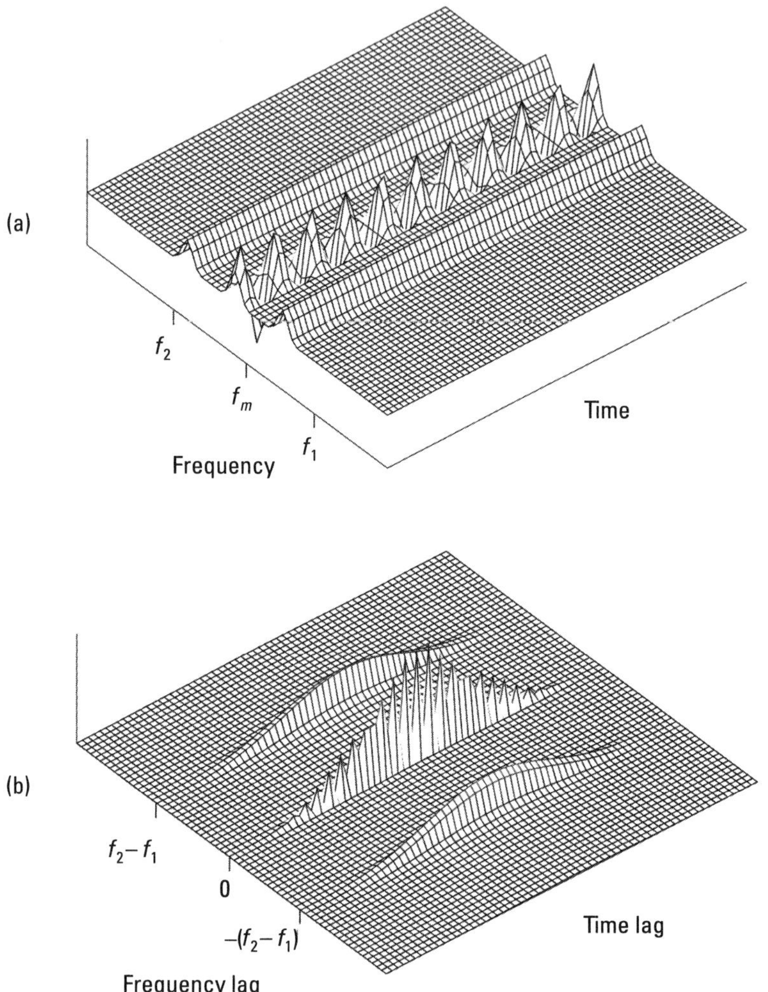

Figure 9.6   Sum of two harmonic exponential signals. (a) WD—autoterms at f_1 and f_2, cross-term at mid-frequency f_m. (b) $|A(\mu,\tau)|$—autoterms at $\mu = 0$, cross-terms at $\mu = \pm(f_2 - f_1)$.

Example 9.2 Sum of Two Harmonic Exponential Signals

Consider a signal $\psi(t) = e^{j2\pi f_1 t} + e^{j2\pi f_2 t}$, $f_2 > f_1$. Its WD is a sum of two autoterms in the form of *delta planes* situated at frequencies $f = f_1$, $f = f_2$ and a *cosine* cross-term in the form of a plane at the mid-frequency $f_m = (f_1 + f_2)/2$. The frequency of the cosine is $f_2 - f_1$. We have

$$W_{\psi}(t,f) = \mathbf{1}(t) \otimes \delta\left(f - f_1\right) + \mathbf{1}(t) \otimes \delta\left(f - f_2\right) + 2\cos\left[2\pi\left(f_2 - f_1\right)t\right]\delta\left(f - f_m\right)$$

The AF of this signal is a sum of two oscillating autoterms mapping the origin of the (μ,τ)-plane and two cross-terms at $\mu = \pm(f_2 - f_1)$, that is,

$$A_{\psi}(\mu,\tau) = \delta(\mu)\left(e^{j2\pi f_1\tau} + e^{j2\pi f_2\tau}\right) + \left[\delta\left(\mu - \left(f_2 - f_1\right)\right) + \delta\left(\mu + \left(f_2 - f_1\right)\right)\right]e^{j2\pi f_m\tau}$$

Figure 9.6(b) shows the magnitude of $A_{\psi}(\mu,\tau)$.

Example 9.3 Sum of Time-Shifted and Modulated Gaussian Signals

Figure 9.7 shows the Wigner distribution of a sum of three $(k = 3)$ time-shifted and modulated Gaussian pulses, $\psi(t) = \sum_{i=1}^{3} A_i / \sqrt{2\pi}\sigma_i \, e^{-(t-t_i)^2/2\sigma_i^2}$, where all $A_i = 1$, $\sigma_i = 1/30$, $t_1 = 0.3$, $t_2 = 0.5$, $t_3 = 0.7$, and $f_1 = f_3 = 20$, $f_2 = -20$. There are three Gaussian autoterms centered at (0.3,20), (0.5,−20), (0.7,20), and three cross-terms between them (compare with Figure 9.3).

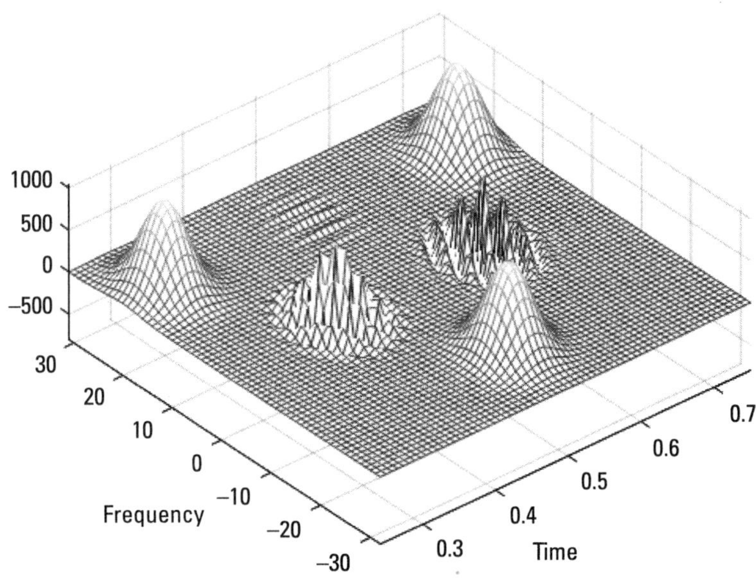

Figure 9.7   The WD of a sum of three time-shifted and modulated Gaussian pulses. Autoterms: three Gaussian pulses, cross-terms: bipolar oscillating components between the autoterms.

9.1.2 WDs and AFs of 2-D Complex Signals

4-D Wigner Distribution

The 4-D Wigner distribution of a 2-D complex signal $\psi(x_1,x_2)$ is a 4-D *real-valued* function given by the 2-D FT with regard to $\pmb{\chi} = (\chi_1,\chi_2)$ (compare with (9.2)):

$$\text{direct 2-D FT:} \quad W_\psi\left(x_1,x_2,f_1,f_2\right) = \iint r(\pmb{x},\pmb{\chi})e^{-j2\pi\left(f_1\chi_1+f_2\chi_2\right)}d\pmb{\chi}$$

$$\text{inverse 2-D FT:} \quad r(\pmb{x},\pmb{\chi}) = \iint W_\psi\left(x_1,x_2,f_1,f_2\right)e^{j2\pi\left(f_1\chi_1+f_2\chi_2\right)}d\pmb{f} \tag{9.23}$$

where the correlation product (compare with (9.1)) is defined as

$$r(\pmb{x},\pmb{\chi}) = \psi\left(x_1 + \frac{\chi_1}{2}, x_2 + \frac{\chi_2}{2}\right)\psi^*\left(x_1 - \frac{\chi_1}{2}, x_2 - \frac{\chi_2}{2}\right) \tag{9.24}$$

Equivalently, the 4-D WD is a 2-D inverse FT with regard to $\pmb{\mu} = (\mu_1,\mu_2)$:

$$\text{inverse 2-D FT:} \quad W_\psi(x_1,x_2,f_1,f_2) = \iint R(\pmb{\mu},\pmb{f})e^{j2\pi\left(\mu_1x_1+\mu_2x_2\right)}d\pmb{\mu}$$

$$\text{direct 2-D FT:} \quad R(\pmb{\mu},\pmb{f}) = \iint W_\psi(x_1,x_2,f_1,f_2)e^{-j2\pi\left(\mu_1x_1+\mu_2x_2\right)}d\pmb{x}$$

$$\tag{9.25}$$

where the 4-D spectrum correlation product is

$$R\left(\mu_1,\mu_2,f_1,f_2\right) = \Psi\left(f_1 + \frac{\mu_1}{2}, f_2 + \frac{\mu_1}{2}\right)\Psi^*\left(f_1 - \frac{\mu_1}{2}, f_2 - \frac{\mu_1}{2}\right) \tag{9.26}$$

The equivalence of definitions (9.23) and (9.25) can be proved in a similar way as for the 2-D WD.

Proof: Let us introduce new variables $x_1 + \chi_1/2 = \alpha_1$, $x_2 + \chi_2/2 = \alpha_2$ into (9.23) giving

$$W_\psi(\pmb{x},\pmb{f})$$

$$= 4\,e^{j4\pi\left(f_1x_1+f_2x_2\right)}\iint \psi(\alpha_1,\alpha_2)\psi^*\left(2x_1 - \alpha_1, 2x_2 - \alpha_2\right)e^{-j4\pi\left(f_1\alpha_1+f_2\alpha_2\right)}d\alpha_1 d\alpha_2$$

The above integral is a 2-D FT of a product of two functions equal to a double convolution of their spectra, that is

$$W_\psi(\pmb{x},\pmb{f}) = 4e^{j4\pi(f_1 x_1 + f_2 x_2)}\left[\Psi\left(2f_1, 2f_2\right) **\Psi^*\left(2f_1, 2f_2\right)e^{-j4\pi(2f_1 x_1 + 2f_2 x_2)}\right]$$

or equivalently

$$W_\psi(\pmb{x},\pmb{f})$$

$$= 4e^{j4\pi(f_1 x_1 + f_2 x_2)}\iint\Psi\left(y_1,y_2\right)\Psi^*\left(2f_1 - y_1, 2f_2 - y_2\right)e^{-j4\pi\left[(2f_1 - y_1)x_1 + (2f_2 - y_2)x_2\right]}dy_1 dy_2$$

A second change of variables $y_1 = f_1 + \mu_1/2$, $y_2 = f_2 + \mu_2/2$ yields (9.25).

4-D Woodward's Ambiguity Function

The 4-D Woodward's ambiguity function [8] is defined as

$$\text{inverse 2-D FT:}\quad A_\psi\left(\mu_1,\mu_2,\chi_1,\chi_2\right) = \iint r(\pmb{x},\pmb{\chi})e^{j2\pi(\mu_1 x_1 + \mu_2 x_2)}d\pmb{x}$$

$$\text{direct 2-D FT:}\quad r(\pmb{x},\pmb{\chi}) = \iint A_\psi\left(\mu_1,\mu_2,\chi_1,\chi_2\right)e^{j2\pi(\mu_1 x_1 + \mu_2 x_2)}d\pmb{\mu} \tag{9.27}$$

or equivalently:

$$\text{direct 2-D FT:}\quad A_\psi\left(\mu_1,\mu_2,\chi_1,\chi_2\right) = \iint R(\pmb{\mu},\pmb{f})e^{-j2\pi(\chi_1 f_1 + \chi_2 f_2)}d\pmb{f}$$

$$\text{inverse 2-D FT:}\quad R(\pmb{\mu},\pmb{f}) = \iint A_\psi\left(\mu_1,\mu_2,\chi_1,\chi_2\right)e^{j2\pi(\chi_1 f_1 + \chi_2 f_2)}d\pmb{\chi} \tag{9.28}$$

Whereas the 4-D WD is real-valued, the 4-D AF can be a *complex-valued* function.

Example 9.4 The 2-D Complex Harmonic Signal

Let us consider the 2-D complex signal $\psi(x_1,x_2) = e^{j2\pi(f_a x_1 + f_b x_2)}$. Its WD is a real 4-D function given by $W_\psi(x_1,x_2,f_1,f_2) = \mathbf{1}(x_1,x_2) \otimes \delta(f_1 - f_a, f_2 - f_b)$ where $\mathbf{1}(x_1,x_2)$ is a unit distribution equal to 1 for all $\pmb{x} = (x_1,x_2)$.

The AF of $\psi(x_1,x_2)$ is $A_\psi(\mu_1,\mu_2,\chi_1,\chi_2) = e^{j2\pi(f_a \chi_1 + f_b \chi_2)} \cdot \delta(\mu_1,\mu_2) \otimes \mathbf{1}(\chi_1,\chi_2)$ where $\mathbf{1}(\chi_1,\chi_2)$ is a unit distribution equal to 1 for all $\pmb{\chi} = (\chi_1,\chi_2)$. We see that the AF is in general a complex function.

The Fourier relations between the 4-D WDs and AFs (compare with (9.9)) are as follows

$$\text{direct 4-D FT:}\quad W_\psi(\pmb{x},\pmb{f}) = \iint\iint A_\psi(\pmb{\mu},\pmb{\chi})\, e^{-j2\pi\left(\chi_1 f_1 + \mu_1 x_1 + \chi_2 f_2 + \mu_2 x_2\right)}d\pmb{\chi}d\pmb{\mu}$$

$$\text{inverse 4-D FT:}\quad A_\psi(\pmb{\mu},\pmb{\chi}) = \iint\iint W_\psi(\pmb{x},\pmb{f})\, e^{j2\pi\left(\chi_1 f_1 + \mu_1 x_1 + \chi_2 f_2 + \mu_2 x_2\right)}d\pmb{x}d\pmb{f}$$

$$\tag{9.29}$$

Chosen Properties of 4-D WDs and AFs

The properties presented in this section are generalizations of the properties of time-frequency distributions collected in Tables 9.1 and 9.2. Let us enumerate some of them:

1. The 4-D WD is always a real valued function, while the corresponding 4-D AF is in general complex.

2. A shift of a 2-D signal in the signal domain by a vector $[x_a, x_b]$ leads to the same shift of its WD:

$$\psi\left(x_1 - x_a, x_2 - x_b\right) \overset{WD}{\leftrightarrow} W_\psi\left(x_1 - x_a, x_2 - x_b, f_1, f_2\right) \tag{9.30}$$

3. A shift of a 2-D signal in the signal domain by a vector $[x_a, x_b]$ leads to the frequency modulation of its AF with regard to (μ_1, μ_2):

$$\psi\left(x_1 - x_a, x_2 - x_b\right) \overset{AF}{\leftrightarrow} A_\psi\left(\mu_1, \mu_2, \chi_1, \chi_2\right) e^{j2\pi(\mu_1 x_a + \mu_2 x_b)} \tag{9.31}$$

4. A modulation of a 2-D signal leads to the frequency-shift of its 4-D WD:

$$\psi\left(x_1, x_2\right) e^{j2\pi(f_a x_1 + f_b x_2)} \overset{WD}{\leftrightarrow} W_\psi\left(x_1, x_2, f_1 - f_a, f_2 - f_b\right) \tag{9.32}$$

5. A modulation of a 2-D signal leads to the frequency-shift of its 4-D AF with regard to (χ_1, χ_2):

$$\psi\left(x_1, x_2\right) e^{j2\pi(f_a x_1 + f_b x_2)} \overset{AF}{\leftrightarrow} A_\psi\left(\mu_1, \mu_2, \chi_1, \chi_2\right) e^{j2\pi(f_a \chi_1 + f_b \chi_2)} \tag{9.33}$$

6. Signal-domain marginal: integration of the WD over f yields the signal energy density of $\psi(x)$:

$$\iint W_\psi\left(x_1, x_2, f_1, f_2\right) df_1\, df_2 = \left|\psi\left(x_1, x_2\right)\right|^2 \tag{9.34}$$

7. Frequency marginal: integration of the WD over x yields the energy density spectrum of $\psi(x)$:

$$\iint W_\psi\left(x_1, x_2, f_1, f_2\right) dx_1\, dx_2 = \left|\Psi\left(f_1, f_2\right)\right|^2 \tag{9.35}$$

8. Parseval's equality:

$$\iint \iint W_\psi\left(x_1,x_2,f_1,f_2\right)dx\,df = \iint |\psi(x)|^2\,dx = \iint |\Psi(f)|^2\,df = E_\psi \quad (9.36)$$

9. The 4-D WD preserves the signal- and frequency-domain supports of a signal. The WD of a single-quadrant analytic signal has a single-quadrant support in the frequency domain.
10. The 4-D WDs and AFs of separable signals $\psi(x_1,x_2) = \psi_a(x_1)\psi_b(x_2)$ are also separable.

9.1.3 WDs and AFs of 2-D Complex Analytic Signals

Since WDs of 2-D signals are 4-D functions, we can study their properties using suitable *cross sections*. For example, for fixed signal coordinates $x_1 = x_{10}$ and $x_2 = x_{20}$, the cross section $W_\psi(x_{10},x_{20},f_1,f_2)$ represents the local spectrum of the signal. Alternatively, choosing $f_1 = f_{10}$ and $f_2 = f_{20}$, we get another cross section $W_\psi(x_1,x_2,f_{10},f_{20})$. It is possible to analyze mixed cross sections of the form $W_\psi(x_{10},x_2,f_{10},f_2)$, $W_\psi(x_1,x_{20},f_1,f_{20})$, and so on. The same applies for 4-D AFs.

9.1.3.1 WDs of 2-D Complex Analytic Signals

Bearing in mind item 9 from the preceding subsection, we can formulate the following property concerning cross sections of the 4-D WD:

Any cross section of the 4-D Wigner distribution of a 2-D analytic signal has a single-quadrant support.

Let us illustrate this property with cross sections $W_{\psi_1}(0,0,f_1,f_2)$ and $W_{\psi_3}(0,0,f_1,f_2)$ of the 4-D WDs of an analytic separable Cauchy signal and a 2-D modulated real band-pass signal with a Gaussian envelope (a quasi-analytic signal).

Example 9.5 Separable 2-D Cauchy Analytic Signal

The 2-D real Cauchy signal is given by

$$u\left(x_1,x_2\right) = \frac{a}{a^2 + x_1^2}\frac{b}{b^2 + x_2^2} \quad (9.37)$$

and its spectrum is

$$U\left(f_1,f_2\right) = \exp\left[-2\pi\left(a|f_1| + b|f_2|\right)\right] \quad (9.38)$$

where $a, b \in \mathbb{R}$ are signal parameters. Applying the formula (5.32) and Table 5.1 (see Chapter 5), we obtain two analytic signals

$$\psi_1(x_1, x_2) = F^{-1}\left\{4 \cdot \mathbf{1}(f_1, f_2) \cdot U(f_1, f_2)\right\} \tag{9.39}$$

$$\psi_3(x_1, x_2) = F^{-1}\left\{4 \cdot \mathbf{1}(f_1, -f_2) \cdot U(f_1, f_2)\right\} \tag{9.40}$$

with $U(f_1, f_2)$ given by (9.38). Figure 9.8 shows cross sections $W_{\psi_1}(0, 0, f_1, f_2)$ and $W_{\psi_3}(0, 0, f_1, f_2)$ for $a = 0.25$, $b = 0.5$. We observe the single-quadrant support of corresponding cross sections and their identical shape (a consequence of separability of the signal). Note that, in the nonseparable case, both cross sections would differ in shape but the single-quadrant support is still conserved.

Example 9.6 2-D Modulated Real Band-Pass Signal with a Gaussian Envelope
The test signal is given by

$$u(x_1, x_2)$$

$$= e^{-\pi\left(x_1^2 + x_2^2\right)}\left[A_1\cos\left(2\pi f_{11}x_1\right)\cos\left(2\pi f_{12}x_2\right) + A_2\cos\left(2\pi f_{21}x_1\right)\cos\left(2\pi f_{22}x_2\right)\right]$$

$$\tag{9.41}$$

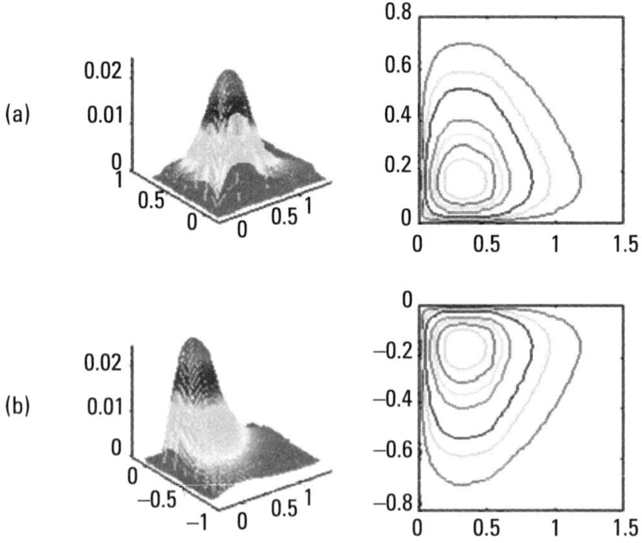

Figure 9.8 Cross sections of WDs of the separable Cauchy signal (9.54): (a) $W_{\psi_1}(0, 0, f_1, f_2)$ and (b) $W_{\psi_3}(0, 0, f_1, f_2)$.

The corresponding 2-D complex analytic signals are

$$\psi_1\left(x_1,x_2\right) \approx e^{-\pi\left(x_1^2+x_2^2\right)}\left[A_1 e^{j2\pi\left(f_{11}x_1+f_{12}x_2\right)} + A_2 e^{j2\pi\left(f_{21}x_1+f_{22}x_2\right)}\right] \quad (9.42)$$

$$\psi_3\left(x_1,x_2\right) \approx e^{-\pi\left(x_1^2+x_2^2\right)}\left[A_1 e^{j2\pi\left(f_{11}x_1-f_{12}x_2\right)} + A_2 e^{j2\pi\left(f_{21}x_1-f_{22}x_2\right)}\right] \quad (9.43)$$

Their spectra are

$$\Psi_1\left(f_1,f_2\right) = A_1 e^{-\pi\left[\left(f_1-f_{11}\right)^2+\left(f_2+f_{12}\right)^2\right]} + A_2 e^{-\pi\left[\left(f_1-f_{21}\right)^2+\left(f_2+f_{22}\right)^2\right]} \quad (9.44)$$

$$\Psi_3\left(f_1,f_2\right) = A_1 e^{-\pi\left[\left(f_1-f_{11}\right)^2+\left(f_2-f_{12}\right)^2\right]} + A_2 e^{-\pi\left[\left(f_1-f_{21}\right)^2+\left(f_2-f_{22}\right)^2\right]} \quad (9.45)$$

Note that if modulation frequencies f_{11}, f_{12}, f_{21} and f_{22} are sufficiently large, the above spectra have approximately single-quadrant supports (signals are quasi-analytic, as discussed in Chapter 8). The Wigner distributions of signals (9.42) and (9.43), respectively, are

$$W_{\psi_1}\left(x_1,x_2,f_1,f_2\right) = e^{-2\pi\left(x_1^2+x_2^2\right)}\, e^{-2\pi\left(f_1-f_{11}\right)^2}\, e^{-2\pi\left(f_2+f_{12}\right)^2} \quad (9.46)$$

$$W_{\psi_3}\left(x_1,x_2,f_1,f_2\right) = e^{-2\pi\left(x_1^2+x_2^2\right)}\, e^{-2\pi\left(f_1-f_{11}\right)^2}\, e^{-2\pi\left(f_2-f_{12}\right)^2} \quad (9.47)$$

Figure 9.9 shows the cross sections $W_{\psi_1}(0,0,f_1,f_2)$ and $W_{\psi_3}(0,0,f_1,f_2)$ of analytic signals corresponding to (9.41), $A_1 = 1$, $A_2 = 0$, $f_{11} = f_{12} = 1.5$. We observe that they are represented by *logons* (called also *atoms* in [17]) with a single-quadrant support in quadrants 1 and 3, respectively.

Example 9.7 Sum of Two Band-Pass Signals with a Gaussian Envelope

Let us substitute in (9.41) $A_1 = A_2 = 1$, $f_{11} = f_{12} = 1.5$, $f_{21} = f_{22} = 3$. The WDs of complex analytic signals are displayed in Figure 9.10. Two logons representing autoterms are separated by a cross-term of doubled amplitude. The distribution W_{ψ_3} shown is a mirror image of W_{ψ_1} with regard to the line $f_2 = 0$. For separable signals, W_{ψ_1} and W_{ψ_3} contain the same spectral information.

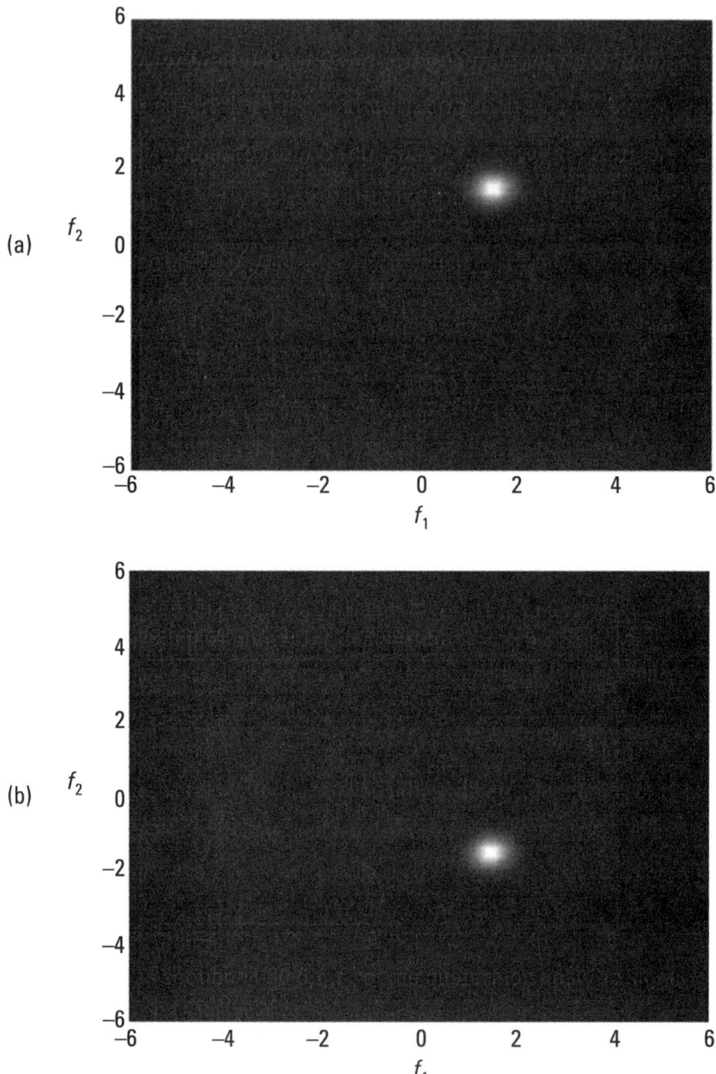

Figure 9.9 Cross sections of (a) $W_{\psi_1}(0,0,f_1,f_2)$ and (b) $W_{\psi_3}(0,0,f_1,f_2)$ of the 2-D modulated real band-pass (quasi-analytic) signal with a Gaussian envelope, $A_1 = 1$, $A_2 = 0$, $f_{11} = f_{12} = 1.5$.

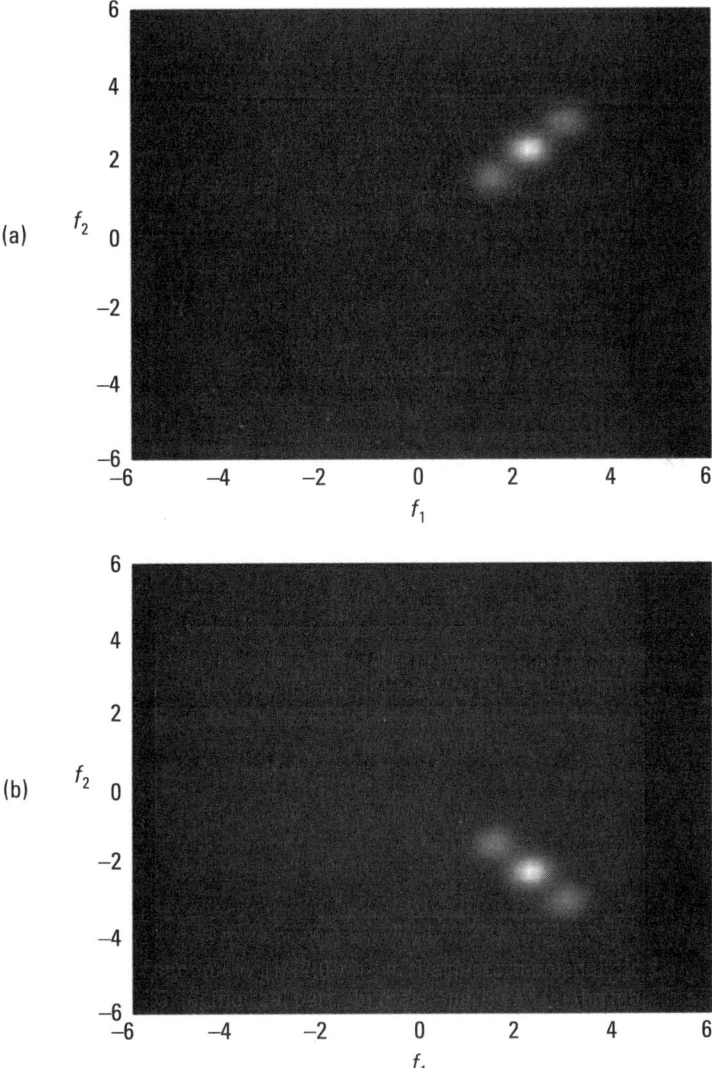

Figure 9.10 Cross sections (a) $W_{\psi_1}(0,0,f_1,f_2)$ and (b) $W_{\psi_3}(0,0,f_1,f_2)$ of analytic signals of (9.42) and (9.43), $A_1 = A_2 = 1$, $f_{11} = f_{12} = 1.5$, $f_{21} = f_{22} = 3$.

9.1.3.2　AFs of 2-D Complex Analytic Signals

The next property concerns cross sections of the AFs.

Any cross section of the 4-D Woodward's ambiguity function of a 2-D analytic signal is an analytic function.

This is a consequence of the fact that 4-D WDs and AFs are pairs of Fourier transforms (see (9.29)). We exploit the property concerning the single-quadrant support of the WD of analytic signals (see Section 9.1.3.1) and rewrite (9.29) as

$$
A_\psi\left(\mu_1,\mu_2,\chi_1,\chi_2\right)
$$

$$
= \int_{-\infty}^{\infty}\int_{-\infty}^{\infty}\left[\int_0^\infty\int_0^\infty W_\psi\left(x_1,x_2,f_1,f_2\right)e^{j2\pi\left(\chi_1 f_1+\chi_2 f_2\right)}\,df_1\,df_2\right]e^{j2\pi\left(\mu_1 x_1+\mu_2 x_2\right)}\,dx_1\,dx_2
$$

$$
(9.48)
$$

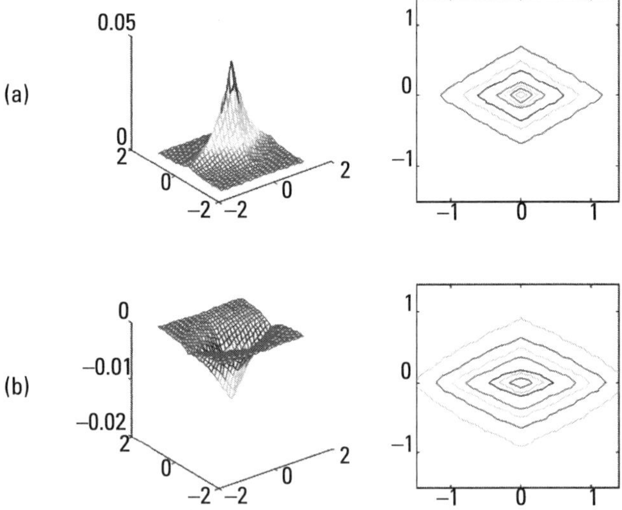

Figure 9.11　(a) Real and (b) imaginary parts of the cross section $A_{\psi_1}(\mu_1,\mu_2,0.1,0.1)$ of a 2-D separable Cauchy signal (9.37).

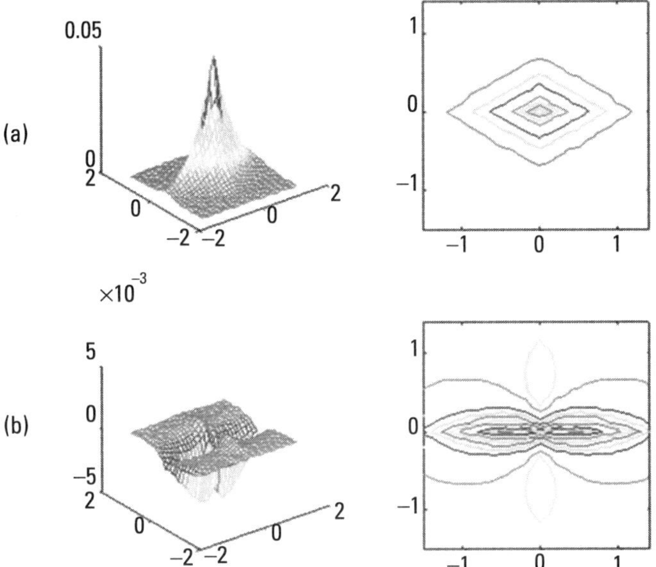

Figure 9.12   (a) Real and (b) imaginary parts of the cross section $A_{\psi_3}(\mu_1,\mu_2,0.1,0.1)$ of a 2-D separable Cauchy signal (9.37).

where

$$\psi_W\left(x_1,x_2,\chi_1,\chi_2\right)=\int_0^\infty\!\!\int_0^\infty W_\psi\left(x_1,x_2,f_1,f_2\right)e^{j2\pi\left(\chi_1 f_1+\chi_2 f_2\right)}\,df_1\,df_2 \qquad (9.49)$$

is a 2-D analytic function with regard to (χ_1,χ_2) (the inverse FT of a single-quadrant WD). More compactly,

$$A_\psi\left(\mu_1,\mu_2,\chi_1,\chi_2\right)=F_x^{-1}\left\{\psi_W\left(x_1,x_2,\chi_1,\chi_2\right)\right\} \qquad (9.50)$$

Figures 9.11 and 9.12 show real and imaginary parts of cross sections $A_{\psi_1}(\mu_1,\mu_2,0.1,0.1)$ and $A_{\psi_3}(\mu_1,\mu_2,0.1,0.1)$ of a separable 2-D Cauchy signal from Example 9.5. Note that, in the nonseparable case, both cross sections would be rotated in the (μ_1,μ_2)-plane.

Figure 9.13 shows the magnitudes of cross sections $A_{\psi_1}(\mu_1,\mu_2,0,0)$ and $A_{\psi_3}(\mu_1,\mu_2,0,0)$ of signals (9.42) and (9.43). Note their four-quadrant support in the domain $\boldsymbol{\mu}=(\mu_1,\mu_2)$ in comparison to the corresponding cross sections of WDs (see Figure 9.10).

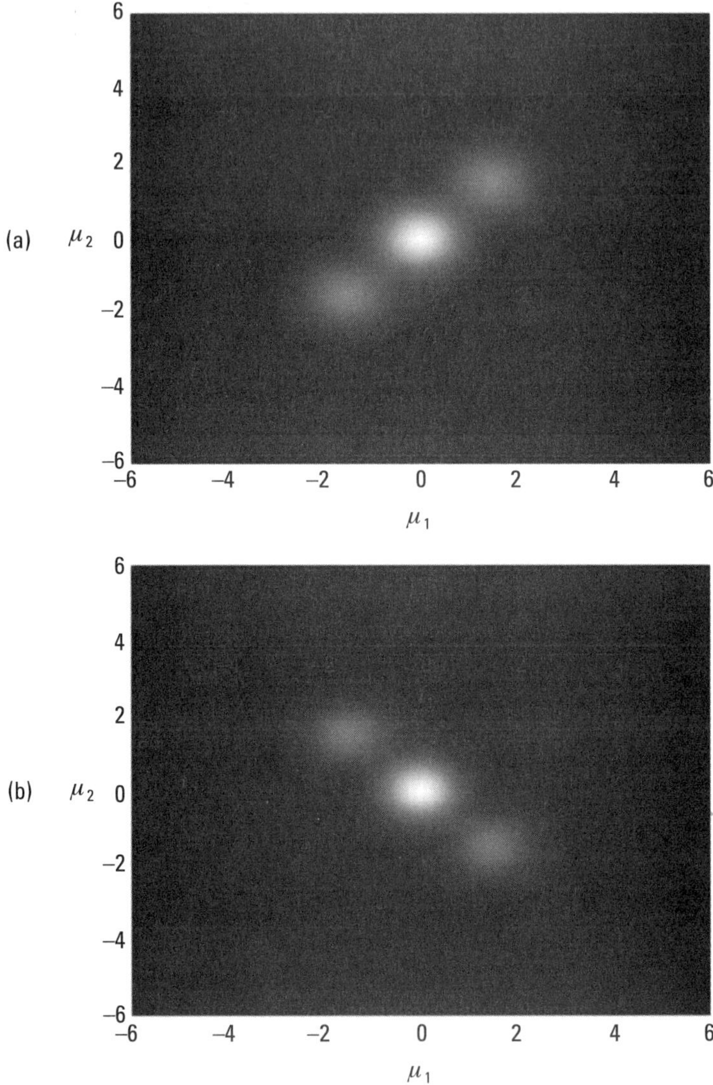

Figure 9.13   Magnitudes (a) $|A_{\psi_1}(\mu_1,\mu_2,0,0)|$ and (b) $|A_{\psi_3}(\mu_1,\mu_2,0,0)|$ of analytic signals of (9.42) and (9.43), $A_1 = A_2 = 1$, $f_{11} = f_{12} = 1.5$, $f_{21} = f_{22} = 3$ [15].

9.2 Wigner Distributions and Woodward Ambiguity Functions of Quaternion and Monogenic Signals

9.2.1 The WDs of Quaternion Signals

In [28], the definition of the 4-D Wigner distribution of 2-D quaternion signals was introduced. The *Quaternion Wigner Distribution* W_q is defined as a two-sided QFT (with regard to $\chi = (\chi_1, \chi_2)$) of a *quaternion correlation product*:

$$\rho_q\left(x_1, x_2, \chi_1, \chi_2\right) = \psi_q\left(x_1 + \frac{\chi_1}{2}, x_2 + \frac{\chi_2}{2}\right)\psi_q^*\left(x_1 - \frac{\chi_1}{2}, x_2 - \frac{\chi_2}{2}\right) \tag{9.51}$$

that is,

$$W_q(\boldsymbol{x}, f) = \iint e^{-e_1 2\pi f_1 \chi_1}\rho_q\left(x_1, x_2, \chi_1, \chi_2\right)e^{-e_2 2\pi f_2 \chi_2}\, d\boldsymbol{\chi} \tag{9.52}$$

It should be noted that it is also possible to define the Quaternion WD using the right-side QFT or left-side QFT of ρ_q (see Chapter 4, (4.50)). However, in the following sections we will keep the original definition [28]. The insertion of $\psi_{q_1}(x_1, x_2) = u + v_1 \cdot e_1 + v_2 \cdot e_2 + v \cdot e_3$ (Chapter 5, (5.78)) and its conjugate $\psi^*_{q_1}(x_1, x_2) = u - v_1 \cdot e_1 - v_2 \cdot e_2 - v \cdot e_3$ into (9.52) yields

$$\rho_q\left(x_1, x_2, \chi_1, \chi_2\right) = A + B \cdot e_1 + C \cdot e_2 + D \cdot e_3 \tag{9.53}$$

where

$$A = u^+u^- + v_1^+v_1^- + v_2^+v_2^- + v^+v^-,\ B = v^+v_2^- - u^+v_1^- + v_1^+u^- - v_2^+v^-$$

$$C = -u^+v_2^- - v^+v_1^- + v_1^+v^- + v_2^+u^-,\ D = v^+u^- - u^+v^- + v_2^+v_1^- - v_1^+v_2^-$$

$$u^+ = u\left(x_1 + \tfrac{\chi_1}{2}, x_2 + \tfrac{\chi_2}{2}\right),\ u^- = u\left(x_1 - \tfrac{\chi_1}{2}, x_2 - \tfrac{\chi_2}{2}\right)$$

and the same shortened notation applies to v, v_1, and v_2. The insertion of (9.53) into (9.52) yields a *quaternion-valued* WD:

$$W_q\left(x_1, x_2, f_1, f_2\right) = W_{q_r} + W_{q_1} \cdot e_1 + W_{q_2} \cdot e_2 + W_{q_3} \cdot e_3 \tag{9.54}$$

where

$$W_{q_r}\left(x_1, x_2, f_1, f_2\right) = \iint\left[Ac_1c_2 + Bs_1c_2 + Cc_1s_2 + Ds_1s_2\right]d\chi_1 d\chi_2 \tag{9.55}$$

$$W_{q_1}\left(x_1,x_2,f_1,f_2\right)=\iint\left[Bc_1c_2-As_1c_2+Dc_1s_2-Cs_1s\alpha_2\right]d\chi_1\,d\chi_2 \qquad (9.56)$$

$$W_{q_2}\left(x_1,x_2,f_1,f_2\right)=\iint\left[Cc_1c_2+D_qs_1c_2-Ac_1s_2-Bs_1s_2\right]d\chi_1\,d\chi_2 \qquad (9.57)$$

$$W_{q_3}\left(x_1,x_2,f_1,f_2\right)=\iint\left[Dc_1c_2-Cs_1c_2-Bc_1s_2+As_1s_2\right]d\chi_1\,d\chi_2 \qquad (9.58)$$

and $c_i=\cos(2\pi f_i x_i)$, $s_i=\sin(2\pi f_i x_i)$. Let us illustrate this part with examples of quaternion WDs of 2-D modulated real band-pass signals with a Gaussian envelope (compare with Examples 9.6 and 9.7).

Example 9.8 A 2-D Modulated Real Band-Pass Signal with a Gaussian Envelope
The quaternion (quasi-analytic) signal corresponding to (9.41) is

$$\psi_q\left(x_1,x_2\right)\approx e^{-\pi\left(x_1^2+x_2^2\right)}\left[A_1e^{e_1\,2\pi f_{11}x_1}e^{e_2\,2\pi f_{12}x_2}+A_2e^{e_1\,2\pi f_{21}x_1}e^{e_2\,2\pi f_{22}x_2}\right] \qquad (9.59)$$

Its W_q [29] is given by

$$\begin{aligned}
W_q\left(x_1,x_2,f_1,f_2\right)&=e^{-2\pi\left(x_1^2+x_2^2\right)}\left\{e^{-2\pi\left(f_1-f_{11}\right)^2}\left[e^{-2\pi\left(f_2-f_{12}\right)^2}+e^{-2\pi\left(f_2+f_{12}\right)^2}\right]\right\}\\
&+e^{-2\pi\left(x_1^2+x_2^2\right)}e^{e_1\,4\pi f_{11}x_1}\left\{e^{-2\pi f_1^2}\left[e^{-2\pi\left(f_2-f_{12}\right)^2}-e^{-2\pi\left(f_2+f_{12}\right)^2}\right]\right\}
\end{aligned} \qquad (9.60)$$

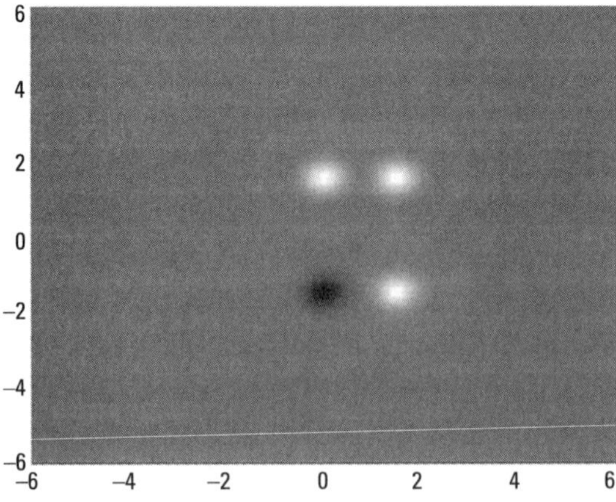

Figure 9.14   The cross section $W_q(0,0,f_1,f_2)$ of the quaternion analytic signal (9.59), $A_1=1$, $A_2=0$, $f_{11}=f_{12}=1.5$.

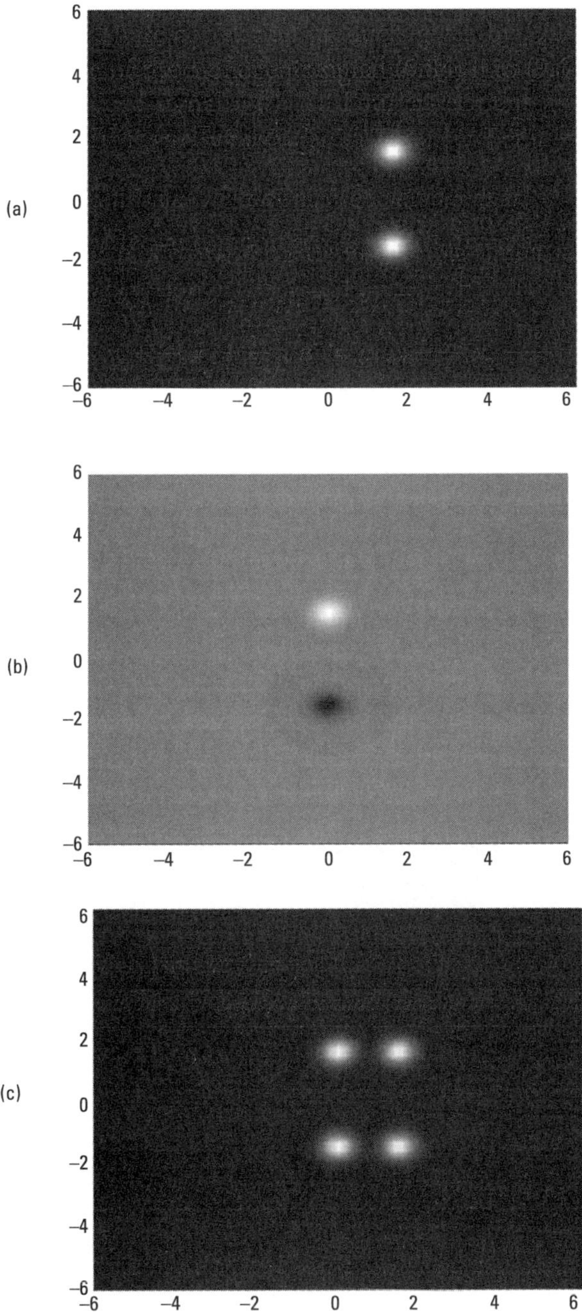

Figure 9.15   The cross section $W_q(1/12, 0, f_1, f_2)$ of the quaternion analytic signal (9.59), $A_1 = 1$, $A_2 = 0$, $f_{11} = f_{12} = 1.5$: (a) the real part, (b) the e_1-part, and (c) the magnitude.

It is known that, in general, the WDs of quaternion signals are quaternion-valued functions. Differently, the cross section $W_q(0,0,f_1,f_2)$ is a real function, since $\exp(e_1 4\pi f_{11} x_{10}) = 1$ (see Figure 9.14). Moreover, it has a two-quadrant support. We also observe a pair of bipolar cross-terms situated along the line $f_1 = 0$. Differently, the cross section $W_q(1/12,0,f_1,f_2)$, $f_1 = 1.5$ is a complex-valued function. Its real parts, the e_1–part and the magnitude, are shown in Figure 9.15(a)–(c).

Example 9.9 Sum of Two Band-Pass Signals with a Gaussian Envelope

Figure 9.16 shows the real-valued cross section $W_q(0,0,f_1,f_2)$ of the quaternion analytic signal (9.59), $A_1 = A_2 = 1$, $f_{11} = f_{12} = 1.5$. We observe a union of the WDs of Figure 9.10 and two groups of cross-terms of opposite polarity.

Figure 9.17(a)–(d) shows that $W_q(1/12,1/24,f_1,f_2)$ is a quaternion-valued function with all terms defined by (9.55)–(9.58). Its magnitude (Figure 9.18) corresponds to the absolute value of the cross section from Figure 9.17(a).

9.2.2   AFs of Quaternion Signals

The quaternion AF is defined as the inverse QFT of the correlation product (9.51) with regard to the signal variables $\boldsymbol{x} = (x_1,x_2)$ [15, 28]:

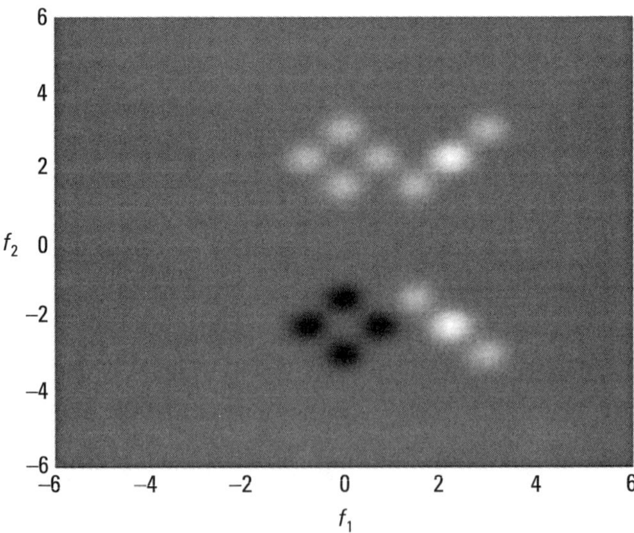

Figure 9.16   The cross section $W_q(0,0,f_1,f_2)$ of the quaternion analytic signal (9.59), $A_1 = A_2 = 1$, $f_{11} = f_{12} = 1.5$, $f_{21} = f_{22} = 3$.

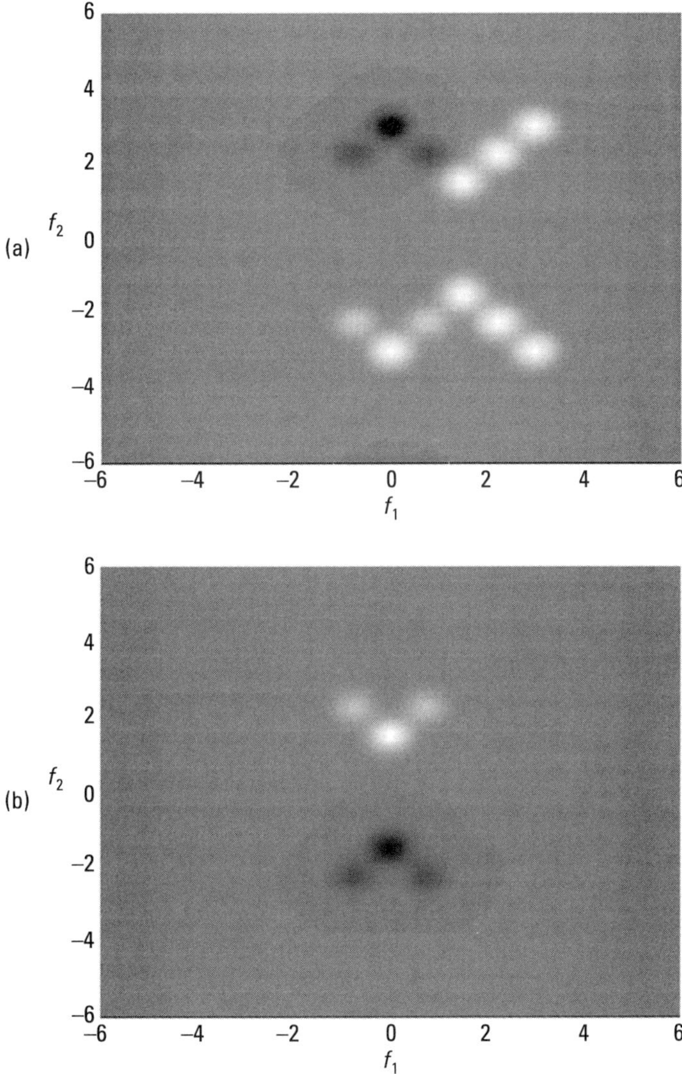

Figure 9.17   Cross sections $W_q(1/12, 1/24, f_1, f_2)$ of the quaternion analytic signal (9.59), $A_1 = A_2 = 1$, $f_{11} = f_{12} = 1.5$, $f_{21} = f_{22} = 3$: (a) the real part, (b) the e_1-part, (c) the e_2-part, and (d) the e_3-part.

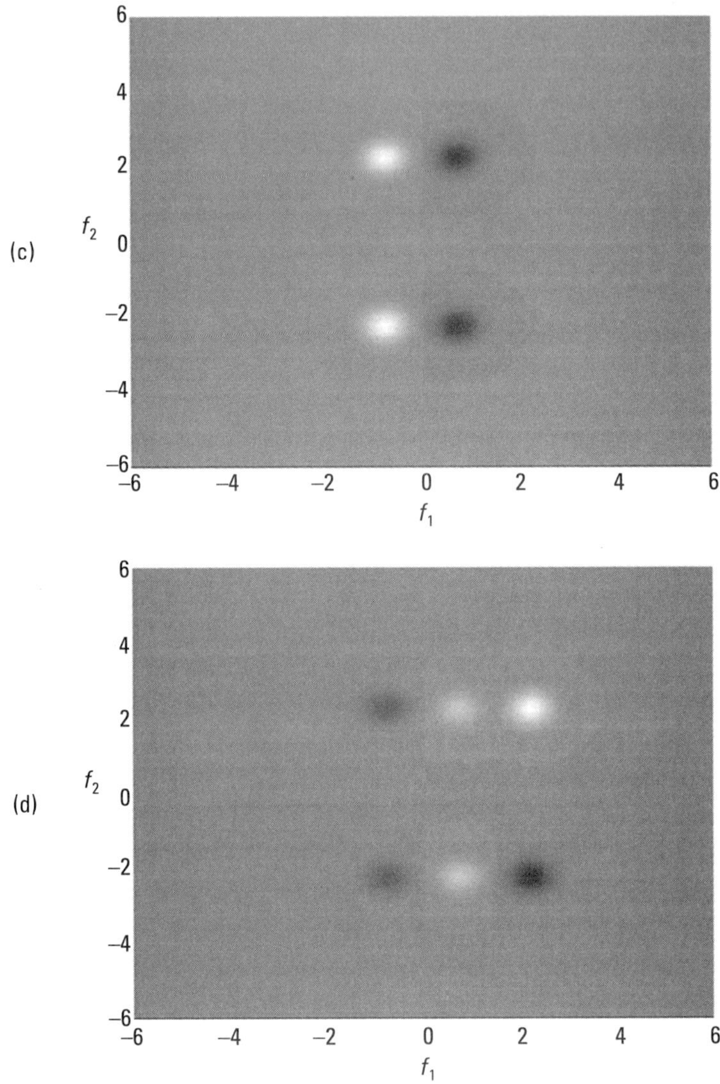

Figure 9.17 Continued

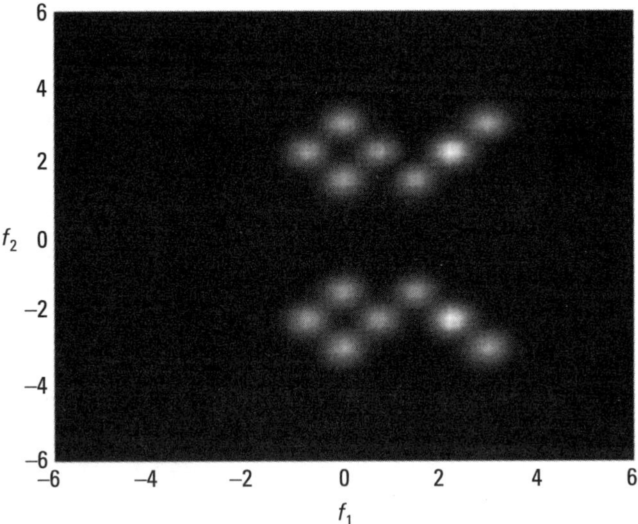

Figure 9.18   The magnitude of the real part of $W_q(1/12,1/24,f_1,f_2)$ of Figure 9.17(a).

$$A_q(\boldsymbol{\mu},\boldsymbol{\chi}) = \iint e^{-e_1 2\pi\mu_1 x_1} \rho_q\left(x_1,x_2,\chi_1,\chi_2\right) e^{-e_2 2\pi\mu_2 x_2}\, d\boldsymbol{x} \tag{9.61}$$

In general, the A_q is *a quaternion-valued* function. As the correlation product ρ_q is a 2-D inverse *QFT* of W_q given by (9.54), we derive the quaternion version of the relation between WD and AF. The substitution of the inverse 2-D *QFT* of (9.54) into (9.61) yields

$$A_q(\boldsymbol{\mu},\boldsymbol{\chi}) = \iint e^{-e_1 2\pi\mu_1 x_1}\left\{\iint e^{e_1 2\pi f_1\chi_1} W_q(\boldsymbol{x},\boldsymbol{f}) e^{e_2 2\pi f_2\chi_2}\, d\boldsymbol{f}\right\} e^{-e_2 2\pi\mu_2 x_2}\, d\boldsymbol{x} \tag{9.62}$$

Of course, other definitions of the quaternion AF are possible. If we change the signs in exponents of (9.61), we get the symmetric form of the two-sided quaternion AF. Other conventions are possible to apply as for example the left-side or right-side QFT (similarly to the quaternion WD described in Section 9.2.1).

Figure 9.19 displays magnitudes of two different cross sections of A_q of a quaternion quasi-analytic signal (9.59). We observe that $|A_q(\mu_1,\mu_2,0,0)|$ is a sum of $|A_1(\mu_1,\mu_2,0,0)|$ and $|A_3(\mu_1,\mu_2,0,0)|$ (see Figure 9.13(a) and (b), respectively). This summation yields the autoterm in the origin of a crest value four times bigger with regard to the cross-terms, which are located in all four quadrants and have the same crest value, as the cross-terms of $|A_1(\mu_1,\mu_2,0,0)|$

and $|A_3(\mu_1,\mu_2,0,0)|$. The previously described summation is expected bearing in mind, that the quaternion signal ψ_q replaces two analytic signals ψ_1 and ψ_3. Note that cross sections with nonzero shift coordinates (χ_1,χ_2) (Figure 9.19(b)) have a more complicated structure than $|A_q(\mu_1,\mu_2,0,0)|$.

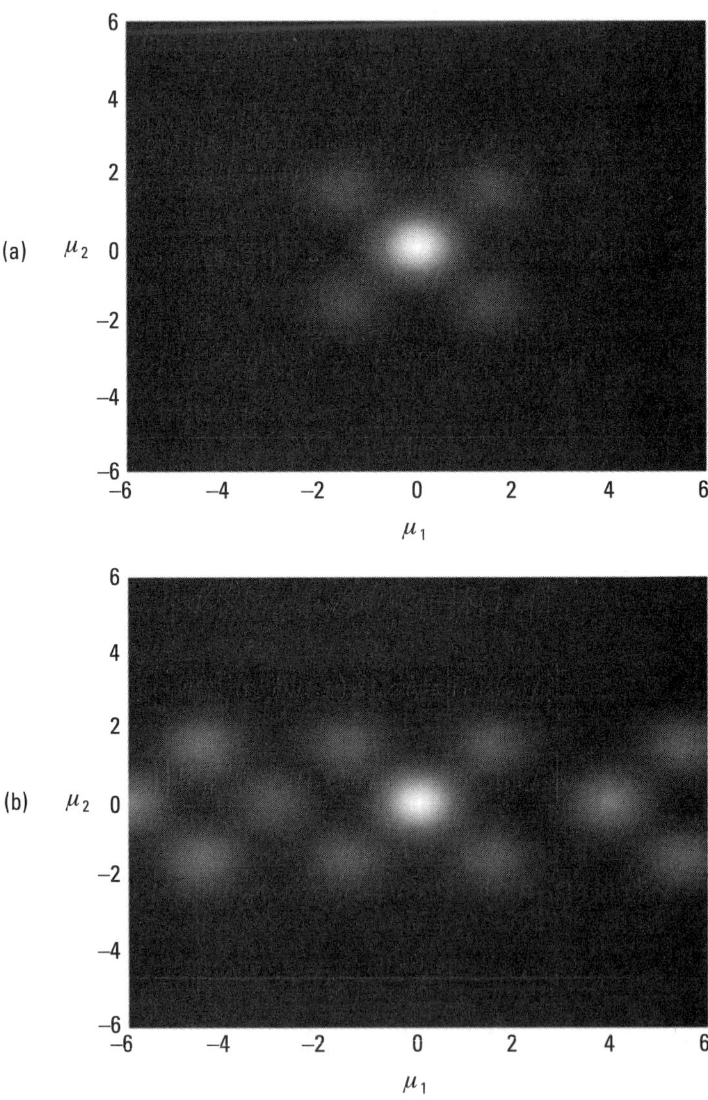

Figure 9.19　Magnitudes (a) $|A_q(\mu_1,\mu_2,0,0)|$ and (b) $|A_q(\mu_1,\mu_2,0.05,0.05)|$ of a quaternion quasi-analytic signal (9.59), $A_1 = A_2 = 1$, $f_{11} = f_{12} = 1.5$, $f_{21} = f_{22} = 3$ [15].

9.2.3 WDs of Monogenic Signals

The monogenic correlation product

$$\rho_M\left(x_1,x_2,\chi_1,\chi_2\right)=\psi_M\left(x_1+\frac{\chi_1}{2},x_2+\frac{\chi_2}{2}\right)\psi_M^*\left(x_1-\frac{\chi_1}{2},x_2-\frac{\chi_2}{2}\right)\qquad(9.63)$$

where the monogenic signal ψ_M and its conjugate are given by (5.84) (see Chapter 5). The Wigner distribution of the monogenic signal is defined as the two-sided *QFT* of (9.63)

$$W_M(\boldsymbol{x},\boldsymbol{f})=\iint e^{-e_1 2\pi f_1\chi_1}\rho_M\left(x_1,x_2,\chi_1,\chi_2\right)e^{-e_2 2\pi f_2\chi_2}\,d\boldsymbol{\chi}\qquad(9.64)$$

The substitution of the monogenic signal (5.84) into (9.64) yields

$$\rho_M\left(x_1,x_2,\chi_1,\chi_2\right)=A_M+B_M\cdot e_1+C_M\cdot e_2+D_M\cdot e_3\qquad(9.65)$$

where

$$A_M=u^+u^-+v_{r1}^+v_{r1}^-+v_{r2}^+v_{r2}^-,\; B_M=u^-v_{r1}^+-u^+v_{r1}^-,$$
$$C_M=u^-v_{r2}^+-u^+v_{r2}^-,\; D_M=v_{r2}^+v_{r1}^--v_{r1}^+v_{r2}^-,$$
$$u^+=u\left(x_1+\tfrac{\chi_1}{2},x_2+\tfrac{\chi_2}{2}\right),\; u^-=u\left(x_1-\tfrac{\chi_1}{2},x_2-\tfrac{\chi_2}{2}\right)$$

and the same applies for v_{r1} and v_{r2}. The W_M is also *a quaternion-valued* function. In contrast to W_q, the imaginary terms of cross sections $W_M(0,0,f_1,f_2)$ may not vanish, except in the case of distributions of selected band-pass signals.

Example 9.10 2-D Modulated Real Band-Pass Signal with a Gaussian Envelope
The monogenic signal corresponding to (9.41) is

$$\psi_M\left(x_1,x_2\right)\approx e^{-\pi\left(x_1^2+x_2^2\right)}\left\{A_1Q_1\left(x_1,x_2\right)+A_2Q_2\left(x_1,x_2\right)\right\}\qquad(9.66)$$

where the quaternion-valued function $Q_1(x_1,x_2)$ is given by

$$Q_1\left(x_1,x_2\right)=$$

$$\cos\left(2\pi f_{11}x_1\right)\cos\left(2\pi f_{12}x_2\right)+e_1\frac{f_{11}}{\sqrt{f_{11}^2+f_{12}^2}}\sin\left(2\pi f_{11}x_1\right)\cos\left(2\pi f_{12}x_2\right)\qquad(9.67)$$

$$+e_2\frac{f_{12}}{\sqrt{f_{11}^2+f_{12}^2}}\cos\left(2\pi f_{11}x_1\right)\sin\left(2\pi f_{12}x_2\right).$$

and $Q_2(x_1,x_2)$ has a form analogous to (9.67) if we change $f_{11} \rightarrow f_{21}$ and $f_{12} \rightarrow f_{22}$. Figure 9.20 shows the cross section $W_M(0,0,f_1,f_2)$ as a real-valued function containing the same spectral information as W_q of Figure 9.16. However, logons exist in all four quadrants and the cross-terms are also duplicated in some places. The cross section $W_M(1/12,1/12,f_1,f_2)$ (Figure 9.21) is a quaternion-valued function with the magnitude corresponding to Figure 9.20.

9.2.4 AFs of Monogenic Signals

The 4-D AF of monogenic signals is a function of frequency-lag and spatial-lag variables $(\mu_1, \mu_2, \chi_1, \chi_2)$. It is given by the inverse QFT of the correlation product (9.63) with regard to spatial variables $\boldsymbol{x} = (x_1,x_2)$. We have

$$A_M(\boldsymbol{\mu},\boldsymbol{\chi}) = \iint e^{-e_1 2\pi\mu_1 x_1} P_M\left(x_1,x_2,\chi_1,\chi_2\right) e^{-e_2 2\pi\mu_2 x_2}\, d\boldsymbol{x} \qquad (9.68)$$

and A_M is *a quaternion-valued* function [15, 28]. The relation (9.63) holds also for A_M and W_M.

Figure 9.22 displays the magnitudes of two different cross sections $|A_M(\mu_1,\mu_2,0,0)|$ and $|A_M(\mu_1,\mu_2,0.1,0.1)|$ of the monogenic signal (9.66). Differently to the magnitude $|A_q(\mu_1,\mu_2,0,0)|$ shown in Figure 9.19(a), the function $|A_M(\mu_1,\mu_2,0,0)|$ contains duplicates of the cross-terms shifted horizontally and

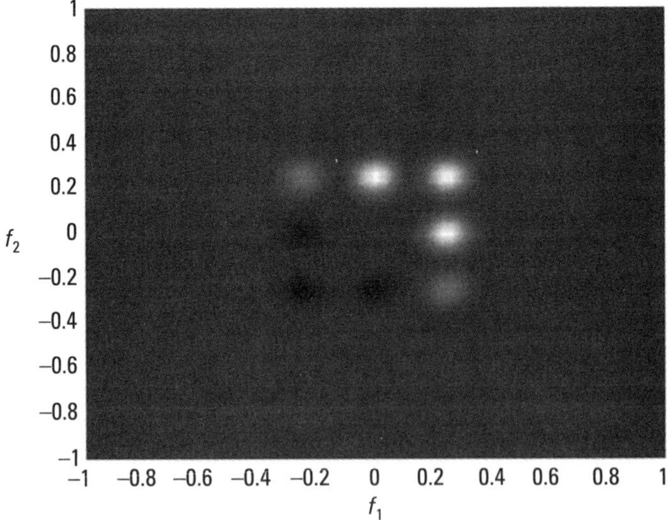

Figure 9.20 The cross section $W_M(0,0,f_1,f_2)$ of the monogenic signal (9.66), $A_1 = 1$, $A_2 = 0$, $f_{11} = f_{12} = 1.5$, $f_{21} = f_{22} = 3$.

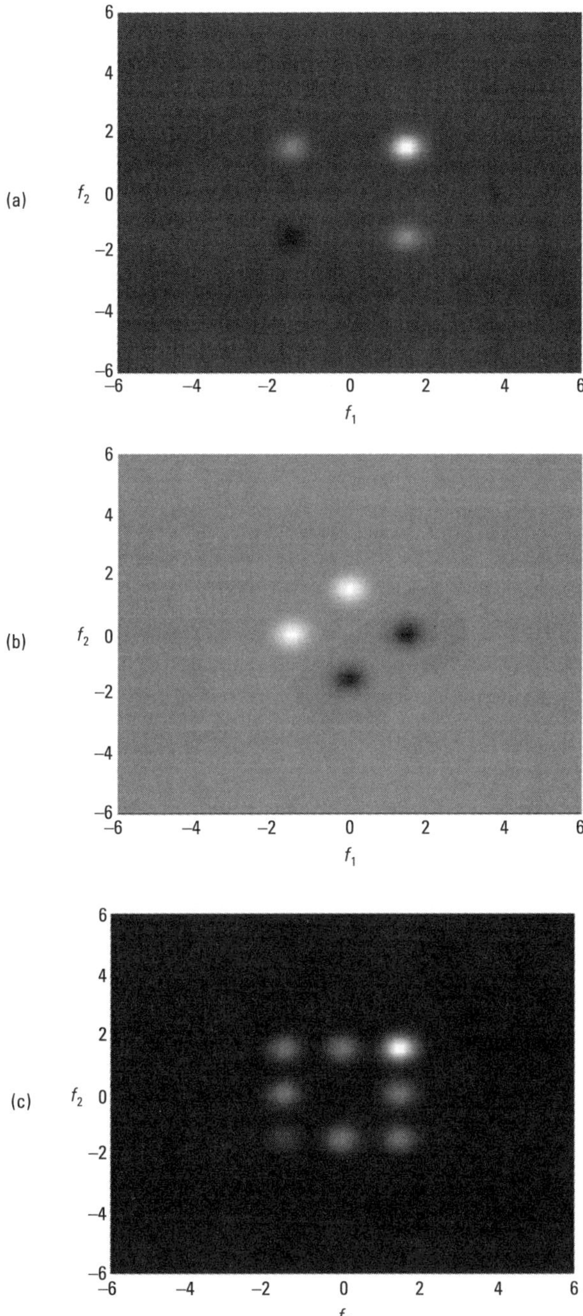

Figure 9.21   The cross section $W_M(1/12,0,f_1,f_2)$ of the monogenic signal (9.66), $A_1 = 1$, $A_2 = 0$, $f_{11} = f_{12} = 1.5$, $f_{21} = f_{22} = 3$: (a) real part, (b) e_1- and e_2-parts together, (c) magnitude.

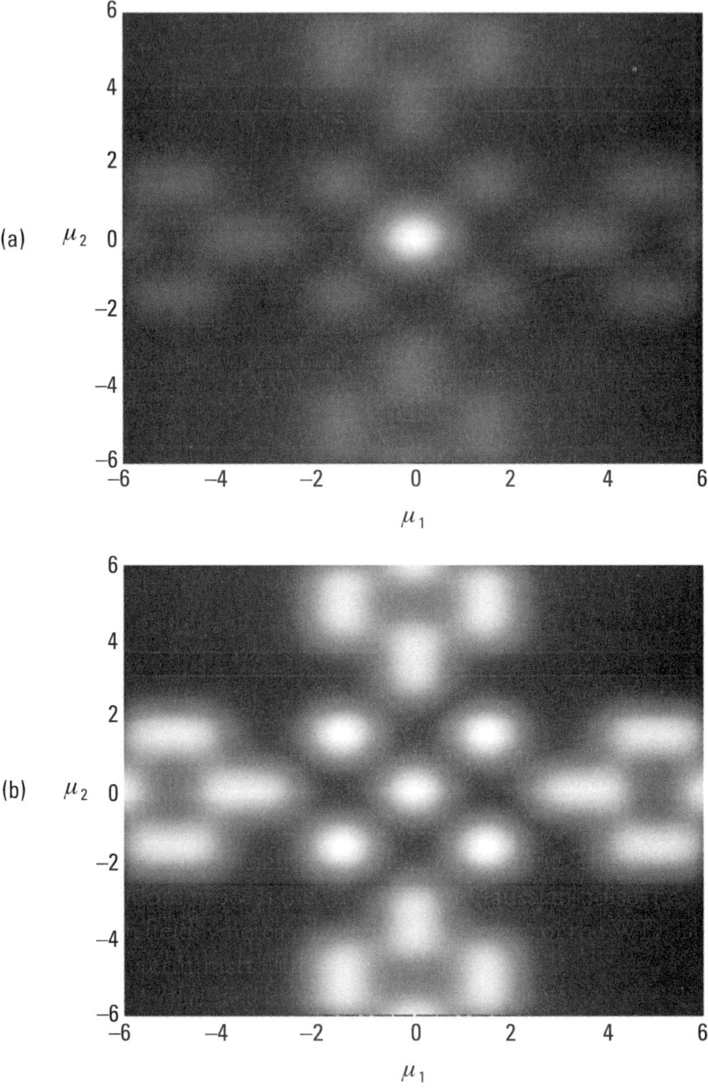

Figure 9.22   Magnitudes (a) $|A_M(\mu_1,\mu_2,0,0)|$ and (b) $|A_M(\mu_1,\mu_2,0.1,0.1)|$ of a monogenic signal (9.66), $A_1 = A_2 = 1$, $f_{11} = f_{12} = 1.5$, $f_{21} = f_{22} = 3$ [15].

vertically in the (μ_1,μ_2)-domain. For the function $|A_M(\mu_1,\mu_2,0.1,0.1)|$ (nonzero values of the shift variables) the crest values of cross-terms are enhanced with regard to the autoterm and change the support from circular to elliptical shape.

9.3 Double-Dimensional Wigner Distributions

In 2002, the new notions of double-dimensional distributions were proposed [30–32]. The idea was simple. First, we calculated the 2-D WD of a given 1-D (analytic) signal as in (9.11). Second, we used the so-obtained WD function as a new 2-D signal and again calculated its 4-D WD or 4-D AF. (Figure 9.23). This operation doubled the dimensionality and that is why these distributions are called *double-dimensional*. This notion is close to the *quartic distributions* defined by O'Neill and Williams in [33] and [34]. Let us recall some basic theoretical results concerning double-dimensional distributions. The details are to be found in [31].

The *double-dimensional Wigner distribution* $W_W^{(2)}(t,f,\mu,\tau)$ is defined in the form

$$W_W^{(2)}(t,f,\mu,\tau)$$

$$= \iint W_\psi\left(t + \frac{\chi_t}{2}, f + \frac{\mu_f}{2}\right) W_\psi\left(t - \frac{\chi_t}{2}, f - \frac{\mu_f}{2}\right) e^{-j2\pi\left(\mu\chi_t + \tau\mu_f\right)}\, d\chi_t d\mu_f$$

$$(9.69)$$

where W_ψ is given by (9.11) or (9.14). The equivalent nonintegral definition known as the *O'Neill-Flandrin formula* is

$$W_W^{(2)}(t,f,\mu,\tau) = W_\psi\left(t + \frac{\tau}{2}, f - \frac{\mu}{2}\right) W_\psi\left(t - \frac{\tau}{2}, f + \frac{\mu}{2}\right) \qquad (9.70)$$

We see that the cross section

$$W_W^{(2)}(t,f,0,0) = \left[W_\psi(t,f)\right]^2 \qquad (9.71)$$

that is, the two distributions differ only in amplitude. The *Wigner distribution of the AF* is given by

$$W_A^{(2)}(\mu,\tau,t,f)$$

$$= \iint A_\psi\left(\mu + \frac{\mu_\mu}{2}, \tau + \frac{\chi_\tau}{2}\right) A_\psi^*\left(\mu - \frac{\mu_\mu}{2}, \tau - \frac{\chi_\tau}{2}\right) e^{-j2\pi\left(\mu_\mu t + \chi_\tau f\right)}\, d\mu_\mu d\chi_\tau$$

$$(9.72)$$

It can be shown that (9.72) is equal to $W_W^{(2)}$ given by (9.69):

$$W_A^{(2)}(\mu,\tau,t,f) = W_W^{(2)}(t,f,\mu,\tau) \tag{9.73}$$

Then, the *ambiguity function of the WD* is defined as

$$
\begin{aligned}
A_W^{(2)}&\left(\mu_\mu,\chi_\tau,\chi_t,\mu_f\right) \\
&= \iint W_\psi\left(t+\frac{\chi_t}{2},f+\frac{\mu_f}{2}\right)W_\psi\left(t-\frac{\chi_t}{2},f-\frac{\mu_f}{2}\right)e^{j2\pi\left(\mu_\mu t+\chi_\tau f\right)}\,dtdf
\end{aligned}
\tag{9.74}
$$

and the *ambiguity function of the AF* called the *double-dimensional ambiguity function* is

$$
\begin{aligned}
A_A^{(2)}&\left(\chi_t,\mu_f,\mu_\mu,\chi_\tau\right) \\
&= \iint A_\psi\left(\mu+\frac{\mu_\mu}{2},\tau+\frac{\chi_\tau}{2}\right)A_\psi^*\left(\mu-\frac{\mu_\mu}{2},\tau-\frac{\chi_\tau}{2}\right)e^{j2\pi\left(\mu\chi_t+\tau\mu_f\right)}\,d\mu d\tau
\end{aligned}
\tag{9.75}
$$

We also have

$$A_A^{(2)}\left(\chi_t,\mu_f,\mu_\mu,\chi_\tau\right) = A_W^{(2)}\left(\mu_\mu,\chi_\tau,\chi_t,\mu_f\right) \tag{9.76}$$

Moreover, it is shown [31] that $A_A^{(2)}$ and $W_W^{(2)}$ form a pair of 4-D Fourier transforms, that is,

$$A_A^{(2)}\left(\chi_t,\mu_f,\mu_\mu,\chi_\tau\right)\overset{4F}{\Longleftrightarrow}W_W^{(2)}(t,f,\mu,\tau) \tag{9.77}$$

and $W_W^{(2)}$ satisfies

$$\iint W_W^{(2)}(t,f,\mu,\tau)\,df\,d\mu = |r(t,\tau)|^2 \tag{9.78}$$

$$\iint W_W^{(2)}(t,f,\mu,\tau)\,dt\,d\tau = |R(f,\mu)|^2 \tag{9.79}$$

$$\iint W_W^{(2)}(t,f,\mu,\tau)\,d\mu d\tau = |W_\psi(t,f)|^2 \tag{9.80}$$

$$\iint W_W^{(2)}(t,f,\mu,\tau)\,d\mu d\tau = |W_\psi(t,f)|^2 \tag{9.81}$$

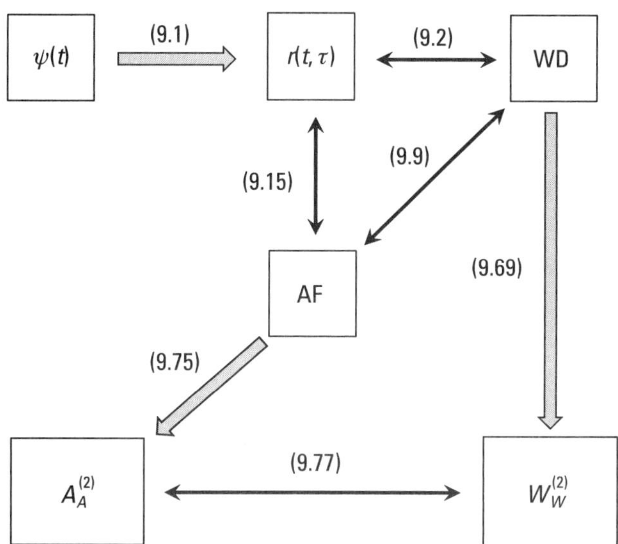

Figure 9.23   The method of calculation of double-dimensional Wigner distributions and ambiguity functions. The corresponding definitions and Fourier relations are marked with numbers of equations.

where $r(t,\tau)$ and $R(f,\mu)$ are correlation products given by (9.10) and (9.12). Figure 9.23 shows the method of derivation of double-dimensional distributions of a complex-valued signal $\psi(t)$. All Fourier relations given by (9.29), (9.74), (9.75), and (9.77) are visualized.

Example 9.11   A Gaussian Time-Shifted and Modulated Signal

The Gaussian time-shifted and modulated signal is given by $\psi(t) = e^{-a(t-t_a)^2} \cdot e^{j2\pi f_a t}$. Its WD is

$$W_\psi(t,f) = \sqrt{\frac{2\pi}{a}}\, e^{-\frac{2\pi^2}{a}(f-f_a)^2}\, e^{-2a(t-t_a)^2}$$

The double-dimensional WD and AF respectively are

$$W_W^{(2)}(t,f,\mu,\tau) = \frac{2\pi}{a} e^{-4a(t-t_a)^2}\, e^{-\frac{4\pi^2}{a}(f-f_a)^2}\, e^{-\frac{\pi^2}{a}\mu^2}\, e^{-a\tau^2}$$

$$A_A^{(2)}\left(\mu_\mu,\chi_\tau,\chi_t,\mu_f\right) = \frac{\pi}{2a} e^{-\frac{\pi^2}{4a}\mu_\mu^2}\, e^{-\frac{a}{4}\chi_\tau^2}\, e^{-a\chi_t^2}\, e^{-\frac{\pi^2}{a}\mu_f^2}\, e^{-j2\pi\left(\mu_\mu t_a + \chi_\tau f_a\right)}$$

Example 9.12 Sum of Time-Shifted and Modulated Gaussian Signals

Figure 9.24(a) shows the cross section of $W_W^{(2)}(t,f,0,0)$ of a sum of two Gaussian pulses $\psi(t) = e^{-a(t-t_1)^2}e^{j2\pi f_1 t} + e^{-a(t-t_2)^2}e^{j2\pi f_2 t}$, $t_1 = 1.5$, $t_2 = -1.5$, $f_1 = 1.5$, $f_2 = 4.5$. The autoterms are well-distinguished in the (t,f)-plane, and we also

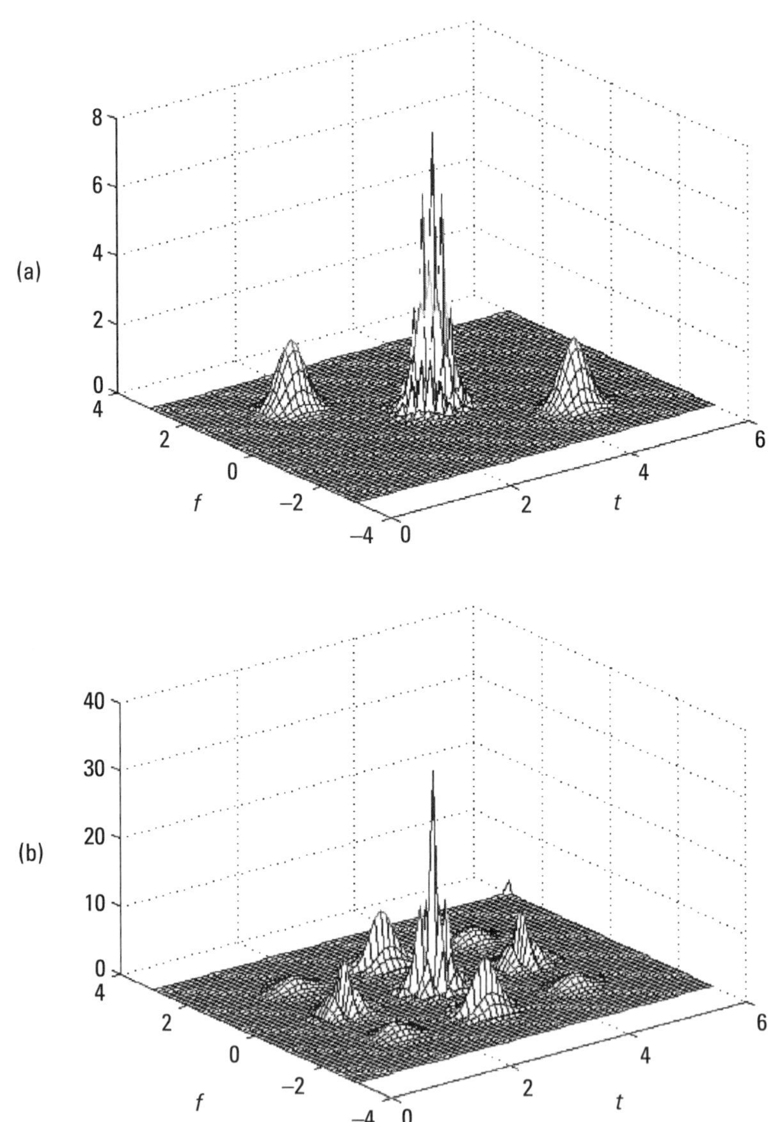

Figure 9.24   Cross section $W_W^{(2)}(t,f,0,0)$ of (a) the sum of two Gaussian pulses, and (b) the sum of four Gaussian pulses.

have a single cross-term situated between the autoterms. Figure 9.24(b) shows the same cross section of $W_W^{(2)}$ for the sum of four Gaussian pulses shifted in time and modulated in frequency:

$$\psi(t) = \sum_{i=1}^{4} \frac{A_i}{\sqrt{2\pi}\sigma_i} \exp\left(-\frac{\left(t - t_i\right)^2}{2\sigma_i^2}\right) \exp\left(j2\pi f_i t\right) ,$$

where all $A_i = 1$, $\sigma_1 = 1/\sqrt{2\pi}$, $t_1 = -2$, $t_2 = 2$, $t_3 = -2$, $t_4 = 2$ and $f_1 = f_2 = 2$, $f_3 = f_4 = 5$. We observe four autocomponents centered at $(-2,2)$, $(2,2)$, $(-2,5)$, and $(2,5)$ of the time-frequency plane and six cross-components (two central cross-terms are superimposed).

The advantage of using $W_W^{(2)}$ instead of the WD can be shown for the real chirp signal $u(t) = \sin(at + bt^2)$, $a = \pi$, $b = \pi$. Figure 9.25(a) presents the cross section $W_W^{(2)}(t,f,0,0)$ of this signal and Figure 9.25(b) shows its Wigner distribution $W_u(t,f)$. We see that both differ only in amplitude, but in $W_W^{(2)}$, the cross-terms are visibly reduced in comparison to the WD.

The methods of reduction of unwanted cross-terms in double-dimensional distributions have been proposed in [32] in the form of so called *double-dimensional pseudo-Wigner distributions*.

9.4 Applications of Space-Frequency Distributions in Signal Processing

The 2-D Wigner distribution and Woodward's ambiguity function are basic tools for time-frequency analysis of nonstationary real and complex signals [17, 20, 21, 25, 35–37]. The WD serves as a basis of the Cohen's class time-frequency distributions [6] largely applied in various fields of signal processing as, for example, in underwater acoustics [38], radar imaging and seismology [39–42], biomedicine [43], mechanical engineering [44], and so on.

Let us briefly present two fields of applications of space-frequency distributions: noise analysis and image processing, in which the authors of this book have made also a small contribution.

9.4.1 Wigner Distribution in Noise Analysis

Noise analysis is a very interesting area of stochastic signal processing, also exploiting the time-frequency approach. It should be noted that—in the case of random signals—we have to distinguish the WD of a single sample function

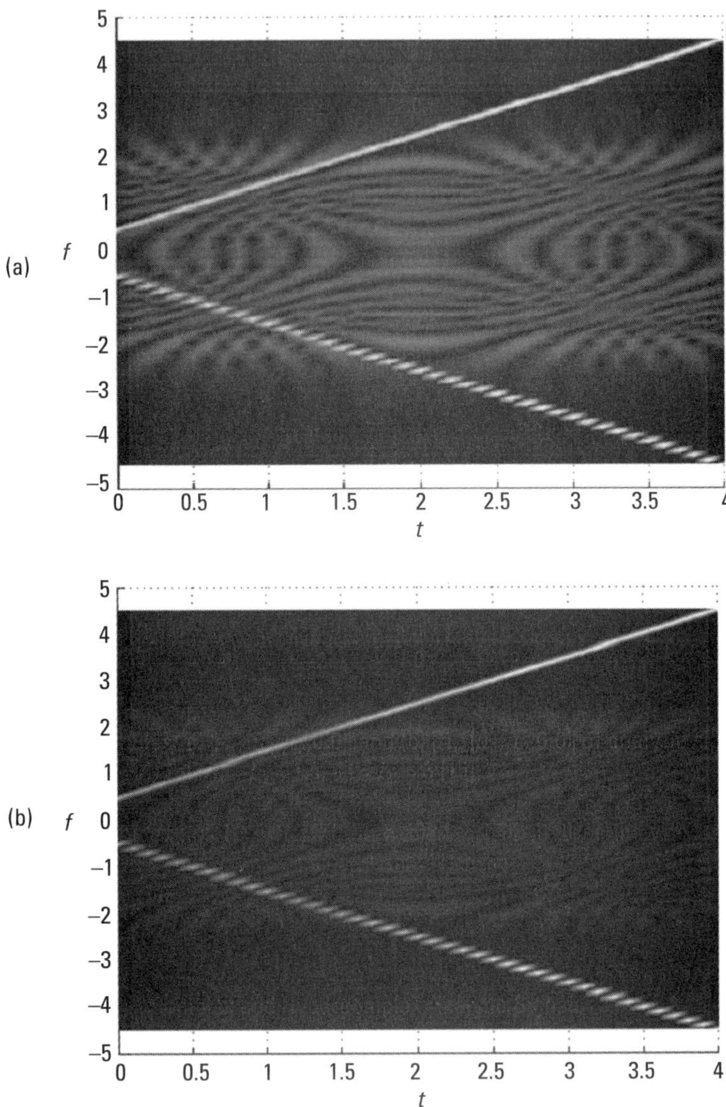

Figure 9.25　The real chirp signal: (a) $W_u(t,f)$, (b) $W_W^{(2)}(t,f,0,0)$.

(realization), $\psi(t)$, from the ensemble averages $\mathrm{E}\{W_\psi(t,f)\}$, where E is the ensemble average operator. Their properties are different. The WD of a single realization of a random process (i.e., a sample of Gaussian noise) has the form of a bipolar random field, whereas the ensemble average of the WD, $\mathrm{E}\{W_\psi(t,f)\}$, is a well-defined deterministic function [45–47].

Example 9.13 WD of a Low-Pass Gaussian Noise

Figure 9.26(a) shows the WD of a single sample of a low-pass stationary Gaussian noise in the form of a random field with a limited (low-pass) frequency support. Figure 9.26(b) displays the slice at $t = 20$ of the ensemble average of two thousand of WDs. The solid line represents the average of autoterms and dotted line represents the average of cross-terms decreasing with the growing number of realizations.

Now, if we consider a deterministic signal $\psi(t)$ contaminated by zero-mean real stationary noise $n(t)$, that is, a sum

$$\xi(t) = \psi(t) + n(t) \tag{9.82}$$

we know that its WD is (compare with (9.20))

$$W_\xi(t, f) = W_\psi(t, f) + W_n(t, f) + W_{\psi n}(t, f) + W_{n\psi}(t, f) \tag{9.83}$$

The ensemble average of (9.83) gives

$$E\left\{W_\xi(t, f)\right\} = W_\psi(t, f) + E\left\{W_n(t, f)\right\} + E\left\{W_{\psi n}(t, f)\right\} + E\left\{W_{n\psi}(t, f)\right\} \tag{9.84}$$

where $E\{W_n(t,f)\} = S_n(f)$ is a noise power density function and $E\{W_{\psi n}(t,f)\} = E\{W_{n\psi}(t,f)\} = 0$ (see, for example, [48]). Finally,

$$E\left\{W_\xi(t, f)\right\} = W_\psi(t, f) + S_n(f) \tag{9.85}$$

We see that the averaging process $E\{W_\xi(t,f)\}$ with a theoretically infinite number of realizations gets the WD of the deterministic signal $\psi(t)$ with a constant component equal to the noise power density.

Example 9.14 Cosine Signal Embedded in Gaussian Noise

Figure 9.27(a) shows the WD of a single realization of the deterministic signal $u(t) = \cos(2\pi f_0 t)$, $f_0 = 20$ embedded in Gaussian noise. Figure 9.27(b) displays the result of an averaging process with 1000 realizations. We see the autoterm at $f = 20$ and a cross-term at zero frequency. Let us mention that we used the analytic signal approach (Wigner-Ville distribution) in simulations.

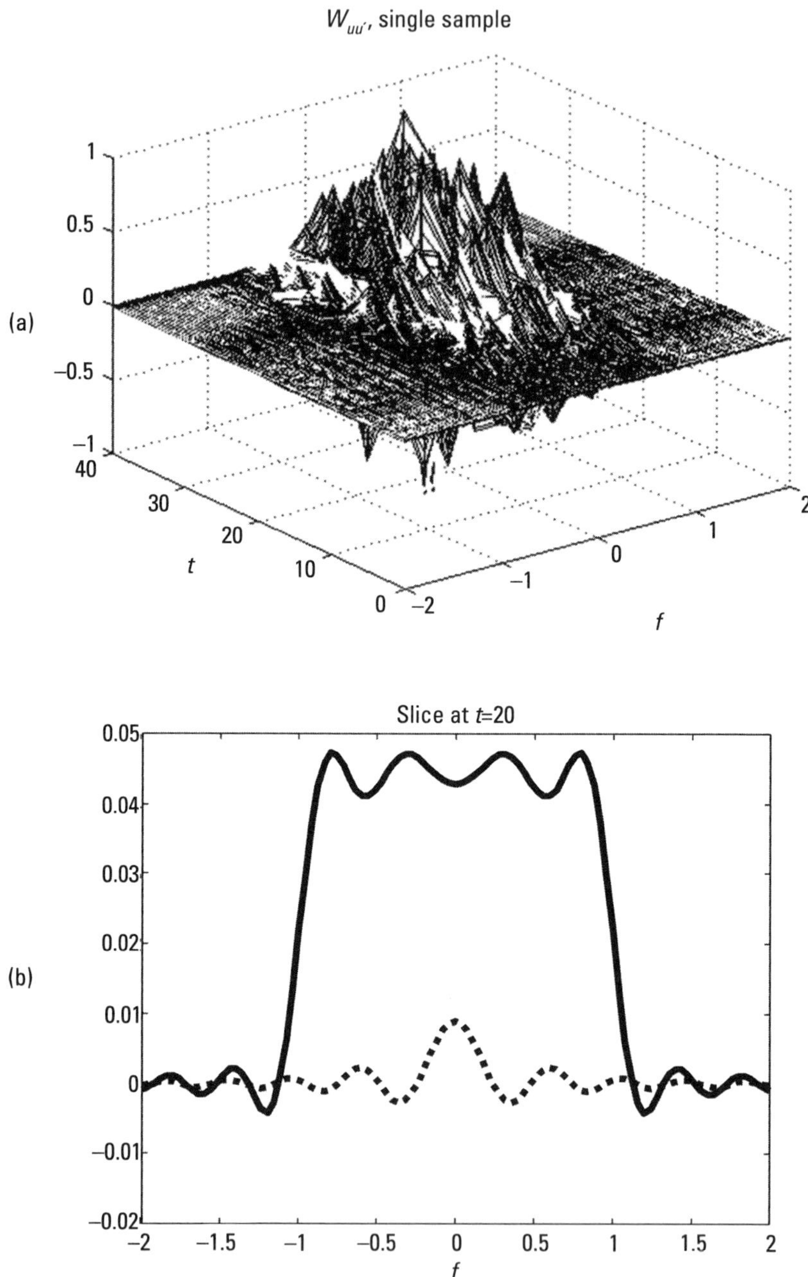

Figure 9.26 The low-pass stationary Gaussian noise. (a) The Wigner distribution of a single realization in the form of a random field, (b) an exemplary slice of the ensemble average of 2000 WDs shows a well-defined deterministic function.

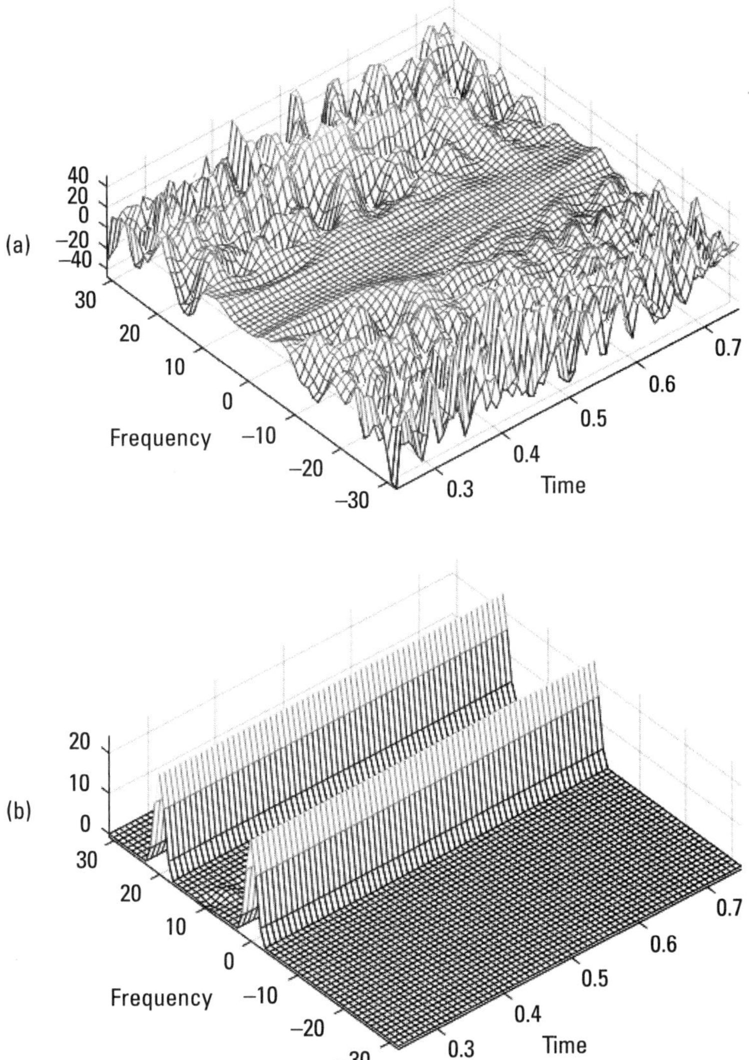

Figure 9.27   The cosine signal contaminated with Gaussian noise. (a) The WD of a single realization. (b) The ensemble average of 1000 WDs.

9.4.2 Wigner Distribution in Image Processing

Real images have a nonstationary structure that makes use of space-frequency distributions very promising. It seems that the first who considered use of the WD in image analysis were Jacobsen and Wechsler [50]. However, in comparison to a wide range of various applications of time-frequency distributions in 1-D signal processing (given in Section 9.4.1), the number of publications is rather limited. A good description of some applications of the WD in image processing is given in [50]. Some examples of the use of spatial-frequency distributions in texture segmentation can be found in [51–54]. In the domain of image watermarking, the Radon-Wigner distribution (a projection of the 2-D WD) is applied [55].

Applications of spatial-frequency distributions of complex and hypercomplex analytic signals in image processing are relatively rare. In [56], Hahn and Snopek analyzed individual cross sections of 4-D WDs of test signals and showed that it is possible to extract a 2-D image from a 2-D random noise field (Figure 9.28). In 2013, Liu and Goutte proposed to apply the quaternion WD in hyperspectral imagery [57].

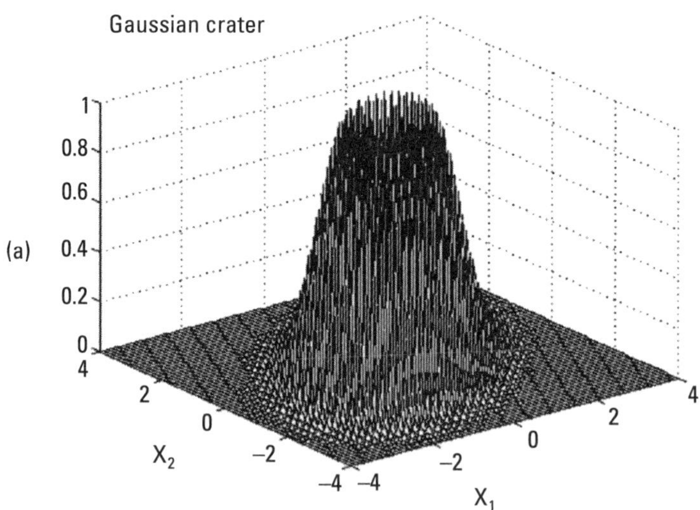

Figure 9.28 2-D Gaussian crater embedded in random noise: (a) 2-D Gaussian crater, (b) 2-D random field, (c) signal embedded in the random field, (d) cross section of the 4-D WD, (e) cross section of the crater (no random field), and (f) cross section of the WD of a sum signal + noise [56].

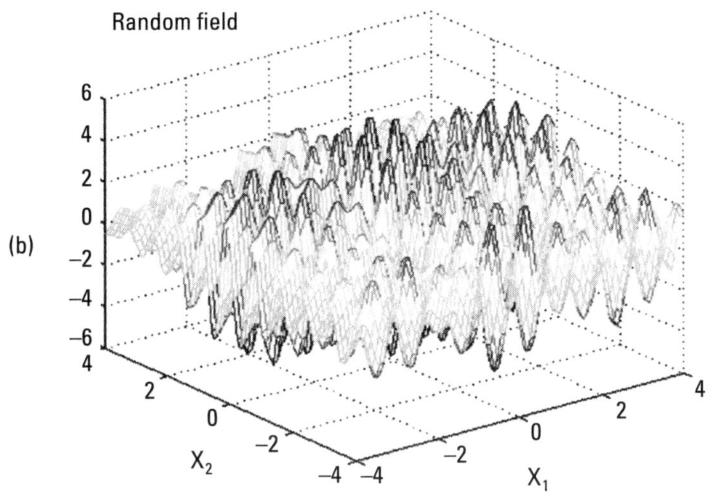

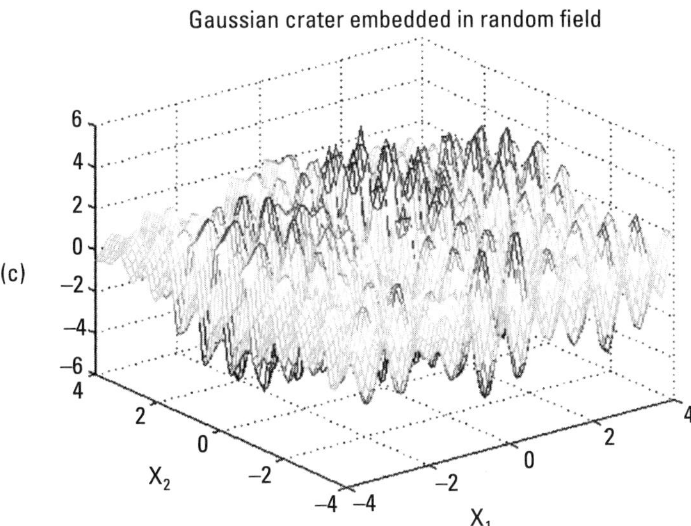

Figure 9.28 Continued

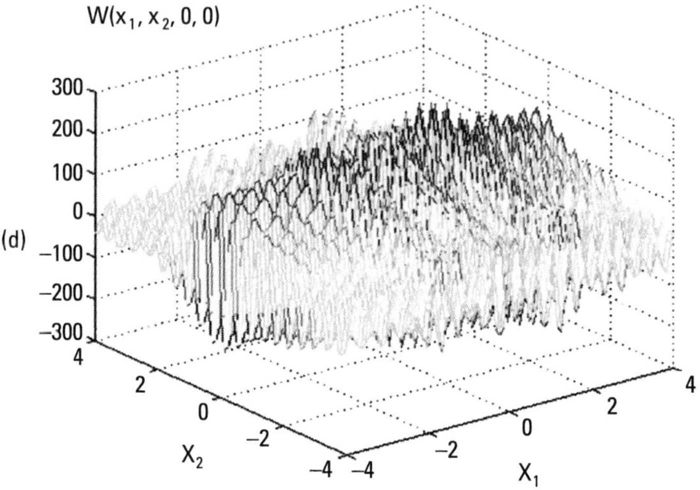

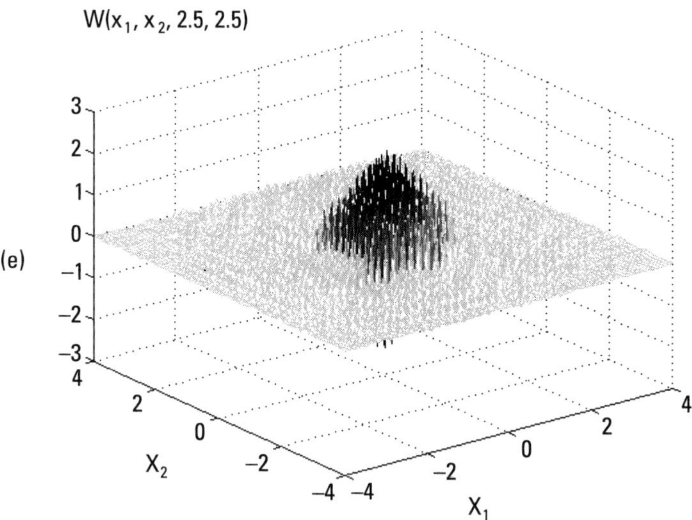

Figure 9.28 Continued

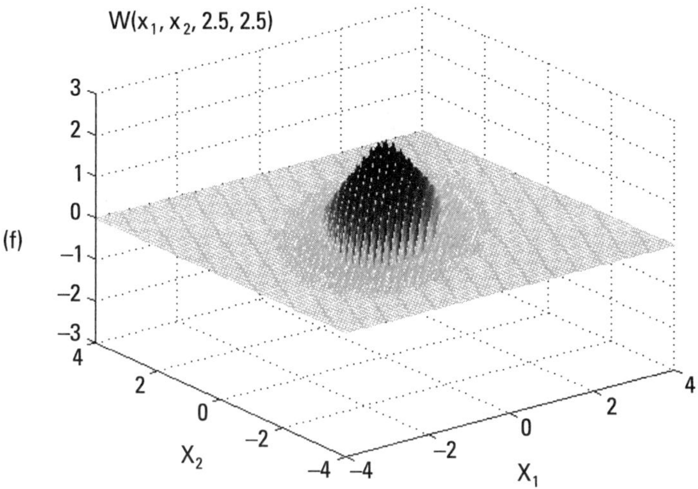

Figure 9.28   Continued

References

[1]   Wigner, E. P., "On the Quantum Correction for Thermodynamic Equilibrium," *Phys. Rev.*, Vol. 40, June 1932, pp. 749–759.

[2]   Ville, J., "Théorie et Applications de la Notion de Signal Analytique," *Câbles et Transmission*, Vol. 2A, 1948, pp. 61–74.

[3]   Claasen, T. C. A. M., and W. F. G Mecklenbräuker, "The Wigner distribution: A Tool for Time-Frequency Signal Analysis, Part I—Continuous-Time Signals," *Philips J. Research*, Vol. 35, 1980, pp. 217–250.

[4]   Claasen, T. C. A. M., and W. F. G. Mecklenbräuker, "The Wigner Distribution: A Tool for Time-Frequency Signal Analysis, Part II—Discrete-Time Signals," *Philips J. Res.*, Vol. 35, 1980, pp. 276–300.

[5]   Claasen, T. C. A. M., and W. F. G Mecklenbräuker, "The Wigner Distribution: A Tool for Time-Frequency Signal Analysis, Part III—Relations with other Time-Frequency Signal Transformations," *Philips J. Res.*, Vol. 35, 1980, pp. 372–389.

[6]   Cohen, L., "Time-Frequency Distributions—A Review," *Proc. IEEE*, Vol. 77, No. 7, 1989, pp. 941–981.

[7]   Hahn, S. L., "Multidimensional Complex Signals with Single-Orthant Spectra," *Proc. IEEE*, Vol. 80, No. 8, 1992, pp. 1287–1300.

[8]   Snopek, K. M., "The Comparison of the 4D Wigner Distributions and the 4D Woodward Ambiguity Functions," *Kleinheubacher Berichte*, Vol. 42, 1999, pp. 237–246.

[9]   Hahn, S. L., "The theory of time-frequency distributions with extension for two-dimensional signals," *Metrology and Measurements Systems*, vol. VIII, no. 2, pp. 113–143, 2000.

[10]   Hahn S. L., "A review of methods of time-frequency analysis with extension for signal plane-frequency plane analysis," *Kleinheubacher Berichte*, Vol. 44, 2001, pp. 163–182.

[11]   Snopek, K. M., "A review of the properties of the Cohen's class time-frequency distributions," *Proc. Applied Electronics 2000*, Západočeská Univerzita v Plzni, Pilsen, September 6–7, 2000, pp. 131–136.

[12]   Snopek, K. M., "A Cohen's class distributions with separable kernels," *Proc. 1st International Conference on Signals and Electronic Systems ICSES 2000*, Ustroń, October 17–20, 2000, pp. 99–104.

[13]   Snopek, K. M., "The Comparison of the 4D Wigner Distributions and the 4D Woodward Ambiguity Functions," *Kleinheubacher Berichte*, Band 42, 1999, pp. 237–246.

[14]   Hahn, S. L., and K. M. Snopek, "Wigner Distributions and Ambiguity Functions in Image Analysis," *Computer Analysis of Images and Patterns, 9th International Conference CAIP 2001*, Warsaw, Poland, September 2001, pp. 537–546.

[15]   Hahn, G., "A survey of ambiguity functions of analytic, quaternionic and monogenic signals," *Image Processing and Communication*, Vol. 10, No. 1, 2005, pp. 13–33.

[16]   Woodward, P.M., *Probability and Information Theory with Applications to Radar*, New York: Pergamon, 1953.

[17]   Flandrin, P., *Time-Frequency/Time-Scale Analysis*, San Diego, CA: Academic Press, 1999.

[18]   Hahn, S. L., and K. M. Snopek, "The Unified Theory of n-Dimensional Complex and Hypercomplex Analytic Signals," *Bull. Polish Ac. Sci., Tech. Sci.*, Vol. 59, No. 2, 2011, pp. 167–181.

[19]   Snopek, K. M. "The Study of Properties of n-D Analytic Signals in Complex and Hypercomplex Domains," *Radioengineering*, Vol. 21, No. 2, April 2012, pp. 29–36.

[20]   Boudreaux-Bartels, G. F., "Mixed Time-Frequency Signals Transformations," in *The Transforms and Applications Handbook*, A. D. Poularikas (Ed.), Boca Raton, FL: CRC Press, Inc., 1996.

[21]   Hlawatsch, F., and G. F. Boudreaux-Bartels, "Linear and Quadratic Time-Frequency Signal Representations," *IEEE Signal Proc. Magazine*, Vol. 9, No. 2, April 1992, pp. 21–67.

[22]   Siebert, W. M., "Studies of Woodward's uncertainty function," *Quart. Progress Rep., Electron. Res. Lab.*, Vol. 15, Cambridge, MA: Mass. Inst. Technol., April 1958, pp. 90–94.

[23]   Stutt, C. A., "Some Results on Real-Part/Imaginary-Part and Magnitude-Phase Relations in Ambiguity Functions," *IEEE Trans. Inform. Theory*, pp. 321–327, October 1964.

[24]   Price R., Hofstetter E.M., "Bounds on the Volume and Height Distributions of the Ambiguity Function," *IEEE Trans. Inform. Theory*, Vol. 11, No.4, 1965, pp. 207–214.

[25]   Mertins, A., *Signal Analysis—Wavelets, Filter Banks, Time-Frequency Transforms, and Applications*, Chichester, England: John Wiley & Sons Ltd., 1999.

[26]   Hlawatsch, F., and G. F. Boudreaux-Bartels, "Linear and Quadratic Time-Frequency Signal Representations," *IEEE Signal Proc. Magazine*, Vol. 9, No. 2, April 1992, pp. 21–67.

[27]   Hlawatsch, F., and P. Flandrin, "The Interference Structure of the Wigner Distribution and Related Time-Frequency Signal Representations" in *The Wigner Distribution. Theory and Applications in Signal Processing*, W. Mecklenbräuker and F. Hlawatsch (eds.), Amsterdam: Elsevier, 1997, pp. 59–133.

[28]   Hahn, S. L., and K. M. Snopek, "Wigner Distributions and Ambiguity Functions of 2D Quaternionic and Monogenic Signals," *IEEE Transactions on Signal Processing*, Vol. 53, No. 8, August 2005, pp. 3111–3128.

[29]   Hahn, S. L., and K. M. Snopek, "The derivation of the Wigner distribution of the quaternionic signal $\psi_q\left(x_1,x_2\right) \approx e^{-\pi\left(x_1^2+x_2^2\right)}\left[A_1 e^{i2\pi f_{11}x_1} e^{j2\pi f_{12}x_2} + A_2 e^{i2\pi f_{21}x_1} e^{j2\pi f_{22}x_2}\right]$," Report No.2, Institute of Radioelectronics, WUT, Warsaw, Poland, 2004. Available: http://www.ire.pw.edu.pl/~ksnopek.

[30]   Hahn, S. L., and K. M. Snopek, "Comparison of selected Cohen's class double-dimensional distributions," *Kleinheubacher Berichte*, Band 45, 2002, pp. 111–115.

[31]   Hahn, S. L., and K. M. Snopek, "Double-dimensional Distributions: Another Approach to Quartic"Distributions," *IEEE Transactions on Signal Processing*, Vol. 50, No. 12, December 2002, pp. 2987–2997.

[32]   Snopek, K. M., "Pseudo-Wigner and double-dimensional pseudo-Wigner distributions with extension for 2-D signals," *Electronics and Telecommun. Quarterly*, Vol. 51, Warsaw, Poland, 2005, pp. 9–21.

[33]   O'Neill, J. C., and W. J. Williams, "A function of time, frequency, lag and Doppler," *IEEE Trans. Signal Processing*, Vol. 47, March 1999, pp. 789–799.

[34]   O'Neill, J. C., and W. J. Williams, "Virtues and vices of quartic time-frequency distributions," *IEEE Trans. Signal Processing*, Vol. 48, September 2000, pp. 2641–2650.

[35]   Boashash, B., "Note on the Use of the Wigner Distribution for Time-Frequency Signal Analysis," *IEEE Trans. Acoustics, Speech, Signal Process*, Vol. 36, No. 9, September 1998, pp. 1518–1521.

[36]   Hammond, J. K., and P. R. White, "The analysis of non-stationary signals using time-frequency methods," *J. Sound and Vibration*, Vol. 190, No. 3, 1996, pp. 419–447.

[37]   D. Eustice, C. Baylis, and R. J. Marks, "Woodward's ambiguity function: From foundations to applications," *Wireless and Microwave Circuits and Systems (WMCS), 2015 Texas Symposium on*, Waco, TX, 2015, pp. 1–17.

[38]   Boashash, B., and P. O'Shea, "A methodology for detection and classification of some underwater acoustic signals using time-frequency analysis techniques," *IEEE Trans. Acoustics, Speech, Signal Process*, Vol. 38, No. 9, November 1990, pp. 1829–1841.

[39]   Angelari, R. D., "The ambiguity function applied to underwater acoustic signal processing: A review," *Ocean Engineering*, Vol. 2, No.1, September 1970, pp. 13–26.

[40]   Gaunaurd, G. C., and H. C. Strifors, "Signal Analysis by Means of Time-Frequency (Wigner-Type) Distributions—Applications to Sonar and Radar Echoes," *Proc. IEEE*, Vol. 84, No. 9, September 1996, pp. 1231–1248.

[41]   Chen, V. C., "Applications of time-frequency processing to radar imaging," *Opt. Eng.*, Vol. 36, No. 4, April 1997, pp. 1151–1161.

[42]   Elizondo, M. A., et al., "Application of the Wigner-Ville distribution to interpret ground-penetrating radar anomalie," *Geofísica Internacional*, Vol. 51, No.2, 2012, pp. 121–127.

[43]   Wood, J. C., and D. T. Barry, "Time-Frequency Analysis of Skeletal Muscle and Cardiac Vibrations," *Proc. IEEE*, Vol. 84, No. 9, September 1996, pp. 1281–1294.

[44]   König, D., and J. F. Böhme, "Wigner-Ville Spectral Analysis of Automotive Signals Captured at Knock," *Applied Signal Processing*, Vol. 3, 1996, pp. 54–64.

[45]   Hahn, S. L., "Stochastic analytic signals and the relation between instantaneous frequency, spectral moments and the Wigner-Ville distribution," *Bull. Pol. Ac. Sci., Tech. Sci.*, Vol.4, 1995, pp. 525–535.

[46]   Matz, G., and F. Hlawatsch, "Wigner distributions (nearly) everywhere: time-frequency analysis of signals, systems, random processes, signal spaces and frames," *Signal Processing*, Vol. 83, 2003, pp. 1355–1378.

[47]   Snopek, K.M., "New Insights into Wigner Distributions of Deterministic and Random Analytic Signals," *Proc. VI International Symposium on Signal Processing and Information Technology ISSPIT'06*, Vancouver, 27–30 August 2006, pp. 374–379.

[48]   Stanković, L., and S. Stanković, "Wigner Distribution of Noisy Signals," *IEEE Trans. Signal Processing*, Vol. 41, No. 2, February 1993, pp. 956–960.

[49]   Beghdadi, A., and R. Iordache, "Image Quality Assessment Using the Joint Spatial/ Spatial-Frequency Representation," *EURASIP Journal on Applied Signal Processing*, Article ID 80537, Hindawi Publishing Corporation, 2006, pp.1–8.

[50]   Jacobsen, L., and H. Wechsler, "Invariant Analogical Image Representation and Pattern Recognition," *Pattern Recognition Letters*, North Holland, Vol. 2, September 1984, pp. 289–299.

[51]   Reed, T. R., and H. Wechsler, "Segmentation of textured images and Gestalt organization using spatial/spatial-frequency representations," *IEEE Trans. Pattern Analysis and Machine Intelligence*, Vol. 12, No. 1, 1990, pp. 1–12.

[52] Zhu, Y. M., R. Goutte, and M. Amiel,"On the use of two-dimensional Wigner-Ville distribution for texture segmentation," *Signal Processing*, Vol. 30, No. 3, February 1993, pp. 329–353.

[53] Cristóbal, G., and J. Hormigo, "Texture segmentation through eigen-analysis of the Pseudo-Wigner distribution," *Pattern Recognition Letters*, Vol. 20, No. 3, 1999, pp. 337–345.

[54] Cristóbal, G., J. Bescós, and J. Santamaria, "Image analysis through the Wigner distribution function," *Appl. Opt.*, Vol. 28, No. 2, Jan. 15, 1989, pp. 262–271.

[55] Stanković, L., I. Djurović, and I. Pitas, "Watermarking in the space/spatial-frequency domain Rusing two-dimensional Radon-Wigner distribution," *IEEE Trans. Image Processing*, Vol. 10, No. 4, 2001, pp. 650–658.

[56] Hahn, S. L., and K. M. Snopek, "Wigner distributions and ambiguity functions in image analysis," *Computer Analysis of Images and Patterns, 9th International Conference CAIP 2001*, Warsaw, Poland, September 2001, *Lecture Notes in Computer Science LNCS 2124*, pp. 537–546.

[57] Liu, Y., and R. Goutte, "Quaternionic Wigner-Ville distribution of analytical signal in hyperspectral imagery," *Int. J. of Advanced Computer Science and Applications (IJACSA)*, Vol. 4, No. 10, 2013, pp. 95–98.

10

Causality of Signals

This chapter presents the definition of causality of multidimensional real signals. Causality of 1-D time signals implies that any response of a system at time t depends only on excitation at an earlier time. A transfer function of a causal and linear time-invariant system is an analytic function of frequency (Kramers-Kronig relations). A multidimensional real signal is called *causal* if its support is limited to $\mathbb{R}^+$. Multidimensional causal signals have analytic spectra. This yields the extension of Kramers-Kronig dispersion relations.

10.1 Kramers-Kronig Relations

All physical systems are causal. The response of the system to an input signal cannot appear earlier than the excitation. For example, the response to a very short input pulse applied at the time t never appears earlier. The response $g(t)$ for the input $\delta(t)$ pulse called the *impulse response* equals zero for $t < 0$. Therefore, the response is a one-sided function in the time domain. Let us recall that 1-D analytic signals have one-sided spectra in the frequency domain. Consequently, due to the duality property of the Fourier transformation, all responses of a physically realizable system have analytic spectra. In order to show details, let us recall the notion of a *transfer function* of a linear time invariant system (LTI). It is the Fourier transform of the impulse response $g(t)$

defined by the quotient of the output steady state harmonic analytic signal $\psi_{\text{output}}(t) = A_2 e^{j(2\pi ft + \varphi_2)}$ to the input steady state signal $\psi_{\text{input}}(t) = A_1 e^{j(2\pi ft + \varphi_1)}$:

$$\Gamma(f) = \frac{A_2 e^{j(2\pi ft) + \varphi_2}}{A_1 e^{j(2\pi ft + \varphi_1)}} = \frac{A_2}{A_1} e^{j(\varphi_2 - \varphi_1)} = A(f) + jB(f) \tag{10.1}$$

In this chapter, we will deal only with complex functions using the notation j and not e_1. The one-sided impulse response $g(t)$ can be decomposed into a sum of noncausal even and odd parts (see Chapter 3, (3.9)): $g(t) = g_e(t) + g_o(t)$. Because $g(t)$ is a real function we have the following Fourier pairs:

$$g_e(t) = 0.5\big[g(t) + g(-t)\big] \overset{F}{\Leftrightarrow} A(f)$$
$$g_o(t) = 0.5\big[g(t) - g(-t)\big] \overset{F}{\Leftrightarrow} B(f) \tag{10.2}$$

The causality of $g(t)$ yields the relations

$$g_o(t) = \operatorname{sgn}(t) g_e(t)$$
$$g_e(t) = \operatorname{sgn}(t) g_o(t) \tag{10.3}$$

Therefore, the impulse response can be written in the form

$$g(t) = g_e(t)\big(1 + \operatorname{sgn}(t)\big) \tag{10.4}$$

that is, it is uniquely defined by the even part. The terms of (10.1) are defined by the Fourier transforms

$$g_e(t) \overset{F}{\Leftrightarrow} A(f), \quad \operatorname{sgn}(t) g_e(t) \overset{F}{\Leftrightarrow} \frac{-j}{\pi f} * A(f) = -j\mathrm{H}\big[A(f)\big] \tag{10.5}$$

The insertion in (10.4) yields the following form of the Kramers-Kronig dispersion relations [4, 5] (the term dispersion is used in optics)

$$\Gamma(f) = A(f) - j\mathrm{H}\big[A(f)\big] \tag{10.6}$$

The real and imaginary parts of the transfer function of a causal system are forming a pair of Hilbert transforms. In the next paragraph, it will be shown that a similar form applies for n-D causal signals. Equation (10.5) can be rewritten in the form

$$\Gamma(f) = A(f) * \left[\delta(f) - j\frac{1}{\pi f} \right] \tag{10.7}$$

The term in parenthesis is the dual form of the complex delta distribution (see Chapter 5, (5.11)). Note the conjugate form with regard to the time domain version.

Example 10.1

Let us present a simple example. The transfer function of a low-pass RC filter is ($\tau = RC$)

$$\Gamma(f) = \frac{1}{1 + j2\pi f\tau} = \frac{1}{1 + (2\pi f\tau)^2} - j\frac{2\pi f\tau}{1 + (2\pi f\tau)^2} = A(f) - jH\left[A(f)\right]$$

Evidently, the real and imaginary parts are forming a pair of Hilbert transforms. The impulse response is $g(t) = \frac{1}{\tau}\left[0.5 + 0.5\,\mathrm{sgn}(t)\right]e^{-t/\tau}$. We have

$$g_e(t) = \frac{1}{2\tau}e^{-\frac{|t|}{\tau}} \overset{F}{\Leftrightarrow} A(f)$$

$$g_o(t) = \frac{1}{2\tau}\mathrm{sgn}(t)e^{-|t/\tau|} \overset{F}{\Leftrightarrow} B(f) = -\frac{1}{2\tau} \cdot \frac{1}{\pi f} * A(f)$$

In this example, the unit step $\mathbf{1}(t) = 0.5 + 0.5\,\mathrm{sgn}(t)$ is also a one-sided function. Its even part equals 0.5 and the odd part is $0.5\,\mathrm{sgn}(t)$. The corresponding Fourier transform of $\mathbf{1}(t)$ is

$$\Gamma(f) = \frac{1}{2}\delta(f) - j\frac{1}{2\pi f} = A(f) - jH\left[A(f)\right]$$

Note that the real and imaginary parts of the complex delta distribution are also a pair of Hilbert transforms.

10.2 Extension of the Notion of Causality to Higher Dimensions

In the previous section, we presented the Kramers-Kronig relations, which show that for causal signals (causal impulse response), the real and imaginary parts of corresponding spectra (transfer functions) are a pair of Hilbert transforms. Let us present the extension of the notion of causality to n-D signals proposed briefly in [2] and later with more details in [3]. Recall that no definition is true or false. The presented definition is based on the duality between n-D analytic signals with single-orthant spectra and signals with single-orthant support in the signal domain. It will be shown that we get the direct extension of Kramers-Kronig dispersion relations. Let us introduce the following definition of causality of n-D signals.

Consider a real signal $g(\boldsymbol{x})$, where $\boldsymbol{x} = (x_1, x_1, \ldots, x_n)$ are Cartesian coordinates of a n-D space. The n-D real signal $g(\boldsymbol{x})$ is called *causal* if its support is limited to $\mathbb{R}^+$, that is, to a single space-orthant $(x_1 > 0, x_2 > 0, \ldots, x_n > 0)$.

10.2.1 Derivation of the Dispersion Relations

A real n-D signal can be decomposed into a sum of 2^n terms with regard to evenness and oddness (see Chapter 3). However, for causal signals all terms are given by the formula (see (10.4))

$$g(\boldsymbol{x}) = g_{ee\ldots e}(\boldsymbol{x}) \prod_{i=1}^{i=n} \left[1 + \mathrm{sgn}(x_i) \right] \tag{10.8}$$

The n terms are all uniquely defined by the even term. For example, in 2-D, we have four terms given by (an example is shown in Figure 10.2).

$$g(x_1, x_2) = g_{ee}(x_1, x_2)(1 + \mathrm{sgn}\, x_1)(1 + \mathrm{sgn}\, x_2) = g_{ee}(1 + s_1 + s_2 + s_1 s_2) \tag{10.9}$$

where $s_i = \mathrm{sgn}\, x_1$. The n-D extension of Kramers-Kronig dispersion relations is given by the Fourier transform of (10.9). In 2-D, we get

$$\Gamma(f_1, f_2) = G(f_1, f_2) - \mathrm{H}[G] - j(\mathrm{H}_1[G] + \mathrm{H}_2[G]) \tag{10.10}$$

This is the dual form of the analytic 2-D signal (see Chapter 5, (5.28)). Notations: H = the total Hilbert transform, H_1 and H_2 = partial Hilbert transforms. Using the iteration property of Hilbert transforms (see Appendix F) we have

$$\mathrm{H}_1[\mathrm{Re}] = \mathrm{H}_1\left[\mathrm{H}_1[G] + \mathrm{H}_2[G] \right] = -G + \mathrm{H}[G] = -\mathrm{Re} \tag{10.11}$$

$$\mathrm{H}_2[\mathrm{Re}] = \mathrm{H}_2\left[\mathrm{H}_1[G] + \mathrm{H}_2[G] \right] = -G + \mathrm{H}[G] = -\mathrm{Re} \tag{10.12}$$

The result of this operation is the same using H_1 or H_2. Using these relations the extension of Kramers-Kronig relations has the form

$$\Gamma\left(f_1, f_2\right) = \text{Re} - j \cdot H_i[\text{Re}]; \quad \text{free choice } i = 1 \text{ or } 2 \tag{10.13}$$

The same relation applies for higher dimensions with free choice of i from 1 to n. In terms of the complex delta distribution, we have the dual form of (5.27) (see Chapter 5):

$$\Gamma(f) = G(f) \prod_{i=1}^{n}\left[\delta\left(f_i\right) - j\frac{1}{\pi f_i} \right] \tag{10.14}$$

The notion of the causality of 2-D and higher dimensional signals does not have the same physical sense as 1-D time signals. It only has the system-theoretical sense. This statement does not exclude applications. For example, an image with a rectangular frame can be made *causal* by a proper choice of the center of the Cartesian coordinates (Figure 10.1) The polar form may define the frequency domain amplitude and phase functions.

Example 10.2

Figure 10.2 shows a causal 2-D rectangular function $g(x_1, x_2) = 1$ inside the support $0 < x_1 < 1$, $0 < x_2 < 1$. The even-even term equals 0.25 for $-1 < x_1 < 1$ and $-1 < x_2 < 1$. The Fourier transform of this function is

$$g_{ee}\left(x_1, x_2\right) \overset{2F}{\Leftrightarrow} G\left(f_1, f_2\right) = 0.25 \frac{\sin\left(2\pi f_1\right)}{2\pi f_1} \frac{\sin\left(2\pi f_2\right)}{2\pi f_2}$$

Figure 10.1 The proper choice of the coordinate system makes the image causal.

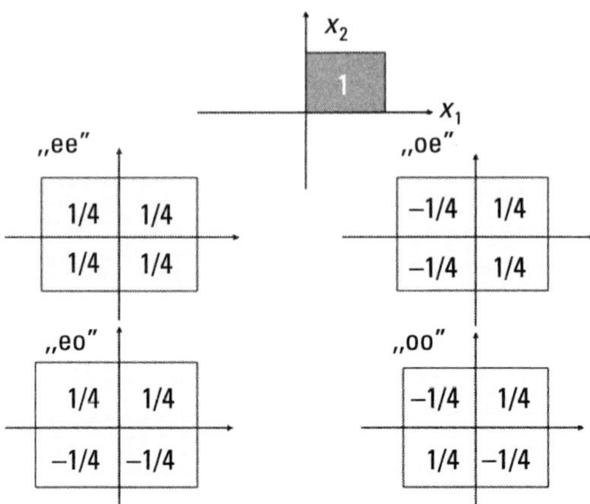

Figure 10.2   The decomposition of the causal cube into even and odd parts.

and

$$\text{sgn}(x_1)\text{sgn}(x_2)g_{ee}(x_1,x_2) \overset{2F}{\Leftrightarrow} -\text{H}\big[G(f_1,f_2)\big] = -0.25\frac{\sin^2(\pi f_1)}{\pi f_1}\frac{\sin^2(\pi f_2)}{\pi f_2}$$

$$\text{sgn}(x_1)g_{ee}(x_1,x_2) \overset{2F}{\Leftrightarrow} -\text{H}_1\big[G(f_1,f_2)\big] = -j0.25\frac{\sin^2(\pi f_1)}{\pi f_1}\frac{\sin(2\pi f_2)}{2\pi f_2}$$

The insertion of the above terms into (10.10) yields the function $\Gamma(f_1,f_2)$.

10.3   Summary

By definition, an n-D signal is called *causal*, if its support is limited to a single orthant in the signal space $\mathbf{x}$ denoted with $\mathbb{R}^+$. Using the duality of the theory of analytic signals with single-orthant spectra, the spectrum of a causal signal is analytic (a boundary distribution of an analytic function). Let us recall that, in the case of signals with single-orthant spectra, the real signal can be reconstructed by knowledge of 2^{n-1} analytic signals. A causal signal is, by definition, a signal with a single-orthant support in the $\mathbf{x}$-domain. Therefore, we have a single Fourier spectrum which defines the dispersion relations. Equations (10.8) and (10.14) define a single Fourier pair

$$g(\mathbf{x}) = g_{ee\ldots e}(\mathbf{x})\prod_{i=1}^{i=n}\left[1+\mathrm{sgn}(x_i)\right] \overset{nF}{\Longleftrightarrow} \Gamma(\mathbf{x}) = \mathrm{Re}(\mathbf{x}) - j \cdot \mathrm{H}_i[\mathrm{Re}(\mathbf{x})]$$

with free choice of i from $1,2,\ldots,n$. Let us write these relations for 1-D, 2-D, and 3-D signals.

1-D:

$$g(t) = g_e(t)(1+\mathrm{sgn}(t)) \overset{F}{\Longleftrightarrow} \Gamma(f) = G(F) - H[G(f)]$$

$$g_e(t) \overset{F}{\Longleftrightarrow} G(f)$$

2-D:

$$g(x_1,x_2) = g_{ee}(x_1,x_2)(1+s_1+s_2+s_1s_2) \overset{2F}{\Longleftrightarrow} \Gamma(f_1,f_2)$$

$$\Gamma(f_1,f_2) = G(f_1,f_2) - H[G] - j \cdot (H_1[G]+H_2[G]),$$

$$g_{ee}(x_1,x_2) \overset{2F}{\Longleftrightarrow} G(f_1,f_2)$$

The imaginary term is the partial Hilbert transform of the real term (free choice of H_1 or H_2).

3-D:

$$g(x_1,x_2,x_3) = g_{eee}\left(1+s_1+s_2+s_3+s_1s_2+s_1s_3+s_2s_3+s_1s_2s_3\right)$$

$$\overset{3F}{\Longleftrightarrow} G(f_1,f_2,f_3) - H_{12}[G] - H_{13}[G] - H_{23}[G]$$

$$- j \cdot (H_1[G]+H_2[G]+H_3[G]-H[G])$$

where $s_i = \mathrm{sgn}\, x_i$, $g_{eee}(x_1,x_2,x_3) \overset{3F}{\Longleftrightarrow} G(f_1,f_2,f_3)$. Again, the imaginary part is given by the partial Hilbert transform H_i of the real part with a free choice of $i = 1$ or 2 or 3.

A Comment about the Paper of Nieto-Vesperinas [6]

Let us present the formulas describing dispersion relations for 2-D signals derived by Nieto-Vesperinas in [6]

$$\operatorname{Re} F\left(x_1, x_2\right) = \frac{1}{\pi^2} \int\limits_{-\infty}^{\infty} \int\limits_{-\infty}^{\infty} \frac{\operatorname{Re} F\left(x_1', x_2'\right)}{\left(x_1' - x_1\right)\left(x_2' - x_2\right)}\, dx_1'\, dx_2'$$

$$+ \frac{1}{\pi} \int\limits_{-\infty}^{\infty} \frac{\operatorname{Im} F\left(x_1, x_2'\right)}{x_2' - x_2}\, dx_2' - \frac{1}{\pi} \int\limits_{-\infty}^{\infty} \frac{\operatorname{Re} F\left(x_1, x_2'\right)}{x_2' - x_2}\, dx_2'$$

$$\operatorname{Im} F\left(x_1, x_2\right) = \frac{1}{\pi^2} \int\limits_{-\infty}^{\infty} \int\limits_{-\infty}^{\infty} \frac{\operatorname{Im} F\left(x_1', x_2'\right)}{\left(x_1' - x_1\right)\left(x_2' - x_2\right)}\, dx_1'\, dx_2'$$

$$- \frac{1}{\pi} \int\limits_{-\infty}^{\infty} \frac{\operatorname{Re} F\left(x_1, x_2'\right)}{x_2' - x_2}\, dx_2' - \frac{1}{\pi} \int\limits_{-\infty}^{\infty} \frac{\operatorname{Re} F\left(x_1, x_2'\right)}{x_2' - x_2}\, dx_2'$$

The authors of this book present the above equations for historical reasons. The paper [6] is dated 1980, while the theory of signals with single-orthant spectra is dated 1992. Nieto-Vesperinas did not use the term *Hilbert transform* and the notion of causality. The above equations have been reprinted by Fiddy [1] (with some typos).

References

[1]   Fiddy, M. A., "The Role of Analyticity in Image Recovery" in *Image Recovery: Theory and Applications*, H. Stark (Ed.), Orlando, FL: Academic Press, 1987, pp.499–529.

[2]   Hahn, S. L., *Hilbert Transforms in Signal Processing*, Norwood, MA: Artech House, 1996.

[3]   Hahn, S. L., "The Definition of Causality of Multidimensional Signals," *Kleinheubacher Berichte*, Band 45, Deutsche Telecom, Darmstadt, 2002, pp. 108–110.

[4]   Kramers, H. A., "La diffusion de la lumière par les atomes," in *Atti del Congresso Internazionale dei Fisici*, Vol. 2, Bologna: Zanichelli, 1927, pp. 545–557.

[5]   Kronig, K., "On the Theory of Dispersion of X-Rays," *J. Optical Soc. Amer.*, Vol. 12, 1926, pp. 547–557.

[6]   Nieto-Vesperinas, M., "Dispersion relations in two-dimensions: Applications in the phase problem," *Optik*, Vol. 56, No. 4, 1980, pp. 377–384.

11

Summary

Complex or hypercomplex? This is the question.

This question can sometimes arise if we analyze or process n-dimensional signals. The authors do not give the direct answer to this question. The content of this book presents both approaches. The reader has a choice to apply in his studies the complex or hypercomplex approach or eventually both, and decide which has more advantages in solving a specific problem.

The starting point in writing this book was Hahn's previous book [1]. In [1], the reader will find a description of properties of n-D complex analytic signals defined in 1992 in the Proceedings of IEEE [2]. In [1], Hahn presented the extension of the notion of Gabor's analytic signal [3] to higher dimensions. However, the content of this book correlates weakly with the content of [1]. Basic knowledge about complex n-D analytic signals is repeated and considerably enlarged in [4]. All descriptions of hypercomplex signals are new, and to a large extent, are included in Snopek's papers [5, 6]. Specifically, the reader will find a large survey of hypercomplex algebra with many tables and a comparison of complex and hypercomplex Fourier transformations. Chapters presenting the notion of ranking of analytic signals and of quasi-analytic signals are new. A large survey of Wigner distributions and ambiguity functions of analytic signals is presented in Chapter 9 in this book. The description of the polar forms of complex and hypercomplex analytic signals are to a large extent new [4]. For 1-D signals, much information included in [1] could not be repeated in this book.

Knowledge about the hypercomplex analytic signals is relatively new. The quaternion analytic signal, including its polar representation, has been described by Bülow and Sommer in [7] and [8]. Shortly thereafter, Thomas Bülow visited Professor Hahn; Bülow believed that the quaternion analytic signal is superior with regard to its complex version because it can be reconstructed from a single quaternion signal in comparison to two 2-D complex signals. Professor Hahn pointed out that the number of functions of the polar representation is the same in the complex case (two amplitudes and two phase functions) and in the quaternion case (a single amplitude and three phase functions). Moreover, there are closed formulae enabling calculation of the quaternion set starting with the complex set (see Chapter 7). In the case of real 2-D separable signals (a product of 1-D signals), in both cases the polar form of the complex and quaternion signals defines a single amplitude and two phase.

The 3-D real signal is represented by four complex analytic signals. The polar form of these signals defines four amplitudes and four phase functions. The corresponding hypercomplex analytic signal has the form of an octonion function. Its polar form defines a single amplitude (the norm of the octonion) and seven phase functions. The exact relations between the four amplitudes and four phase functions of the complex case and the single amplitude and seven phase functions of the octonion are unknown. However, approximate relations are described in [4] (details in Chapter 7).

Again, for separable 3-D real signals the polar form in both cases defines a single amplitude and three phase functions. The complex signals with single-orthant spectra can be defined for any dimension. The hypercomplex analytic signal representing a 4-D real signal has the form of a sedenion. Its polar form is unknown. Readers will find the definition of the monogenic signal described in the paper of Felsberg and Sommer [9] and in Chapter 5 of this book. Its Wigner distributions are described in Chapter 9. Evidently, the support of the spectrum of the monogenic signal is not restricted to a single quadrant.

In writing this book, the authors have been unable to include everything they might have wished due to the constraints on the number of pages available for print. Many applications can be found in the list of references of successive chapters, such as the controversy about the uniqueness of the notion of the instantaneous amplitude and phase defined by Gabor's analytic signal described in [10]. In conclusion, we believe that this book is the first presenting in a single-volume theoretical background of n-D complex and hypercomplex analytic signals.

This book is a common work of both authors. The content has been designed by both authors. Stefan Hahn is the author of Chapters 1, 3, 5, 7, 8, and 10. Kajetana Snopek is the author of Chapters 2, 4, 6, and 9. Each author served as a reviewer for chapters written by the other. The authors believe that their intellectual contributions are equal.

References

[1] Hahn, S. L., *Hilbert Transforms in Signal Processing*, Norwood, MA: Artech House, 1996.

[2] Hahn, S. L., "Multidimensional Complex Signals with Single-Orthant Spectra," *Proc. IEEE*, Vol. 80, No. 8, 1992, pp. 1287–1300.

[3] Gabor, D., "Theory of Communications," *Trans. Inst. Electr. Eng.*, Vol. 3, 1946, pp. 429–456.

[4] Hahn, S. L., and K. M. Snopek, "The Unified Theory of n-Dimensional Complex and Hypercomplex Analytic Signals," *Bull. Polish Ac. Sci., Tech. Sci.*, Vol. 59, No. 2, 2011, pp. 167–181.

[5] Snopek, K. M., "New Hypercomplex Analytic Signals and Fourier Transforms in Cayley-Dickson Algebras," *Electr. Tel. Quarterly*, Vol. 55, No. 3, 2009, pp. 403–419.

[6] Snopek, K.M., *Studies of Complex and Hypercomplex Analytic Signals*, Dissertation, Warsaw University of Technology, Oficyna Wydawnicza, Prace Naukowe, Elektronika, No. 190, 2013, p. 190.

[7] Bülow, T., "Hypercomplex spectral signal representation for the processing and analysis of images," in Bericht Nr. 99–3, Institut für Informatik und Praktische Mathematik, Christian-Albrechts-Universität Kiel, August 1999.

[8] Bülow, T., and G. Sommer, "Hypercomplex Signals—A Novel Extension of the Analytic Signal to the Multidimensional Case," *IEEE Trans. Sign. Proc.*, Vol. 49, No. 11, 2001, pp. 2844–2852.

[9] Felsberg, M., and G. Sommer, "The Monogenic Signal," *IEEE Trans. Sign. Proc.*, Vol. 49, No. 12, 2001, pp. 3136–3144.

[10] Hahn, S. L., "On the uniqueness of the definition of the amplitude and phase of the analytic signal," *Signal Processing*, Vol. 83, 2003, pp. 1815–1820.

Appendix A

Table of Properties of 1-D Fourier Transformation

No.	Name	Signal	Fourier transform
1	Notation	$u(t)$ or $F^{-1}\{U(f)\}$	$U(f)$ or $F\{u(t)\}$
2	Integral definition	$u(t) = \int_{-\infty}^{\infty} U(f)e^{j2\pi ft}\,df$	$U(f) = \int_{-\infty}^{\infty} u(t)e^{-j2\pi ft}\,dt$
3	Duality	$U(t)$	$u(-f)$
4	Linearity	$a \cdot u_1(t) + b \cdot u_2(t);\ a,b \in \mathbb{R}$	$a \cdot U_1(f) + b \cdot U_2(f)$
5	Scaling	$u(a \cdot t);\ a \in \mathbb{R}^+$	$\dfrac{1}{|a|}U\left(\dfrac{f}{a}\right)$
6	Time reversal	$u(-t)$	$U(-f)$
7	Conjugation	$u^*(t)$	$U^*(-f)$
8	Time shift	$u(t-a);\ a \in \mathbb{R}$	$U(f)e^{-j2\pi fa}$
9	Modulation	$u(t)e^{j2\pi f_0 t};\ f_0 \in \mathbb{R}$ $u(t)\cos(2\pi f_0 t)$ $u(t)\sin(2\pi f_0 t)$	$U(f - f_0)$ $\dfrac{1}{2}\left[U(f - f_0) + U(f + f_0)\right]$ $\dfrac{1}{2j}\left[U(f - f_0) - U(f + f_0)\right]$
10	Time differentiation	$\dfrac{d^n u(t)}{dt^n}$	$(j2\pi f)^n U(f)$
11	Frequency differentiation	$(-jt)^n u(t)$	$\dfrac{d^n U(f)}{df^n}$
12	Integration	$\int_{-\infty}^{t} u(\tau)\,d\tau$	$\dfrac{1}{j2\pi f}U(f)$
13	Product	$u_1(t) \cdot u_2(t)$	$U_1(f) * U_2(f)$

No.	Name	Signal	Fourier transform
14	Convolution	$u_1(t) * u_2(t) = \int_{-\infty}^{\infty} u_1(\tau)u_2(t-\tau)\,d\tau$	$U_1(f) \cdot U_2(f)$
15	Autocorrelation	$R_u(\tau) = \int_{-\infty}^{\infty} u(t)u^*(t-\tau)\,dt$	$\|U(f)\|^2$
16	Cross-correlation	$R_{12}(\tau) = \int_{-\infty}^{\infty} u_1(t)u_2^*(t-\tau)\,dt$	$U_1(f)U_2^*(f)$
		$R_{21}(\tau) = \int_{-\infty}^{\infty} u_2(t)u_1^*(t-\tau)\,dt$	$U_2(f)U_1^*(f)$
17	Energy equality (Raileigh's theorem)	$\int_{-\infty}^{\infty} \|u(t)\|^2 dt = \int_{-\infty}^{\infty} \|U(f)\|^2 df$	
18	Scalar product (Parseval's theorem)	$\int_{-\infty}^{\infty} u_1(t)u_2^*(t)\,dt = \int_{-\infty}^{\infty} U_1(f)U_2^*(f)\,df$	
19	Time integral	$\int_{-\infty}^{\infty} u(t)\,dt$	$U(0)$
20	First moment	$\int_{-\infty}^{\infty} t \cdot u(t)\,dt$	$jU'(0)$
21	Normalized first moment	$\dfrac{\int_{-\infty}^{\infty} t \cdot u(t)\,dt}{\int_{-\infty}^{\infty} u(t)\,dt}$	$j\dfrac{U'(0)}{U(0)}$
22	Second-order moment	$\int_{-\infty}^{\infty} t^2 \cdot u(t)\,dt$	$-U''(0)$
23	Normalized second-order moment	$\dfrac{\int_{-\infty}^{\infty} t^2 \cdot u(t)\,dt}{\int_{-\infty}^{\infty} u(t)\,dt}$	$-\dfrac{U''(0)}{U(0)}$
24	nth moment	$\int_{-\infty}^{\infty} t^n \cdot u(t)\,dt$	$j^n U^{(n)}(0)$

Selected Bibliography

Oppenheim, A.V., A. S. Willsky, and I. T. Young, *Signals and Systems*, Englewood Cliffs, NJ: Prentice Hall Signal Processing Series, 1984.

Poularikas, A. D. (ed.), *The Transforms and Applications Handbook, Second Edition*, Boca Raton, FL: CRC Press, Inc., 2000.

Allen, R. L., and D. W. Mills, *Signal Analysis: Time, Frequency, Scale and Structure*, Piscataway, NJ: Wiley-Interscience, IEEE Press, 2004.

Appendix B

Table of Chosen 1-D Fourier Pairs

No.	Name	Signal	Fourier Transform
1	Notation	$u(t)$	$U(f) = \mathrm{Re}\,U(f) + j\mathrm{Im}\,U(f)$
2	Real	$\mathrm{Re}\,u(t) = u(t)$	$U(f) = U^*(-f),\ \mathrm{Re}\,U(f) = \mathrm{Re}\,U(-f),\ \mathrm{Im}\,U(f) = -\mathrm{Im}\,U(-f),\ \lvert U(f)\rvert = \lvert U(-f)\rvert,\ \arg U(f) = -\arg U(-f)$
3	Even part	$u_e(t)$	$\mathrm{Re}\,U(f)$
4	Odd part	$u_o(t)$	$j\mathrm{Im}\,U(f)$
5	Even signal	$u(t) = u(-t)$	$U(f) = U(-f)$
6	Odd signal	$u(t) = -u(-t)$	$U(f) = -U(-f)$
7	Analytic signal	$u(t) + j \cdot v(t)$	$(1 + \mathrm{sgn}\,f)U(f)$
8	Conjugate analytic signal	$u(t) - j \cdot v(t)$	$(1 - \mathrm{sgn}\,f)U(f)$
9	Dirac's delta pulse	$\delta(t)$ $\delta(t - a)$	1 $\exp(-j2\pi f a)$
10	Constant	1	$\delta(f)$
11	Complex exponential	$\exp(j2\pi f_0 t);\ f_0 \in \mathbb{R}$	$\delta(f - f_0)$
12	Signum	$\mathrm{sgn}\,t$	$\dfrac{1}{j\pi f}$
13	Hyperbolic	$\dfrac{1}{\pi t}$	$-j\,\mathrm{sgn}\,f$
14	Unit step	$\mathbf{1}(t)$	$\dfrac{1}{2}\delta(f) + \dfrac{1}{j2\pi f}$
15	Cosine	$\cos(2\pi f_0 t);\ f_0 > 0$	$\dfrac{1}{2}\delta(f - f_0) + \dfrac{1}{2}\delta(f + f_0)$

No.	Name	Signal	Fourier Transform
16	Sine	$\sin(2\pi f_0 t);\ f_0 > 0$	$\dfrac{1}{2j}\delta(f - f_0) - \dfrac{1}{2j}\delta(f + f_0)$
17	One-sided exponential	$\exp(-at)\cdot 1(t);\ \mathrm{Re}\ a > 0$	$\dfrac{1}{a + j2\pi f}$
18	Two-sided exponential	$\exp(-a\lvert t\rvert);\ \mathrm{Re}\ a > 0$	$\dfrac{2a}{a^2 + 4\pi^2 f^2}$
19	Cauchy signal	$\dfrac{1}{t^2 + a^2};\ a > 0$	$\dfrac{\pi}{a}\exp(-2\pi a\lvert f\rvert)$
20	Product of a polynomial and an exponential	$t^n \cdot \exp(-at)\cdot 1(t);$ $\mathrm{Re}\ a > 0, n = 1,2,\dots$	$\dfrac{n!}{(a + j2\pi f)^{n+1}}$
21	Rectangle pulse	$\mathrm{rect}_a(t) = \begin{cases} 1 & \lvert t\rvert < \dfrac{a}{2} \\ 0.5 & t = \pm\dfrac{a}{2} \\ 0 & \text{elsewhere} \end{cases}$	$\dfrac{\sin(\pi f a)}{\pi f}$
22	Sinc	$\dfrac{\sin(at)}{t};\ a > 0$	$\pi\,\mathrm{rect}_a(\pi f)$
23	Triangle pulse	$\mathrm{tri}_a(t) = \begin{cases} \dfrac{1}{a}(a - \lvert t\rvert) & \lvert t\rvert \le a \\ 0 & \text{elsewhere} \end{cases}$	$\dfrac{\sin^2(\pi f a)}{\pi^2 f^2 a}$
24	Sinc squared	$\dfrac{\sin^2(at)}{t^2};\ a > 0$	$a\pi \cdot \mathrm{tri}_a(\pi f)$
25	Gaussian	$\exp\!\left(-\dfrac{t^2}{2a^2}\right);\ a > 0$	$\sqrt{2\pi}\,\lvert a\rvert\exp(-8\pi^2 a^2 f^2)$
26	Fourier series	$x(t) = x(t + T_0);\ T_0 > 0$ $x(t) = \displaystyle\sum_{n\in\mathbb{Z}} c_n \exp\!\left(j2\pi\dfrac{n}{T_0}t\right)$	$\displaystyle\sum_{n\in\mathbb{Z}} c_n\delta\!\left(f - \dfrac{n}{T_0}\right)$
27	Comb	$\displaystyle\sum_{n\in\mathbb{Z}} \delta(t - nT_0)$	$\dfrac{1}{T_0}\displaystyle\sum_{n\in\mathbb{Z}} \delta\!\left(f - \dfrac{n}{T_0}\right)$

Selected Bibliography

Allen, R. L., and D. W. Mills, *Signal Analysis: Time, Frequency, Scale and Structure*, Piscataway, NJ: Wiley-Interscience,, IEEE Press, 2004.

Oppenheim, A. V., A. S. Willsky, and I. T. Young, *Signals and Systems*, Englewood Cliffs, NJ: Prentice Hall Signal Processing Series, 1984.

Poularikas, A. D. (ed.), *The Transforms and Applications Handbook, Second Edition*, Boca Raton, FL: CRC Press, Inc., 2000.

Appendix C

Table of Properties of 2-D Fourier Transformations

No.	Name	Signal	Fourier Transform
1	Notation	$u(x_1, x_2)$ or $F^{-1}\{U(f_1, f_2)\}$	$U(f_1, f_2)$ or $F\{u(x_1, x_2)\}$
2	Integral definition	$u(x_1, x_2)$ $$= \int\limits_{-\infty}^{\infty}\int\limits_{-\infty}^{\infty} U(f_1, f_2) e^{j2\pi(f_1 x_1 + f_2 x_2)} \, df_1 \, df_2$$	$U(f_1, f_2)$ $$= \int\limits_{-\infty}^{\infty}\int\limits_{-\infty}^{\infty} u(x_1, x_2) e^{-j2\pi(f_1 x_1 + f_2 x_2)} \, dx_1 \, dx_2$$ $$U(f_1, f_2) = F_{x_2}\{F_{x_1}\{u(x_1, x_2)\}\}$$ F_{x_i} are 1-D FTs w.r.t. x_i
3	Identity	$F\{F\{u(x_1, x_2)\}\} = u(-x_1, -x_2)$ $F^{-1}\{F\{u(x_1, x_2)\}\} = u(x_1, x_2)$	
4	Duality	$U(x_1, x_2)$	$u(-f_1, -f_2)$
5	Hermitian symmetry	$U(f_1, f_2) = U^*(-f_1, -f_2)$ $U(f_1, -f_2) = U^*(-f_1, f_2)$ $U(-f_1, f_2) = U^*(f_1, -f_2)$	
6	Linearity	$a \cdot u_1(x_1, x_2) + b \cdot u_2(x_1, x_2); \; a, b \in \mathbb{R}$	$a \cdot U_1(f_1, f_2) + b \cdot U_2(f_1, f_2)$
7	Scaling	$u(ax_1, bx_2); \; a, b \in \mathbb{R}^+$	$\dfrac{1}{|ab|} U\left(\dfrac{f_1}{a}, \dfrac{f_2}{b}\right)$
8	Signal-domain reversal	$u(-x_1, -x_2)$	$U(-f_1, -f_2)$
9	Conjugation	$u^*(f_1, f_2)$	$U^*(-f_1, -f_2)$

No.	Name	Signal	Fourier Transform
10	Signal-domain shift	$u(x_1 - a, x_2 - b);\ a, b \in \mathbb{R}$	$U(f_1, f_2) e^{-j2\pi(f_1 a + f_2 b)}$
11	Modulation	$u(x_1, x_2) e^{j2\pi(f_a x_1 + f_b x_2)};\ f_a, f_b \in \mathbb{R}$	$U(f_1 - f_a, f_2 - f_b)$
12	Signal-domain differentiation	$\partial^2 u(x_1, x_2)/\partial x_1 \partial x_2$ $\partial u(x_1, x_2)/\partial x_i;\ i = 1,2$ $\partial^2 u(x_1, x_2)/\partial x_i^2;\ i = 1,2$	$-4\pi^2 f_1 f_2 \cdot U(f_1, f_2)$ $j2\pi f_i \cdot U(f_1, f_2)$ $-4\pi^2 f_i^2 \cdot U(f_1, f_2)$
13	Separability	$u_1(x_1) \cdot u_2(x_2)$	$U_1(f_1) \cdot U_2(f_2)$
14	Product	$u_1(x_1, x_2) \cdot u_2(x_1, x_2)$	$U_1(f_1, f_2) ** U_2(f_1, f_2)$
15	Convolution	$u_1(x_1, x_2) ** u_2(x_1, x_2)$	$U_1(f_1, f_2) \cdot U_2(f_1, f_2)$
16	Auto-correlation	$R_u(\chi_1, \chi_2)$ $= \int\limits_{-\infty}^{\infty} \int\limits_{-\infty}^{\infty} u(x_1, x_2) u^*(x_1 - \chi_1, x_2 - \chi_2)\, dx_1\, dx_2$	$\left\| U(f_1, f_2) \right\|^2$
17	Cross-correlation	$R_{12}(\chi_1, \chi_2)$ $= \int\limits_{-\infty}^{\infty} \int\limits_{-\infty}^{\infty} u_1(x_1, x_2) u_2^*(x_1 - \chi_1, x_2 - \chi_2)\, dx_1\, dx_2$ $R_{21}(\chi_1, \chi_2)$ $= \int\limits_{-\infty}^{\infty} \int\limits_{-\infty}^{\infty} u_2(x_1, x_2) u_1^*(x_1 - \chi_1, x_2 - \chi_2)\, dx_1\, dx_2$	$U_1(f_1, f_2) U_2^*(f_1, f_2)$ $U_2(f_1, f_2) U_1^*(f_1, f_2)$
18	Energy equality (Rayleigh's theorem)	$\int\limits_{-\infty}^{\infty} \int\limits_{-\infty}^{\infty} \left\| u(x_1, x_2) \right\|^2 dx_1\, dx_2 = \int\limits_{-\infty}^{\infty} \int\limits_{-\infty}^{\infty} \left\| U(f_1, f_2) \right\|^2 df_1\, df_2$	
19	Scalar product (Parseval's theorem)	$\int\limits_{-\infty}^{\infty} \int\limits_{-\infty}^{\infty} u_1(x_1, x_2) u_2^*(x_1, x_2)\, dx_1\, dx_2 = \int\limits_{-\infty}^{\infty} \int\limits_{-\infty}^{\infty} U_1(f_1, f_2) U_2^*(f_1, f_2)\, df_1 df_2$	
20	Circular symmetry	$u(x_1, x_2) = u(r);\ r = \left(x_1^2 + x_2^2\right)^{1/2}$	$U(f_1, f_2) = U(\rho\cos\phi, \rho\sin\phi)$ $= 2\pi \int\limits_0^{\infty} r u(r) J_0(2\pi r \rho)\, dr = \tilde{U}(\rho);$ $\tilde{U}(\rho) -$ Hankel transform, $J_0(\cdot) -$ 0-order Bessel function $\rho = \left(f_1^2 + f_2^2\right)^{1/2}$

No.	Name	Signal	Fourier Transform
21	Signal-domain integral	$\displaystyle\int\limits_{-\infty}^{\infty}\int\limits_{-\infty}^{\infty} u\left(x_1, x_2\right) dx_1\, dx_2$	$U(0,0)$

Selected Bibliography

Gonzalez, R. C., and P. Wintz, *Digital Image Processing, Second Edition*, Upper Saddle River, NJ: Addison-Wesley Publishing Company, Inc., 1987.

Poularikas, A.D. (ed.), *The Transforms and Applications Handbook, Second Edition*, Boca Raton, FL: CRC Press, Inc., 2000.

Stark, H., and F. B. Tuteur, *Modern Electrical Communications: Theory and Systems*, Englewood Cliffs, NJ: Prentice-Hall, Inc., 1979.

Appendix D

Chosen 2-D Fourier Pairs

No.	Name	Signal	Fourier Transform
1	Notation	$u(x_1, x_2)$	$U_1(f_1, f_2) = \mathrm{Re}\, U_1(f_1, f_2) + j\, \mathrm{Im}\, U_1(f_1, f_2)$
2	Constant	1	$\delta(f_1, f_2)$
3	Dirac's delta pulse	$\delta(ax_1, bx_2);\ a,b \in \mathbb{R}\backslash\{0\}$ $\delta(x_1 - a, x_2 - b)$	$\dfrac{1}{\|ab\|}\delta(x_1, x_2)$ $\exp\left(-j2\pi(f_1 a + f_2 b)\right)$
4	Complex exponential	$\exp\left[\,j2\pi(ax_1 + bx_2)\right];\ a,b \in \mathbb{R}$	$\delta(f_1 - a, f_2 - b)$
5	Signum	$\mathrm{sgn}(x_1)\mathrm{sgn}(x_2)$	$\dfrac{1}{j\pi f_1}\dfrac{1}{j\pi f_2}$
6	Cauchy signal	$\dfrac{1}{x_1^2 + a^2}\dfrac{1}{x_2^2 + b^2};\ a,b > 0$	$\dfrac{\pi^2}{ab}\exp\left[-2\pi\left(a\|f_1\| + b\|f_2\|\right)\right]$
7	Circle	$u(r) = \begin{cases} 1, & \|r\| < a \\ 0, & \|r\| \geq a \end{cases}$; $r = x_1^2 + x_2^2,\ a > 0$	$U(\rho) = a\dfrac{J_1(\pi a \rho)}{\rho};\ \rho^2 = f_1^2 + f_2^2$
8	Rectangle	$\mathrm{rect}_{a,b}(x_1, x_2) = \begin{cases} 1 & \|x_1\| < a/2, \|x_2\| < b/2 \\ 0.5 & x_1 = \pm a/2, x_2 = \pm b/2 \\ 0 & \text{elsewhere} \end{cases}$	$\dfrac{1}{\pi^2}\dfrac{\sin(\pi f_1 a)}{f_1}\dfrac{\sin(\pi f_2 b)}{f_2}$
9	Triangle	$\mathrm{tri}_a(x_1)\mathrm{tri}_b(x_2)$	$\dfrac{1}{\pi^4}\dfrac{\sin^2(\pi f_1 a)}{f_1^2 a}\dfrac{\sin^2(\pi f_2 b)}{f_2^2 b}$
10	Pyramid	$\begin{cases} \dfrac{h}{a}\left(a - \|x_1\| - \|x_2\|\right), & (x_1, x_2) \in S \\ 0, & (x_1, x_2) \notin S \end{cases}$ h – height, S – support	$\dfrac{4h}{4\pi^2\left(f_2^2 - f_1^2\right)}\left[\dfrac{\sin(2\pi f_1 a)}{2\pi f_1 a} - \dfrac{\sin(2\pi f_2 a)}{2\pi f_2 a}\right]$

No.	Name	Signal	Fourier Transform
11	Gaussian	$\exp\left[-\pi\left(a^2 x_1^2 + b^2 x_2^2\right)\right];\ a,b > 0$	$\dfrac{1}{\lvert ab \rvert}\exp\left[-\pi\left(\dfrac{f_1^2}{a^2} + \dfrac{f_2^2}{b^2}\right)\right]$
		$\dfrac{1}{2\pi\sigma_1\sigma_2\sqrt{1-\rho^2}}e^{\frac{-1}{2(1-\rho^2)}\left(\frac{x_1^2}{\sigma_1^2} + \frac{x_2^2}{\sigma_2^2} - \frac{2\rho x_1 x_2}{\sigma_1\sigma_2}\right)}$	$e^{-2\pi^2\left(\sigma_1^2 f_1^2 + \sigma_2^2 f_2^2 + 2\rho\sigma_1\sigma_2 f_1 f_2\right)}$
		$u(r) = \dfrac{1}{2\pi\sigma^2}\exp\left(-\dfrac{r^2}{2\sigma^2}\right);\ \sigma > 0,\ r^2 = x_1^2 + x_2^2$	$U(\rho) = \exp\left(-2\pi^2\rho^2\sigma^2\right);\ \rho^2 = f_1^2 + f_2^2$
12	Comb	$\mathrm{comb}(x_1)\mathrm{comb}(x_2) =$ $\displaystyle\sum_{n,m\in\mathbb{Z}} \delta(x_1 - nT_a)\delta(x_2 - mT_b)$	$\dfrac{1}{T_a T_b}\displaystyle\sum_{n,m\in\mathbb{Z}} \delta\left(f_1 - \dfrac{n}{T_a}\right)\delta\left(f_2 - \dfrac{m}{T_b}\right)$

Selected Bibliography

Poularikas, A. D. (Ed.), *The Transforms and Applications Handbook, Second Edition*, Boca Raton, FL: CRC Press, Inc., 2000.

Stark, H., and F. B. Tuteur, *Modern Electrical Communications: Theory and Systems*, Englewood Cliffs, NJ: Prentice-Hall, Inc., 1979.

Appendix E

Table of Properties of Quaternion Fourier Transformation of Real Signals

No.	Name	Real Signal	Quaternion Fourier Transform
1	Notation	$u(x_1, x_2)$	$U_q(f_1, f_2)$
2	Right-sided QFT*	$u(x_1, x_2)$ $= \int_{\mathbb{R}^2} U_q(f_1, f_2) e^{e_2 2\pi f_2 x_2} e^{e_1 2\pi f_1 x_1}\, df_1 df_2$	$U_q(f_1, f_2)$ $= \int_{\mathbb{R}^2} u(x_1, x_2) e^{-e_1 2\pi f_1 x_1} e^{-e_2 2\pi f_2 x_2}\, dx_1 dx_2$
3	Two-sided QFT*	$u(x_1, x_2)$ $= \int_{\mathbb{R}^2} e^{e_2 2\pi f_2 x_2} U_q(f_1, f_2) e^{e_1 2\pi f_1 x_1}\, df_1 df_2$	$U_q(f_1, f_2)$ $= \int_{\mathbb{R}^2} e^{-e_1 2\pi f_1 x_1} u(x_1, x_2) e^{-e_2 2\pi f_2 x_2}\, dx_1 dx_2$
4	Relation to 2-D FT	$U_q(f_1, f_2) = U(f_1, f_2) \dfrac{1 - e_3}{2} + U(f_1, -f_2) \dfrac{1 + e_3}{2}$	
5	Duality	$U_q(x_1, x_2)$	$u(-f_1, -f_2)$
6	Involutions	$U_q(-f_1, f_2) = -e_2 \cdot U_q(f_1, f_2) \cdot e_2$ $U_q(f_1, -f_2) = -e_1 \cdot U_q(f_1, f_2) \cdot e_1$ $U_q(-f_1, -f_2) = -e_3 \cdot U_q(f_1, f_2) \cdot e_3$	
7	Linearity	$a \cdot u(x_1, x_2) + b \cdot w(x_1, x_2);\ a, b \in \mathbb{R}$	$a \cdot U_q(f_1, f_2) + b \cdot W_q(f_1, f_2)$
8	Scaling	$u(ax_1, bx_2);\ a, b \in \mathbb{R}^+$	$\dfrac{1}{|ab|} U_q\left(\dfrac{f_1}{a}, \dfrac{f_2}{b} \right)$
9	Signal-domain reversal	$u(-x_1, -x_2)$	$U_q(-f_1, -f_2)$

No.	Name	Real Signal	Quaternion Fourier Transform
10	Signal-domain shift	$u(x_1 - a, x_2 - b);\ a,b \in \mathbb{R}$	$e^{-e_1 2\pi f_1 a} \cdot U_q(f_1, f_2) \cdot e^{-e_2 2\pi f_2 b}$
11	Modulation	$e^{e_1 2\pi a x_1} \cdot u(x_1, x_2) \cdot e^{e_2 2\pi b x_2};\ a,b \in \mathbb{R}$	$U_q(f_1 - a, f_2 - b)$
12	Signal-domain differentiation	$\partial^2 u(x_1, x_2)/\partial x_1 \partial x_2$ $\partial u(x_1, x_2)/\partial x_1$ $\partial u(x_1, x_2)/\partial x_2$ $\partial^2 u(x_1, x_2)/\partial x_i^2;\ i = 1,2$	$4\pi^2 e_1 f_1 \cdot U_q(f_1, f_2) \cdot e_2 f_2$ $2\pi e_1 f_1 \cdot U_q(f_1, f_2)$ $2\pi \cdot U_q(f_1, f_2) \cdot e_2 f_2$ $-4\pi^2 f_i^2 \cdot U(f_1, f_2)$
13	Separability	$u_1(x_1) \cdot u_2(x_2)$	$U_1(f_1) \cdot U_2(f_2)$
14	Energy equality (Rayleigh's theorem)	$\displaystyle\int_{-\infty}^{\infty}\int_{-\infty}^{\infty} \left|u(x_1, x_2)\right|^2 dx_1\, dx_2 = \int_{-\infty}^{\infty}\int_{-\infty}^{\infty} \left|U_q(f_1, f_2)\right|^2 df_1\, df_2$	
15	Signal-domain integral	$\displaystyle\int_{-\infty}^{\infty}\int_{-\infty}^{\infty} u(x_1, x_2)\, dx_1\, dx_2$	$U_q(0,0)$

*For real signals both definitions are equivalent.

Selected Bibliography

Bülow T., M. Felsberg, and G. Sommer, "Noncommutative Hypercomplex Fourier Transforms of Multidimensional Signals," in *Geometric Computing with Clifford Algebra*, G. Sommer (ed.), Berlin: Springer-Verlag, 2001, pp. 187–207.

Appendix F

Properties of 1-D Hilbert Transformations

No.	Name	Signal or Inverse Transform	Hilbert Transform
1	Notations	$u(t)$ or $H^{-1}[v(t)]$ $\quad u(t)\overset{H}{\Leftrightarrow}v(t)$	$v(t)$ or $H[u(t)]$
2	Integral definition	$u(t) = \dfrac{1}{\pi}P\displaystyle\int_{-\infty}^{\infty}\dfrac{v(\eta)}{\eta-t}d\eta$	$v(t) = \dfrac{-1}{\pi}P\displaystyle\int_{-\infty}^{\infty}\dfrac{u(\eta)}{\eta-t}d\eta$
3	Convolution notation	$u(t) = -\dfrac{1}{\pi t}*v(t)$	$v(t) = \dfrac{1}{\pi t}*u(t)$
		(by numerical convolution no sample at $t=0$)	
4	Fourier spectra	$u(t)\overset{F}{\Leftrightarrow}U(f)$ $U(f) = j\,\mathrm{sgn}(f)V(f)$	$v(t)\overset{F}{\Leftrightarrow}V(f)$ $V(f) = -j\,\mathrm{sgn}(f)U(f)$
5	Linearity	$au_1(t) + bu_2(t)$	$av_1(t) + bv_2(t)$
6	Scaling	$u(at);\ a>0$	$v(at)$
7	Time reversal	$u(-at)$	$-v(-at)$
8	Time shift	$u(t-a)$	$v(t-a)$
9	Parity	$u_e(t) + v_o(t)$; e-even; o-odd	$v_o(t-a) + v_e(t)$
		The Hilbert transform of an even function $u_e(t) = u(-t)$ is an odd function and vice versa.	
10		Time derivatives: Two options.	
10a	First option	$\dfrac{du(t)}{dt} = \dfrac{-1}{\pi t}*\dfrac{dv[t]}{dt}$	$\dfrac{dv(t)}{dt} = \dfrac{1}{\pi t}*\dfrac{du[t]}{dt}$
10b	Second option	$\dfrac{du(t)}{dt} = \dfrac{d}{dt}\!\left(\dfrac{-1}{\pi t}\right)*v(t)$	$\dfrac{dv(t)}{dt} = \dfrac{d}{dt}\!\left(\dfrac{1}{\pi t}\right)*u(t)$

No.	Name	Signal or Inverse Transform	Hilbert Transform
11	Convolutions	$u_1(t)*u_2(t) = -v_1(t)*v_2(t)$	$u_1(t)*v_2(t) = v_1(t)*u_2(t)$
12	Autoconvolution equality	$u(t)*u(t) = -v(t)*v(t)$	
13	Energy equality	$\int_{-\infty}^{\infty} u^2(t)\,dt = \int_{-\infty}^{\infty} v^2(t)\,dt$; The energy of the DC term is rejected.	
14	Multiplication by t	$tu(t)$	$tv(t) - \dfrac{1}{\pi}\int_{-\infty}^{\infty} u(\eta)\,d\eta$
15	Multiplication of signals with nonoverlapping spectra	$u_1(t) \overset{F}{\Longleftrightarrow} U_1(f)$; low-pass spectrum $u_2(t) \overset{F}{\Longleftrightarrow} U_2(f)$; high-pass spectrum $u_1(t)u_2(t)$	$u_1(t)v_2(t)$
14	Analytic signal	$\psi(t) = u(t) + jv(t)$	$H[\psi(t)] = -j\psi(t)$
15	Product of anal. sign.	$\psi(t) = \psi_1(t)\psi_2(t)$; $H[\psi(t)] = \psi_1(t)H[\psi_2(t)] = H[\psi_1(t)]\psi_2(t)$	
16	Iterations	$H[u(t)] = v(t)$; $H\{H[u(t)]\} = -u(t)$ $H\{H[H(u(t))]\} = -v[t]$; $H\{H\{H[H(u(t))]\}\} = u(t)$	

Appendix G

1-D Hilbert Pairs

Name	Function (Signal)	Hilbert Transform
Notations	$u(t)$	$u(t)\overset{H}{\Longleftrightarrow}v(t)\qquad v(t)$
Integral definition	$u(t)=\dfrac{1}{\pi}\mathrm{P}\displaystyle\int_{-\infty}^{\infty}\dfrac{v(\eta)}{\eta-t}\,d\eta=\mathrm{H}^{-1}[v(t)]$	$v(t)=\dfrac{-1}{\pi}\mathrm{P}\displaystyle\int_{-\infty}^{\infty}\dfrac{u(\eta)}{\eta-t}\,d\eta=\mathrm{H}[u(t)]$

(by numerical calculation of the convolution no sample at $t=0$, symmetric samples around zero)

Name	Function (Signal)	Hilbert Transform
Cosine	$\cos(\omega_0 t)$	$\sin(\omega_0 t)$
Sine	$\sin(\omega_0 t)$	$-\cos(\omega_0 t)$
Delta pulse	$\delta(t)$	$\dfrac{1}{\pi t}$
Approximation of delta	$\delta(t)=\lim_{a\to 0}\left[\dfrac{a}{\pi\left(a^2+t^2\right)}\right]$	$\dfrac{1}{\pi t}=\lim_{a\to 0}\left[\dfrac{t}{\pi\left(a^2+t^2\right)}\right]$
Square pulse equals $1,\,-a<t<a$.	$-a\Pi a(t)$	$\dfrac{1}{\pi}\ln\left|\dfrac{t+a}{t-a}\right|$
Triangle pulse	$1-\left|\dfrac{t}{a}\right|;\;-a<t<a$	$\dfrac{1}{\pi}\left\{\ln\left|\dfrac{t+a}{t-a}\right|+\dfrac{t}{a}\ln\left|\dfrac{t^2+a^2}{t^2}\right|\right\}$
Gaussian pulse	$\dfrac{1}{\sqrt{2\pi}\sigma}\exp\left(\dfrac{-t^2}{2\sigma^2}\right)$	$2\displaystyle\int_{0}^{\infty}\exp\left(-2\pi^2\sigma^2 f^2\right)\sin(2\pi f t)\,df$

Name	Function (Signal)	Hilbert Transform
Sinc pulse	$\dfrac{\sin(at)}{at}$	$\dfrac{\sin^2(at/2)}{(at/2)} = \dfrac{1-\cos(at)}{at}$
Video test pulse	$\cos^2[\pi t/(2a)];\ |t| \le a$	$2\displaystyle\int_0^\infty \dfrac{2a^2}{4a^2-\omega^2}\dfrac{\sin[\pi\omega/(2a)]}{\omega}\sin(\omega t)df;$ $\omega = 2\pi f$
A constant	a	zero
Parabolic pulse	$1-(t/a)^2;\ |t| \le a$	$\dfrac{-1}{\pi}\left\{\left[1-(t/a)^2\ln\left(\dfrac{t-a}{t+a}\right)-\dfrac{2t}{a}\right]\right\}$
Symmetric expotential	$\exp(-a|t|)$	$2\displaystyle\int_0^\infty \dfrac{2a}{a^2-(2\pi f)^2}\sin(2\pi ft)\,df$
Antisymmetric exponential	$\mathrm{sgn}(t)\exp(-a|t|)$	$-2\displaystyle\int_0^\infty \dfrac{2a}{a^2-(2\pi f)^2}\cos(2\pi ft)\,df$
One-sided exponential	$[0.5+0.5\,\mathrm{sgn}(t)]\exp(-at)$	$\displaystyle\int_0^\infty \dfrac{a\sin(\omega t)-\omega\cos(\omega t)}{a^2-\omega^2}df;$ $\omega = 2\pi f$
Sampling sequence	$u(t) = \displaystyle\sum_{n=-\infty}^{\infty} \delta(t-nT)$	$v(t) = \dfrac{1}{T}\displaystyle\sum_{n=-\infty}^{\infty}\cos[(\pi/T)(t-nT)]$
Fourier series	$U_0 + \displaystyle\sum_{n=1}^{\infty} A_n\cos(n\omega_0 t + \varphi_n)$	$\displaystyle\sum_{n=1}^{\infty} A_n\sin(n\omega_0 t + \varphi_n)$
Even square wave	$\mathrm{sgn}\left[\cos(\omega_0 t)\right]$	$\dfrac{2}{\pi}\ln\left[\tan(\omega_0 t/2 + \pi/4)\right]$
Odd square wave	$\mathrm{sgn}\left[\sin(\omega_0 t)\right]$	$\dfrac{2}{\pi}\ln\left[\tan(\omega_0 t/2)\right]$
Squared cosine	$\cos^2(\omega_0 t)$	$0.5\sin(2\omega_0 t)$
Squared sine	$\sin^2(\omega_0 t)$	$-0.5\sin(2\omega_0 t)$

Hilbert transform of the interpolatory expansion of a series of samples $u\left(\dfrac{k}{2W}\right)$, sampling frequency $f_s = 2W$.

Name	Function (Signal)	Hilbert Transform
	$$u(t) = \sum_{k=-\infty}^{\infty} u\left(\frac{k}{2W}\right)\frac{\sin(y)}{y};$$ $$y = 2\pi W\left(t - \frac{k}{2W}\right)$$	$$v(t) = \sum_{k=-\infty}^{\infty} u\left(\frac{k}{2W}\right)\frac{1-\cos(y)}{y}$$
Using random values of samples enables the generation of a random signal and its Hilbert transform		
Analytic signal	$\psi(t) = u(t) + jv(t)$	$H[\psi(t)] = -j\psi(t)$
Example	$e^{j\omega t}$	$H(e^{j\omega t}) = -je^{j\omega t}$
Product of analytic signals	$\psi(t) = \psi_1(t)\,\psi_2(t)$	$H[\psi_1(t)\psi_2(t)] = -j\psi_1(t)\psi_2(t)$

Appendix H

2-D Hilbert Quadruples

The Hilbert quadruple is formed by the real signal $u(x_1,x_2)$, the partial Hilbert transform with regard to x_1 denoted $v_1(x_1,x_2)$, the partial Hilbert transform with regard to x_2 denoted $v_2(x_1,x_2)$, and the total Hilbert transform wiht regard to both variables denoted $v(x_1,x_2)$. All are functions of (x_1,x_2). Below, they are denoted as u, v_1, v_2 and v.

Integral Definitions

$$v_1 = H_1(u) = \frac{1}{\pi}P\int_{-\infty}^{\infty} \frac{u(\eta_1,x_2)}{(x_1-\eta_1)}\,d\eta_1;$$

$$v_2 = H_2(u) = \frac{1}{\pi}P\int_{-\infty}^{\infty} \frac{u(\eta_2,x_1)}{(x_2-\eta_2)}\,d\eta_2;$$

$$v = H(u) = \frac{1}{\pi^2}P\int_{-\infty}^{\infty}\int_{-\infty}^{\infty} \frac{u(\eta_1,\eta_2)}{(x_1-\eta_1)(x_2-\eta_2)}\,d\eta_1\,d\eta_2$$

P denotes the Cauchy principal value (symmetrical sampling around the orign, no sample at the origin).

Convolution Definitions

$$u = \delta(x_1,x_2)**u; \quad v_1 = H_1(u) = \frac{\delta(x_2)}{\pi x_1}**u;$$

$$v_2 = H_2(u) = \frac{\delta(x_1)}{\pi x_2}**u; \quad v = H(u) = \frac{1}{\pi^2 x_1 x_2}**u$$

Separable 2-D Functions

A separable 2-D real functions is a product of 1-D functions $u(x_1,x_2) = g_1(x_1)$ $g_2(x_2)$. Let us define 1-D Hilbert pairs $g_1(x_1)\overset{H}{\Leftrightarrow}h_1(x_1)$ and $g_2(x_2)\overset{H}{\Leftrightarrow}h_2(x_2)$, where H denotes 1-D Hilbert transformation. We have

$$u = g_1(x_1)g_2(x_2); \quad v_1 = h_1(x_1)g_2(x_2); \quad v_2 = g_1(x_1)h_2(x_2); \quad v = h_1(x_1)h_2(x_2)$$

Table H.1
2-D Hilbert Pairs

Name
2-D delta $u = \delta(x_1)\delta(x_2); \quad v_1 = \dfrac{\delta(x_2)}{\pi x_1}; \quad v_2 = \dfrac{\delta(x_1)}{\pi x_2}; \quad v = \dfrac{1}{\pi^2 x_1 x_2}$
2-D Gaussian pulse $u = \dfrac{1}{\sqrt{2\pi}\sigma_1}\exp\left(\dfrac{-x_1^2}{2\sigma_1^2}\right)\dfrac{1}{\sqrt{2\pi}\sigma_2}\exp\left(\dfrac{-x_2^2}{2\sigma_2^2}\right);$ $v_1 = \left\{2\displaystyle\int_0^{\infty}\exp\left(-2\pi^2\sigma_1^2 f_1^2\right)\sin\left(2\pi f_1 x_1\right)df_1\right\}\dfrac{1}{\sqrt{2\pi}\sigma_2}\exp\left(\dfrac{-x_2^2}{2\sigma_2^2}\right)$ $v_2 = \dfrac{1}{\sqrt{2\pi}\sigma_1}\exp\left(\dfrac{-x_1^2}{2\sigma_1^2}\right)\left\{2\displaystyle\int_0^{\infty}\exp\left(-2\pi^2\sigma_2^2 f_2^2\right)\sin\left(2\pi f_2 x_2\right)df_2\right\}$ $v = \left\{2\displaystyle\int_0^{\infty}\exp\left(-2\pi^2\sigma_1^2 f_1^2\right)\sin\left(2\pi f_1 x_1\right)df_1\right\}\left\{2\displaystyle\int_0^{\infty}\exp\left(-2\pi^2\sigma_2^2 f_2^2\right)\sin\left(2\pi f_2 x_2\right)df_2\right\}$

    Quadruples can be analogously derived using Table H.1.

    Hilbert quadruples of interpolatory expansion of a raster of 2-D samples $a(i_1,i_2)$.

$$-N_1 \le i_1 \le N_1, -N_2 \le i_2 \le N_2$$

The expansion assumes sampling in a rectangular raster with periods Δ_1 along the axis x_1 and Δ_2 along the axis x_2. Notations:

$$X_1 = 2\pi W_1(x_1 - i_1\Delta_1), X_2 = 2\pi W_2(x_2 - i_2\Delta_2)$$

W_1 and W_2 are the cutoff frequencies of the 2-D lowpass spectral window.

$$u(x_1,x_2) = \sum_{i_1=-N_1}^{N_1} \sum_{i_2=-N_2}^{N_2} a(i_1,i_2) \frac{\sin(X_1)}{X_1} \frac{\sin(X_2)}{X_2}$$

$$v_1(x_1,x_2) = \sum_{i_1=-N_1}^{N_1} \sum_{i_2=-N_2}^{N_2} a(i_1,i_2) \frac{1-\cos(X_1)}{X_1} \frac{\sin(X_2)}{X_2}$$

$$v_2(x_1,x_2) = \sum_{i_1=-N_1}^{N_1} \sum_{i_2=-N_2}^{N_2} a(i_1,i_2) \frac{\sin(X_1)}{X_1} \frac{1-\cos(X_2)}{X_2}$$

$$v(x_1,x_2) = \sum_{i_1=-N_1}^{N_1} \sum_{i_2=-N_2}^{N_2} a(i_1,i_2) \frac{1-\cos(X_1)}{X_1} \frac{1-\cos(X_2)}{X_2}$$

Periodic Functions

$$u = \cos(2\pi f_1 x_1)\cos(2\pi f_2 x_2)$$

$$v_1 = \sin(2\pi f_1 x_1)\cos(2\pi f_2 x_2)$$

$$v_2 = \cos(2\pi f_1 x_1)\sin(2\pi f_2 x_2)$$

$$v = \sin(2\pi f_1 x_1)\sin(2\pi f_2 x_2)$$

$$u = \sin(2\pi f_1 x_1)\sin(2\pi f_2 x_2)$$

$$v_1 = -\cos(2\pi f_1 x_1)\sin(2\pi f_2 x_2)$$

$$v_2 = -\sin(2\pi f_1 x_1)\cos(2\pi f_2 x_2)$$

$$v = \cos(2\pi f_1 x_1)\cos(2\pi f_2 x_2)$$

$$u = \cos\left[2\pi\left(f_1 x_1 + f_2 x_2\right)\right]$$

$$v_1 = v_2 = \sin\left[2\pi\left(f_1 x_1 + f_2 x_2\right)\right]$$

$$v = -\cos\left[2\pi\left(f_1 x_1 + f_2 x_2\right)\right]$$

$$u = \sin\left[2\pi\left(f_1 x_1 + f_2 x_2\right)\right]$$

$$v_1 = v_2 = -\cos\left[2\pi\left(f_1 x_1 + f_2 x_2\right)\right]$$

$$v = -\sin\left[2\pi\left(f_1 x_1 + f_2 x_2\right)\right]$$

2-D Square Wave

$$u = \operatorname{sgn}\left[\cos\left(2\pi f_1 x_1\right)\right]\operatorname{sgn}\left[\cos\left(2\pi f_2 x_2\right)\right]$$

$$v_1 = \left(\frac{2}{\pi}\right)\ln\left[\tan\left(2\pi f_1 x_1 + \pi/4\right)\right]\operatorname{sgn}\left[\cos\left(2\pi f_2 x_2\right)\right]$$

$$v_2 = \left(\frac{2}{\pi}\right)\operatorname{sgn}\left[\cos\left(2\pi f_1 x_1\right)\right]\ln\left[\tan\left(2\pi f_2 x_2 + \pi/4\right)\right]$$

$$v = \left(\frac{4}{\pi^2}\right)\ln\left[\tan\left(2\pi f_1 x_1 + \pi/4\right)\right]\ln\left[\tan\left(2\pi f_2 x_2 + \pi/4\right)\right]$$

2-D Delta Samping Sequence

$$u = \sum_{n_1=-\infty}^{\infty}\sum_{n_2=-\infty}^{\infty}\delta\left[\left(x_1 - n_1 a\right),\left(x_2 - n_2 b\right)\right]$$

$$v_1 = \frac{1}{a}\sum_{n_1=-\infty}^{\infty}\sum_{n_2=-\infty}^{\infty}\cot\left[\left(\frac{\pi}{a}\right)\left(x_1 - n_1 a\right)\right]\delta\left(x_2 - n_2 b\right)$$

$$v_2 = \frac{1}{b}\sum_{n_1=-\infty}^{\infty}\sum_{n_2=-\infty}^{\infty}\delta\left(x_1 - n_1 a\right)\cot\left[\left(\frac{\pi}{b}\right)\left(x_2 - n_2 b\right)\right]$$

$$v = \frac{1}{ab}\sum_{n_1=-\infty}^{\infty}\sum_{n_2=-\infty}^{\infty}\cot\left[\left(\frac{\pi}{a}\right)\left(x_1 - n_1 a\right)\right]\cot\left[\left(\frac{\pi}{b}\right)\left(x_2 - n_2 b\right)\right]$$

List of Symbols

$\times$	Cartesian product
$*$	convolution
$\underbrace{* \ \cdots \ *}_{k \ \text{times}}$	k–fold convolution
$(\cdot) \overset{F}{\Leftrightarrow} (\cdot)$	Fourier transform pair
$(\cdot) \overset{H}{\Leftrightarrow} (\cdot)$	Hilbert transform pair
$\lVert \cdot \rVert$	norm
$(\cdot)^*$	conjugate of a number or a function
$(\cdot)^{-1}$	reciprocal of a number or a function
$(\cdot)_e$	even component of a 1-D function (e.g. $u_e(t)$)
$(\cdot)_o$	odd component of a 1-D function (e.g., $u_o(t)$)
$(\cdot)_{ee}$	even-even component of a 2-D function (e.g., $u_{ee}(x_1,x_2)$)
$(\cdot)_{eo}$	even-odd component of a 2-D function (e.g., $u_{eo}(x_1,x_2)$)
$(\cdot)_{oe}$	odd-even component of a 2-D function (e.g., $u_{oe}(x_1,x_2)$)

$(\cdot)_{oo}$	odd-odd component of a 2-D function (e.g., $u_{oo}(x_1,x_2)$)
$(\cdot)_{eee}$	even-even-even component of a 3-D function (e.g., $u_{eee}(x_1,x_2,x_3)$)
$(\cdot)_{eeo}$	even-even-odd component of a 3-D function (e.g., $u_{eeo}(x_1,x_2,x_3)$)
$(\cdot)_{eoe}$	even-odd-even component of a 3-D function (e.g., $u_{eoe}(x_1,x_2,x_3)$)
$(\cdot)_{eoo}$	even-odd-odd component of a 3-D function (e.g., $u_{eoo}(x_1,x_2,x_3)$)
$(\cdot)_{ooo}$	odd-odd-odd component of a 3-D function (e.g., $u_{ooo}(x_1,x_2,x_3)$)
$(\cdot)_{ooe}$	odd-odd-even component of a 3-D function (e.g., $u_{ooe}(x_1,x_2,x_3)$)
$(\cdot)_{oeo}$	odd-even-odd component of a 3-D function (e.g., $u_{oeo}(x_1,x_2,x_3)$)
$(\cdot)_{ooo}$	odd-odd-odd component of a 3-D function (e.g., $u_{ooo}(x_1,x_2,x_3)$)
∂D_n	closed contour of the region D_n in $\mathbb{R}^n$
$\mathbf{1}(f)$	n-D unit step (single-orthant operator)
$\arg(\cdot)$	argument of a complex number or a function
$A(t)$	instantaneous amplitude of $u(t)$
$A_i(x_1,x_2)$	2-D local amplitude of $u(x_1,x_2)$
$A_q(x_1,x_2)$	2-D local amplitude of a quaternion signal
$A_M(x_1,x_2)$	2-D local amplitude of a monogenic signal
$A_q(x_1,x_2,\chi_1,\chi_2)$	4-D quaternion ambiguity function
$A_M(x_1,x_2,\chi_1,\chi_2)$	4-D ambiguity function of a monogenic signal

$A_\psi(\cdot)$	Woodward's ambiguity function of $\psi(\boldsymbol{x})$
$A_A^{(2)}(\cdot)$	ambiguity function of A_ψ
$A_W^{(2)}(\cdot)$	ambiguity function of W_ψ
$\mathbb{C}$	algebra of complex numbers
$Cl_{p,q}(\mathbb{R})$	Clifford algebra over $\mathbb{R}$
$Cl_{0,1}(\mathbb{R})$	Clifford algebra of complex numbers
$Cl_{1,0}(\mathbb{R})$	Clifford algebra of double numbers
$Cl_{0,2}(\mathbb{R})$	Clifford algebra of quaternions
$Cl_{1,1}(\mathbb{R})$	Clifford algebra of coquaternions
$Cl_{0,3}(\mathbb{R})$	algebra of Clifford biquaternions
$Cl_{0,4}(\mathbb{R})$	algebra of Clifford bioctonions
e_{ijk}	product of imaginary units e_i, e_j and e_k (e.g., e_{123})
E_ψ	energy of the signal ψ
f	frequency variable (1-D)
$f(t)$	instantaneous frequency of $u(t)$
$\boldsymbol{f} = (f_1, f_1, \ldots, f_n)$	n-D frequency domain variable
$F\{\cdot\}$	Fourier transformation of a signal
$F^{-1}\{\cdot\}$	inverse Fourier transformation of a signal
F_x	Fourier transformation with respect to the x-variable
F_x^{-1}	inverse Fourier transformation with respect to the x-variable
$\mathbb{H}$	algebra of quaternions
$H[\cdot]$	Hilbert transformation of a signal

$\mathbf{H}[\cdot]$	Hilbert operator
i, j, k	imaginary units
$\mathbf{I}[\cdot]$	identity operator
$\mathrm{Im}(\cdot)$	imaginary part of a complex number or a function
o, o_i	octonion number
$o_{Cl}(\boldsymbol{x})$	Clifford n-D bioctonion signal
$\mathbb{O}$	algebra of octonions
$\mathrm{OFT}(\cdot)$	Octonion Fourier transform
P.V.	Cauchy principal value
q, q_i	quaternion number
$q_{Cl}(\boldsymbol{x})$	Clifford n-D biquaternion signal
r_i	real number
$r(\boldsymbol{x},\boldsymbol{\chi})$	$2n$-D correlation product in $\boldsymbol{x}$-domain
$\mathbb{R}$	set of real numbers
$R(\boldsymbol{\mu},\boldsymbol{f})$	$2n$-D correlation product in $\boldsymbol{f}$-domain
$\mathrm{Re}(\cdot)$	real part of a complex number or a function
s	sedenion number
$\mathrm{sgn}(\cdot)$	signum function
$\mathbb{S}$	algebra of sedenions
t	time variable (1-D)
$u(t)$	1-D real signal
$u(\boldsymbol{x})$	n-D real signal
$u_{\mathrm{rec}}(\boldsymbol{x})$	reconstructed signal of u
$U(\cdot)$	complex Fourier transform (spectrum) of u

$U_{CD}(\cdot)$	Cayley-Dickson Fourier transform (spectrum) of u
$U_q(\cdot)$	quaternion Fourier transform (spectrum) of u
$v(\boldsymbol{x})$	Total Hilbert transform of $u(\boldsymbol{x})$
$v_i(\cdot)$	first-order partial Hilbert transform of u
$v_{ij}(\cdot),\ i < j$	second-order partial Hilbert transform of u
$V(\cdot)$	Complex Fourier spectrum of v
$W_i(\cdot)$	auto-term of the Wigner distribution of ψ
$W_{ij}(\cdot)$	cross-term of the Wigner distribution of ψ
$W_q(x_1,x_2,f_1,f_2)$	4-D Quaternion Wigner distribution
$W_M(x_1,x_2,f_1,f_2)$	4-D Wigner distribution of a monogenic signal
$W_\psi(\cdot)$	Wigner distribution of ψ
$W_A^{(2)}(\cdot)$	Wigner distribution of A_ψ
$W_W^{(2)}(\cdot)$	Wigner distribution of W_ψ
$\boldsymbol{x} = (x_1,x_1,\ldots,x_n)$	n-D signal domain variable
$z,\ z_i$	complex number
$\delta(\cdot)$	Dirac delta (delta distribution)
$\boldsymbol{\mu} = (\mu_1,\mu_1,\ldots,\mu_n)$	n-D frequency-domain shift variable
$\Psi(f)$	Fourier spectrum of $\psi(t)$
ρ,ρ_{ij}	correlation coefficient of the n-D Gaussian signal
$\rho_q(\cdot)$	quaternion correlation product
$\rho_M(\cdot)$	monogenic correlation product
σ_i	standard deviation of the n-D Gaussian signal
$\phi(t)$	instantaneous phase of $u(t)$

$\phi_i(\boldsymbol{x})$	(local) phase function of the n-D complex/hypercomplex analytic signal
$\boldsymbol{\chi} = (\chi_1, \chi_1, \ldots, \chi_n)$	n-D signal-domain shift variable
$\psi(t)$	1-D analytic signal
$\psi_{(f_1>0)}(\boldsymbol{x})$	n-D complex analytic signal of rank 1
$\psi^{\mathbb{CD}}_{(f_1>0)}(\boldsymbol{x})$	n-D Cayley-Dickson hypercomplex analytic signal of rank 1
$\psi_i(\boldsymbol{x})$	n-D complex analytic signal with the spectrum in the ith orthant
$\psi_{i,j}(\boldsymbol{x})$	n-D lower rank complex analytic signal with the spectrum in orthants labelled with i and j
ψ_{CD}	Cayley-Dickson hypercomplex number
$\psi_o(\boldsymbol{x})$	n-D octonion analytic signal
$\psi_q(\boldsymbol{x})$	n-D quaternion analytic signal
$\psi(\boldsymbol{z})$	generalized Cauchy integral
$\psi_\delta(t)$	1-D complex delta distribution
$\psi_\delta(\boldsymbol{x})$	n-D complex delta distribution
$\psi_{hyp}(\boldsymbol{x})$	n-D Cayley-Dickson delta distribution
ω	product of elements of the basis of the Clifford algebra (e.g., $\omega = e_{123}$)
$\omega(t)$	angular frequency of $u(t)$

About the Authors

Stefan L. Hahn was born in Poznań, Poland, on February 20, 1921. He received a Ph.D. and D.Sc. in radioengineering from the Warsaw University of Technology (Politechnika Warszawska), Warsaw, Poland, in 1958 and 1962, respectively. He served at the Warsaw University of Technology as a professor and as head of the Institute of Radioelectronics of the Faculty of Electronics and Information Technology from 1949–1991. He is still at the Institute of Radioelectronics and Multimedia Technology as a professor emeritus.

His research interests include signal and modulation theory, quantum frequency standards, and radiocommunication systems. He is the author of over 100 scientific papers printed in Poland, the United States, England, and Germany. Professor Hahn was vice-chairman and chairman of the URSI Commission A on Electromagnetic Metrology from 1981–2010. He is a member of the Polish Academy of Sciences, the New York Academy of Sciences, and is a life senior member of IEEE.

Kajetana M. Snopek was born in Sosnowiec, Poland, in 1966. She received an M.Sc. in applied mathematics from the Warsaw University of Technology (WUT) in 1991. Since 1991, she has been with the Institute of Radioelectronics and Multimedia Technology (formerly the Institute of Radioelectronics), Faculty of Electronics and Information Technology, WUT. In 2002, she received her Ph.D. and in 2014, she received her D.Sc. in electronics. She holds the position of associate professor at the Institute and a head of Radiocommunication Signals and Systems Division.

She is the author of several papers presented at international symposia and conferences and published in scientific journals. Her research activities include complex and hypercomplex signal theory and time-frequency analysis and its extension to the n-D case. During the last 10 years, she has served as a reviewer of several prestigious international scientific journals. Her teaching areas include signals and systems and biomedical signal and image analysis. She is a member of IEEE.

Index

1-D analytic signals, 88, 117–19
  ambiguity functions (AFs), 190–99
  as boundary distributions, 88–93, 94
  extension of, 117–19
  polar representation of, 144–48
  spectrum of the modulated signal, 176–77
  Wigner distributions, 190–99
1-D Fourier pairs, 255–56
1-D Fourier transformation properties,
  253–54
1-D Gaussian signals
  illustrated, 55
  spectrum of, 56
1-D Hilbert pairs, 269–71
1-D quasi-analytic signals
  defined, 176
  frequency modulation, 180
  modulated Gaussian pulse, 177, 178
  modulated rectangular pulse, 177, 179
  modulation function, 176
  phase, 180
  See also Quasi-analytic signals
1-D real signals
  amplitude spectrum of, 60
  phase spectrum of, 60
  spectrum in terms of even/odd
    components, 59–61
2-D analytic signals, 94–99
  developed form, 94
  examples, 97–99
  four options of the spectrum, 96
  Gaussian signal, 100
  Hilbert transforms recovery, 96–97
  partial Hilbert transforms, 100
  polar representation of, 148–65
  second term of the real part, 95

  single quadrant spectrum, 95, 96
  total Hilbert transform, 100
2-D Cayley-Dickson analytic signals
  example, 137–38
  ranking, 135–38
  ranking related to real signal, 140
  single-domain definitions of, 135
2-D complex analytic signals
  ambiguity functions (AFs), 208–10
  amplitude functions, 149
  amplitudes, 154
  phase functions, 149, 156
  polar representation, 148–50
  ranking, 129–31
  ranking in frequency domain, 130
  ranking in signal domain, 130
  reconstruction formula, 149
  reconstruction of 2-D Gaussian signal, 156
  separable real signals, 150
  with single-orthant spectra, 148–50
  Wigner distributions, 203–7
2-D complex harmonic signals, 201
2-D complex signals
  ambiguity functions (AFs), 200–203
  Wigner distributions, 200–203
2-D delta sampling sequence, 276
2-D Fourier pairs, 263–64
2-D Fourier transforms
  Hermitian symmetry, 64
  mixed (sine-cosine), 64
  properties, 259–61
  QFT relation, 78–79
2-D Gaussian signals
  spectrum illustration, 58
  spectrum of, 56–59
2-D Hilbert quadruples, 273–76

2-D hypercomplex quaternion analytic signals
   amplitude, 150, 158–59
   phase functions, 150–51, 158–59
   polar representation, 150–51
   reconstruction formula, 151
   reconstruction of 2-D Gaussian signal, 160
2-D quaternion signals, 110–12
   example, 112
   Hermitian symmetry, 111
   inverse QFT, 111
   in successive quadrants, 111
2-D real signals
   amplitude spectrum, 64
   components, 61
   frequency content, 64
   phase spectrum, 64
   pyramid signal example, 64–67
   QFT of, 70
   spectrum of, 61–67
2-D signals
   nonseparable Gaussian, 50–51
   terms, 47–48
2-D square wave, 276
$2n$-D Wigner distribution, 188, 189
3-D analytic signals, 99–109
   examples, 107–9
   extension to, 104–5
   Fourier spectrum, 109
   Gaussian signal, 107
   Hermitian symmetry, 104
   inverse Fourier transform, 105, 106
   partial Hilbert transforms, 105
   polar representation, 165–73
   total Hilbert transform, 106
   union of terms, 108
3-D Cayley-Dickson analytic signals
   example, 139–41
   ranking, 138–41
   ranking related to real signal, 140
3-D complex analytic signals
   amplitudes, 166
   form, 165
   phases, 166
   polar representation, 165–67
   ranking, 131–35
   ranking in frequency domain, 133
   ranking in signal domain, 132
   separable real signals, 166–67
3-D Fourier transform
   Hermitian symmetric, 68–69
   OFT relation, 79–81
3-D hypercomplex analytic signals, 112–13

3-D octonion analytic signals
   amplitudes, 168
   examples, 169–73
   phase functions, 168
   polar representation, 167–73
   real signal reconstruction, 168–69
   separable real signals, 169
3-D real signals
   frequency content, 69, 76
   half-space support of Fourier spectrum, 69
   octonion Fourier transform (OFT), 70, 73
   representation of, 250
   separable, 250
   spectrum of, 67–69
   symmetry properties of even-odd
      components, 68
3-D signals
   single-orthant operators, 47
   terms, 48
4-D Wigner distributions, 200–201, 202–3
4-D Woodward's ambiguity functions (AFs),
   201, 202–3

Alternativity, 73
Ambiguity functions (AFs)
   1-D signals, 190–99
   2-D, 189, 192
   2-D complex analytic signals, 208–10
   2-D complex signals, 200–203
   4-D, 201, 202–3
   autoterms of, 194–99
   as complex valued function, 192
   cross-terms of, 194–99
   monogenic signals, 220–22
   properties of 1-D signals, 194
   as quadratic transformations of signals, 194
   of quaternion signals, 214–18
   in symmetrical form, 192
   Woodward, 191
Amplitude
   complex, 144
   defined, 144
   instantaneous, 145
   local, 35
Amplitude spectrum
   of 1-D real signal, 60
   2-D pyramid signal, 65–67
   2-D real signals, 64
   defined, 59
   one-sided exponential signal, 60–61, 62
Analytic functions
   1-D, 88–93

analytic signals and, 87
Analytic signals
  analytic functions and, 87
  as boundary distributions, 1
  Cauchy, 91
  defined, 88, 90
  hypercomplex n-D, 109–13
  with identity operator, 90
  magnitudes of, 210
  monogenic 2-D signals, 113
  n-D, 94–109
  one-sided spectrum, 91
  partial, 129
  polar representation of, 143–73
  product of, 268
  quasi, 99, 175–85
  ranking of, 125–41
  with single orthant spectra, 117–20
  Vakman's conditions and, 93
Applications
  of Clifford Algebras, 33
  of complex and hypercomplex Fourier
    transforms, 81–82
  in domain of analytic signals, *83*
  of n-D analytic signals, 120–21
  in signal processing, 31–36
  of space-frequency distributions, 227–35
Associativity, 23
Autoconvolution equality, 268
Autocorrelation, 254, 260
Autoterms, 194–99

Band-pass signals
  modulated real, 204–5, 212–14, 219–20
  sum of two, 205–7, 214
Biassociativity, 23
Boundary distributions, 1, 88–93, 94, 116

Cauchy Hilbert transforms, 181
Cauchy-Riemann equations, 181
Causality of signals
  coordinate systems and, 245
  decomposition, 246
  defined, 241, 246
  derivation of dispersion relations, 244–46
  extension to higher dimension, 243–46
  formula, 244
  Kramers-Kronig relations and, 241–43, 244
  summary, 246–48
Cayley-Dickson algebras
  octonions, 19–23
  operations in, 12

overview, 9–11
  of quaternions, 15
  See also Hypercomplex algebras
Cayley-Dickson construction
  continuation of, 14
  defined, 11
Cayley-Dickson Fourier transformation, 53,
    69–77
  defined, 69
  formulas, 69–71
  general form, 69–70
  inverse, 71
  theoretical basics, 53
Cayley-Dickson spectrum, 70
Circular symmetry, 260
Clifford algebras
  application of, 33
  of bioctonions, 28–31
  of biquaternions, 24–27
  defined, 23–24
  defining, 24
  general properties of, 24
  notation, 24
  See also Hypercomplex algebras
Clifford bioctonions
  Clifford algebra for, 28–31
  defined, 28
  expression as hypercomplex sum, 28
  multiplication rules in, 29
  properties of, 30–31
  relations of commutation, 30
  relations of noncommutation, 30–31
Clifford biquaternions
  Clifford algebra of, 24–27
  defined, 24
  multiplication of imaginary units, 27
  multiplication properties, 27
  properties of, 25–27
Common hyperplane, 127
Commutativity, 78
Complex algebras, in multidimensional signal
    theory, 33–36
Complex amplitude, 144
Complex analytic signals
  defined in signal domain, 119
  polar representation of, 143–73
  ranking of, 129–35
  Wigner distributions of, 187–210
  Woodward ambiguity function (AF) of,
    187–210
  See also 3-D complex analytic signals; 2-D
    complex analytic signals

Complex delta distribution
approximating functions of, 90, 91
defined, 6, 91
n-D, 94
real and imaginary parts of, 243
Complex Fourier spectrum, 44
Complex Fourier transforms
applications of, 81–82
hypercomplex relations, 77–81
Complex numbers, polar representation of, 144
Conjugation
1 D Fourier transformation property, 253
2-D Fourier transformation property, 259
octonions, 20–21
quaternions, 16
Contour integration, 89
Convolution, 254, 260, 267, 268
Cosine Fourier transform
1-D, 59
2-D, 63
Cosine signals embedded in Gaussian noise, 229, 231
Cross-correlation, 254, 260
Cross sections
of 2-D modulated real band-pass signal, 206
of analytic signals, 207
defined, 203
of monogenic signal, 220, 221
of quaternion analytic signal, 212, 213, 214, 215–16
of sum of four Gaussian pulses, 226
of sum of two Gaussian pulses, 226
of WDs of separable Cauchy signal, 204
Cross-terms
of 2-D separable Cauchy signal, 208, 209
of AF of bicomponent signals, 196
cosine, 198
number of, 195
reduction, 194–95
of WD of bicomponent signals, 196
Cylindrical 2-D analytic signal, 162–65

Decomposition
2-D signals, 47–48
3-D signals, 48
nonseparable 2-D Gaussian signal, 50–51
real functions into even and odd terms, 47–51
single-quadrant cube signal, 49
Delta planes, 196, 198

Discrete QFT, 81
Double-dimensional pseudo-Wigner distributions, 227
Double-dimensional Wigner distributions, 223–27
of the AF, 223–24
ambiguity function of the AF, 224
defined, 223
examples, 225–27
form, 223
method of calculation, 225
O'Neill-Flandrin formula, 223
Duality, 253, 259, 265
Duality of the Fourier transformation, 57

Energy equality (Rayleigh's theorem), 254, 260, 266, 268
Euler's formula, 5

Fano scheme, 20
Fast Fourier Transform (FFT), 81
Finite energy part, 181
First Cauchy integral, 88
First moment, 254
Fourier pairs
1-D, 255–56
2-D, 263–64
Fourier relations between correlation products, 189, 190, 192
Fourier spectrum
complex, 44
as Hilbert transformation property, 267
n-D signal, 43
Fourier transformations
1-D properties, 253–54
2-D properties, 259–61
in analysis of n-dimensional signals, 53–83
Cayley-Dickson, 53, 69–77
complex n-D, 54–69
in complex notation, 5
duality of, 57
mixed (sine-cosine), 63
octonion Fourier transform (OFT), 3, 53, 70, 73–77
quaternion Fourier transform (QFT), 3, 53, 70, 71–73
relations between complex and hypercomplex, 77–81
Frequency differentiation, 253
Frequency domain, 1
Frequency domain method, 3
Frequency domain space, 175

Frequency modulated harmonic carrier, 148
Frequency modulation, 180
Frequency spectrum
    1-D Gaussian signal, 54–56
    1-D real signal, 59–61
    2-D pyramid signal, 64–67
    2-D real signal, 61–67
    3-D real signal, 67–69
    amplitude spectrum, 59, 60–61
    Cayley-Dickson, 70
    defined, 54
    imaginary spectrum, 59
    of the modulated signal, 176–77
    phase spectrum, 59, 60–61
    real spectrum, 59

Gabor's 1-D analytic signal, 4, 5, 6, 88, 117, 146
Gaussian crater embedded in random noise,
    232–35

Hamiltonian rule, 12–13
Harmonic exponential signals, 195–96,
    198–99
Harmonic independence, 93
Hermitian symmetry, 59, 64, 68–69, 104,
    111, 259
Hilbert pairs
    1-D, 269–71
    2-D, 274
Hilbert quadruples
    2-D delta sampling sequence, 276
    2-D square wave, 276
    convolution definitions, 273
    defined, 273
    integral definitions, 273
    periodic functions, 275–76
    separable 2-D functions, 274–75
Hilbert transforms, 5, 44, 87, 89, 92–93
    1-D properties of, 267–68
    Cauchy, 181
    partial, 6, 95, 98, 244, 247
    real and imaginary part formation of, 242,
        243
    total, 98, 100, 106
    Vakman's conditions and, 93, 144–45
Historical survey, 4–6
Hypercomplex algebras
    applications in signal processing, 31–36
    Cayley-Dickson, 9–23
    Clifford, 23–31
    comparison of, 31
    in multidimensional signal theory, 33–36
    as normed division algebras, 10
    relations between, 32
    summary, 36–37
    survey of, 9–37
Hypercomplex analytic signals, 109–13
    2-D quaternion signals, 110–12
    3-D, 112–13
    Cayley-Dickson version, 110
    Clifford algebra version, 110
    defined, 109
    defined in signal domain, 120
    delta distribution, 110
    Fourier transform notation, 109
    knowledge about, 250
    polar representation of, 143–73
    ranking of, 135–41
Hypercomplex Fourier transforms
    applications of, 81–82
    complex relations, 77–81

Identity, 259
Identity operator, 90
Image processing, Wigner distribution in,
    232–35
Imaginary part, complex number, 10
Imaginary spectrum, 59
Impulse response, 241, 242
Instantaneous amplitude, 145
Instantaneous angular velocity, 145
Instantaneous complex frequency, 6, 147
    defined, 145
    representation with Wigner distribution,
        148
Instantaneous complex phase, 145, 147
Instantaneous frequency, 146
Instantaneous radial velocity, 145
Integral definition, 253, 259, 267
Integration, 253
Interpolating function, 148
Inverse Fourier transform
    3-D analytic signals, 105, 106
    defined, 57
    of enlarged spectrum, 141
    n-D Cayley-Dickson, 71
    octonion, 71
    quaternion, 111
Involutions, 265
Iterations, 268

Jacobi's elliptic functions, 5

Kramers-Kronig relations, 241–43, 244

Leakage coefficient, 175
Linearity, 253, 259, 265, 267
Low-pass Gaussian noise, 229, 230

Mixed (sine-cosine) FTs, 63
Modified modulation function, 176
Modulated Gaussian pulse, 177, 178
Modulated real band-pass signals, 204–5,
    212–14, 219–20
Modulated rectangular pulse, 177, 179
Modulation, 253, 260, 266
Modulus of octonions, 21
Monogenic 2-D signals, 113–17
    amplitude, 160–61
    boundary distributions, 116
    examples, 116–17
    Gaussian signal, 118
    operators, 114, 115
    orientation angle, 152
    phase, 153
    phase functions, 160–61
    polar representation, 114, 152–53
    quaternion Fourier transform (QFT), 114
    reconstruction of 2-D Gaussian signal,
        162
    Riesz kernels, 113
Monogenic signals
    4-D AF of, 220
    ambiguity functions (AFs) of, 220–22
    cross sections, 220, 221
    defined, 88
    magnitudes of, 222
    Wigner distributions of, 219–20
Multiplication
    alternative, 23
    of imaginary units, 15, 20, 27
    in quasi-analytic signal definition, 183
    quaternions, 18–19
    of signals, 268
    by t, 268
Multiplication rules, 13, 14
    Clifford bioctonions, 29
    Clifford biquaternions, 27
Multiplicative inverse
    octonions, 22
    quaternions, 17

n-D analytic signals
    2-D complex signals, 94–99
    3-D complex signals, 99–109
    applications of, 120–21
    complex delta distribution, 94

definitions, 43
    lower rank, 141
    with single-orthant spectra, 43
n-D Cayley-Dickson hypercomplex signals, 35
n-D Clifford bioctonion signals, 36
n-D Clifford biquaternion signals, 35
n-D complex analytic signals, hierarchy of,
    136
n-D complex signals
    defined, 34
    frequency analysis of, 53–83
    local amplitude of, 35
    with single-orthant spectra, 34
n-D Fourier spectrum, 1
n-D hypercomplex delta distribution, 6
n-D hypercomplex signals
    frequency analysis of, 53–83
    general form, 34
n-D octonion signals, 34
n-D quasi-analytic signals
    complex carrier, 183
    defined, 183
    example, 185
    shifted spectrum definition, 184
n-D quaternion signals, 34
Nieto-Vesperinas, 247–48
Noise analysis
    as area of stochastic signal processing, 227
    cosine signal embedded in Gaussian noise,
        229, 231
    low-pass Gaussian noise, 229, 230
    Wigner distribution in, 227–31
Nonassociativity, 23
Norm
    n-D Cayley-Dickson hypercomplex signals,
        35
    n-D Clifford bioctonion signals, 36
    octonions, 21–22
    quaternions, 16–17
Normalized first moment, 254
Normalized second-order moment, 254
Normed division algebras, 10
nth moment, 254

Octants, 126
Octonion Fourier transform (OFT)
    3-D FT relation, 79–81
    3-D real signal, 70, 73
    defined, 3
    single-octant support, 77
    spectrum, 80
    superposition of involutions, 77

symmetry properties of even-odd
components, 74
symmetry relations, 75–76
theoretical basics, 53
Octonions
alternativity, 73
Cayley-Dickson algebras of, 19–23
conjugation, 20–21
modulus of, 21
multiplicative inverse, 22
norm, 21–22
properties of, 20–23
summation, 22–23
O'Neill-Flandrin formula, 223
One sided exponential signal, 60–61
Orientation angle, 152
Orthants
2-D, 127
3-D, 127
defined, 126
labeling of, 45, 127
n D-Cartesian space, 44
1-D, 127
in ranking, 125

Parity, 267
Parseval's equality, 203
Partial Hilbert transforms, 6, 95, 98, 105, 244
Periodic functions, 275–76
Phase signals, 181–83
finite energy part, 181
instantaneous angular frequency, 182
Phase spectrum
1-D real signals, 60
2-D pyramid signal, 65–67
2-D real signals, 64
defined, 59
as odd function, 59
one-sided exponential signal, 60–61, 62
Polar representation
of 1-D analytic signals, 144–48
of 2-D analytic signals, 148–65
of 2-D complex analytic signals with
single-quadrant spectra, 148–50
of 2-D hypercomplex quaternion analytic
signals, 150–51
of 3-D analytic signals, 165–73
of 3-D complex analytic signals, 165–67
of 3-D octonion analytic signals, 167–73
of analytic signals, 143–73
of complex analytic signals, 143–73
of complex numbers, 144

examples, 153–65
formula, 145
of hypercomplex analytic signals, 143–73
introduction to, 143–44
of monogenic 2-D signals, 152–53
Product, 253, 260

Quasi-analytic signals
1-D, 176–80
defined, 99, 175–76
n-D, 183–85
phase signals, 181–83
Quaternion AFs, 214
Quaternion correlation product, 211
Quaternion Fourier transform (QFT)
2-D, first-quadrant support, 73
2-D FT relation, 78–79
2-D real signal, 70
defined, 70
discrete, 81
even/odd components, 71–73
image watermarking detection scheme, 81
inverse, 111
monogenic 2-D signals, 114
properties of real signals, 265–66
right-sided, 265
in successive quadrants, 72
terms, 71
theoretical basics, 53
two-sided, 265
Quaternions
Cayley-Dickson algebras of, 15–19
conjugation, 16
defined, 12
general form, 4
multiplication, 18–19
multiplicative inverse, 17
norm, 16–17
polar form of, 19
properties of, 15–19
in representing rotations, 33
reversibility, 15–16
summation, 17
Quaternion signals
ambiguity functions (AFs) of, 214–18
magnitudes of, 218
Wigner distributions of, 211–14
Quaternion Singular Value Decomposition,
81
Quaternion-valued function, 217
Quaternion-valued WDs, 211
Quaternion Wigner distribution, 211

Ranking of analytic signals
   2-D Cayley-Dickson analytic signals,
     135–38
   2-D complex signals, 129–35
   3-D Cayley-Dickson analytic signals,
     138–41
   3-D complex signals, 131–35
   complex, 129–35
   defined, 125
   hypercomplex, 135–41
   process, 141
   summary, 141
Real part, complex number, 10
Real spectrum, 59
Reversibility, quaternions, 15–16
Riesz transforms, 6

Scalar product (Parseval's theorem), 254, 260
Scaling, 253, 259, 265, 267
Second Cauchy integral, 88
Second-order moment, 254
Sedenions, 13
Seminorm, 25
Separability, 260, 266
Separable real signals, 151–52
   3-D complex analytic signals, 166–67
   3-D octonion analytic signals, 169
   polar form of, 250
Signal domain, 1
Signal-domain differentiation, 260, 266
Signal-domain integral, 261, 266
Signal domain method, 3
Signal-domain reversal, 259, 265
Signal-domain shift, 260, 266
Sine Fourier transform
   1-D, 59
   2-D, 63
Single-orthant operators, 45–47
Space-frequency distributions, applications,
   227–35
Spectrum. *See* Frequency spectrum
Spectrum of the modulated signal, 176–77
Spiral phase quadrature transform, 116
Suborthants
   defined, 126–29
   in ranking, 125
Subquadrants
   in 2-D, 128
   in 3-D, 128–29

Summary, this book, 249–51
Summation
   octonions, 22–23
   quaternions, 17
Symbols list, 277–82

Time derivatives, 267
Time differentiation, 253
Time-frequency Wigner distribution, 190
Time integral, 254
Time reversal, 253, 267
Time shift, 253, 267
Total Hilbert transform
   calculation in single domain, 98
   3-D analytic signals, 106
   2-D analytic signals, 100
Transfer function, 241, 242

Vakman's conditions, 93, 144–45

Watermark detection scheme, 81–82
Watermark embedding algorithm, 81
Wigner distributions, 148, 187–235
   1-D signals, 190–99
   2-D, 188, 189, 192
   2-D complex analytic signals, 203–7
   2-D complex signals, 200–203
   4-D, 200–201, 202–3
   autoterms of, 194–99
   calculation of, 188
   cross-terms of, 194–99
   defined, 188
   double-dimensional, 223–27
   introduction of, 187–88
   low-pass Gaussian noise, 229, 230
   monogenic signals, 219–20
   in noise analysis, 227–31
   properties of 1-D signals, 193
   quaternion signals, 211–14
   quaternion-valued, 211
   sum of three time-shifted and Gaussian
     pulses, 199
   time-frequency, 190
Wigner-Ville distribution (WVD), 188, 229
Woodward ambiguity function
   4-D, 201, 202–3
   defined, 191
   See also Ambiguity functions (AFs)